Principles of
Celestial Mechanics

Principles of Celestial Mechanics

PHILIP M. FITZPATRICK
AUBURN UNIVERSITY

ACADEMIC PRESS New York and London

ACADEMIC PRESS, INC.
111 Fifth Avenue, New York, New York 10003

United Kingdom Edition published by
ACADEMIC PRESS, INC. (LONDON) LTD.
Berkeley Square House, London W1X 6BA

Library of Congress Catalog Card Number: 77-119612
AMS 1968 Subject Classification No. 8510

PRINTED IN THE UNITED STATES OF AMERICA

FOR

ROSALIND

Contents

11. Hamilton–Jacobi Theory with Examples

12. The Gravitational Potential

13. Gravity and Drag Effects on the Orbit of an Artificial Earth Satellite

14. Gravity Effects on the Rotational Motion of an Artificial Earth Satellite

Preface

This text is an outgrowth of lectures in celestial mechanics which I have given at Auburn University over the last seven years. It is designed to introduce the reader to the principles and concepts which underlie the most commonly used general perturbations methods of celestial mechanics and to demonstrate the application of these methods to several important satellite problems. My aim has been to provide a coherent treatment of certain fundamental information which is needed for the understanding of current analytical investigations into the motions of artificial satellites. Established ideas which are basic to the subject as a whole are treated in the first twelve chapters and four appendixes. The remaining two chapters contain illustrative applications of the basic ideas to outstanding unsolved problems of satellite orbital and attitude dynamics. In particular, the last chapter contains results on the rotational motions of satellites which, to my knowledge, have not been previously published in textbook form.

The text is designed not only for astronomy majors but also for undergraduates and beginning graduate students in mathematics, physics, and engineering who have no background in astronomy but, nevertheless, wish to acquire a grasp of the basic concepts and theoretical methods of celestial mechanics. Although this book is intended primarily for use in a two-semester sequence at either the senior level or a combination senior level–beginning graduate level, the order in which the topics are developed is such that the material in the first seven chapters and the four appendixes will meet the requirements of most one-semester courses on the subject.

Reasonable prerequisites for undergraduates who take this course are (1) a sound understanding of the elementary calculus, and (2) a mastery of first courses in ordinary differential equations, vector analysis, mechanics, and complex variables. For gifted undergraduates and graduate students, the cited prerequisites could be relaxed somewhat. Particle dynamics is developed from first principles. Therefore prior work in mechanics, while desirable, is not a must in every case. The use of complex variables is pretty well restricted to Chapter 4 and Appendix 3. This subject as a requirement might be dropped at the discretion of the instructor.

The choice and synthesis of material are considerably different from that usually found in introductory treatments of celestial mechanics. Topics such as special perturbations methods, the three-body problem, the techniques and methods of observational astronomy, and astronomical applications are ordinarily part of an introduction to the subject. These items are slighted in this text in favor of material on Hamilton–Jacobi theory, canonical transformations, analytical procedures of satellite surveillance, gravitational potential, and satellite orbit and attitude dynamics. Such material in the opinion of the author, more closely fits the needs of people whose main interest is in the formulation of analytical solutions to the problems of satellite motions.

A somewhat unusual feature of the book is that it contains a description of the basic ideas which underlie the procedures used at control centers to maintain surveillance of orbiting satellites. This puts the reader in a position to appreciate the type of analytical support which is needed in the area. There is an especially thorough treatment of gravitational potential in Chapter 12. Material of this type is important to people who are concerned with formulating theories for the translational and rotational motions of artificial satellites. A specifically distinct feature of the book is that its last chapter is concerned with applications of principles discussed in earlier chapters to the problem of describing the rotational motions of an artificial satellite about its center of mass. This material is designed to introduce the reader to a fascinating area of research and at the same time challenge him with some of its unsolved problems.

The working of a significant number of problems is obligatory if the reader is to obtain maximum benefit from the text. The concepts of celestial mechanics oftentimes appear deceptively easy and it is only by intensive testing of his understanding that the reader can learn to think correctly in terms of basic ideas and definitions and develop skill in applying them to new situations. A goodly number of the exercises have been drawn from the vast reservoir of problems which appear in the classical literature on the subject. These problems, some of which are easy and some difficult, have been carefully selected to provide the reader, who is willing to expend the effort, with a deeper insight into the full content of the major concepts discussed in the text. Except for a few problems whose original sources I can no longer identify, the remainder of the problems have been specially devised and vary in their degrees of difficulty. Many of this latter group of problems serve either to complete the proofs and developments begun in the text or to introduce new concepts and techniques not touched upon in the body of the text. In most cases a problem will contain a statement of the specific result which the reader is required to establish.

This textbook has been written from the viewpoint of one who has been working in the field of celestial mechanics but who lays no claim to being a specialist. My purpose has not been to teach all about celestial mechanics but rather to build a framework of principles and methods that can readily be added to as it is required. In the process of building this framework I hope that I have been able to communicate to the reader some of my own enthusiasm for this intriguing, sometimes frustrating, but always challenging area of study.

Acknowledgments

More than to any other source of knowledge which I can identify, I am indebted to certain unpublished lecture material on celestial mechanics developed and made available to me by Leland E. Cunningham of the University of California, Berkeley, California. I am further indebted to Dr. Cunningham for graciously extending his permission to present the account of his theory of the motion of a satellite of an oblate planet given in Chapter 13. Also I would like to render grateful thanks to my son, Philip M. Fitzpatrick, Jr., of Auburn University for his painstaking efforts in preparing all of the drawings and the book jacket design.

The research which led to the results described in Chapter 14 was sponsored by the National Aeronautics and Space Administration.

Chapter 1

Fundamental Concepts

1.1. INTRODUCTION

Celestial mechanics applies the concepts of mathematics and mechanics to the study of the motion of celestial objects. In the present work, we will restrict our considerations to the motions of objects within our solar system. We presuppose a working knowledge of the elementary principles of mechanics, of differential and integral calculus, and of ordinary and partial differential equations. We also assume a reasonable familiarity with the elements of functional analysis for both vector and scalar functions.

Throughout this book we will be concerned with particle dynamics. In fact, the central theme will be woven about the motion (i.e., the trajectory) of a single particle moving under the influence either of one other particle or under the influence of one other dominant particle, plus small perturbing influences from other particles in the dynamical system. We treat particles as mass-points and since we will be dealing with the solar system, one may well inquire how it is possible to consider bodies in the solar system, the Sun, for example, as particles. The justification for this procedure is based upon the following theorem:

Theorem 1.1.1. A sphere composed of uniformly dense shells attracts other bodies as though the entire mass of the sphere were concentrated at its center.

To establish a proof of Theorem 1.1.1, one need invoke only a suitable definition of force and some elementary integral calculus. It is more suitable for our purpose here, however, to use an alternative method of proof in which an intermediate result known as Gauss's theorem is first established. Before stating the theorem, we review briefly some well-known results from vector analysis.

Consider a vector field **F** (boldface notation indicates a vector quantity) of the form $\mathbf{F} = f(r)\,\mathbf{r}$, where $\mathbf{r} = x\mathbf{i} + y\mathbf{j} + z\mathbf{k}$ is the position vector measured from the origin of the rectangular reference system $Oxyz$ and $r = (x^2 + y^2 + z^2)^{1/2} = |\mathbf{r}|$. The divergence $\nabla \cdot \mathbf{F}$ may be written

$$\begin{aligned}\nabla \cdot f(r)\,\mathbf{r} &= \frac{\partial}{\partial x}[f(r)\,x] + \frac{\partial}{\partial y}[f(r)\,y] + \frac{\partial}{\partial z}[f(r)\,z] \\ &= 3f(r) + x\frac{\partial f}{\partial x} + y\frac{\partial f}{\partial y} + z\frac{\partial f}{\partial z} \\ &= 3f(r) + f'(r)\frac{(x^2 + y^2 + z^2)}{r} \\ &= 3f(r) + rf'(r)\end{aligned}$$

In particular, if $\mathbf{F} = (a/r^3)\,\mathbf{r}$, where a is a constant and $r \neq 0$, so that $f(r) = a/r^3$ and $f'(r) = -3a/r^4$, we have

$$\nabla \cdot f(r)\,\mathbf{r} = \frac{3a}{r^3} + r\left(\frac{-3a}{r^4}\right) = 0$$

The preceding result can be combined with the divergence theorem to establish Gauss's theorem. Recall that the divergence theorem, under certain restrictions that will always be satisfied in our applications here, relates a surface and volume integral through the formula

$$\iint_S \mathbf{F} \cdot d\mathbf{S} = \iiint_\tau \nabla \cdot \mathbf{F}\, d\tau$$

where **F** is a vector function of position and τ is the volume bounded by the closed surface S. The notation $d\mathbf{S} = \mathbf{n}\, dS$ associates with the element of surface area dS a vector $d\mathbf{S}$ whose magnitude is dS and whose direction is that of **n**, the outwardly drawn unit normal to the surface.

Gauss's Theorem. For S a closed surface and **r** the position vector of any point measured from an origin O,

(1) $\displaystyle\iint_S \frac{a}{r^3}\mathbf{r} \cdot d\mathbf{S} = 0$ if O is exterior to S;

(2) $\displaystyle\iint_S \frac{a}{r^3}\mathbf{r} \cdot d\mathbf{S} = 4\pi a$ if O is interior to S.

Proof. If O is outside of S, $r \neq 0$, and $\nabla \cdot (a/r^3)\,\mathbf{r} = 0$ everywhere inside of τ. From the divergence theorem it follows that

$$\iint_S \frac{a}{r^3}\,\mathbf{r} \cdot d\mathbf{S} = 0$$

If O is interior to S, let $\boldsymbol{\rho}$ describe a small sphere of radius ρ centered at O and contained entirely within S. If τ' is the region bounded by S and S', where S' is the surface area of the small sphere, we have, $\nabla \cdot (a/r^3)\,\mathbf{r} = 0$ everywhere in τ' since $r \neq 0$.

Reapplying the divergence theorem, we write

$$\begin{aligned}\iint_{S+S'} \frac{a}{r^3}\,\mathbf{r} \cdot d\mathbf{S} &= \iint_S \frac{a}{r^3}\,\mathbf{r} \cdot d\mathbf{S} + \iint_{S'} \frac{a}{\rho^3}\,\boldsymbol{\rho} \cdot d\mathbf{S}' \\ &= \iiint_{\tau'} \nabla \cdot \frac{a}{r^3}\,\mathbf{r}\,d\tau' = 0\end{aligned}$$

Thus

$$\iint_S \frac{a}{r^3}\,\mathbf{r} \cdot d\mathbf{S} = -\iint_{S'} \frac{a}{\rho^3}\,\boldsymbol{\rho} \cdot d\mathbf{S}' = a \iint_{S'} \frac{dS'}{\rho^2} = 4\pi a \qquad (1.1.1)$$

since on S' the outwardly directed normal is oppositely directed to $\boldsymbol{\rho}$ and the quantity dS'/ρ^2 represents an increment of solid angle that, when integrated over the entire sphere, is equal to 4π.

We are now in a position to give a proof of Theorem 1.1.1.

Proof. Consider the closed surface indicated by S in Figure 1.1.1. Let the element of mass dm_i be centered at a point O that lies within the surface S. Let $\mathbf{r}$ be a position vector from O to any point P on S. The gravity field $\mathbf{g}_i$ at P on the surface S, due to dm_i, is assumed to be given by Newton's law of attraction. Thus

$$\mathbf{g}_i = \frac{-k^2\,dm_i}{r^3}\,\mathbf{r} \qquad (k^2 = \text{gravitational constant})$$

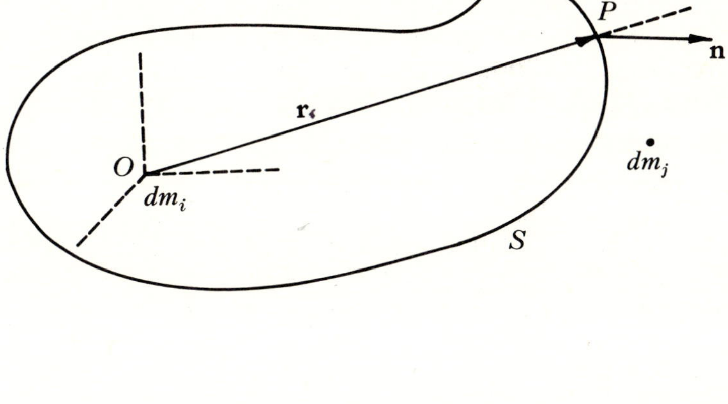

FIGURE 1.1.1

From Gauss's theorem, with $a = -k^2\, dm_i$, we have

$$\iint_S \mathbf{g}_i \cdot d\mathbf{S} = -4\pi k^2\, dm_i$$

By experiment, it is found that when several masses, m_1 , m_2 ,..., m_n , act simultaneously on a test body, the forces on the test body satisfy the law of vector addition. This is an example of the "principle of superposition of fields." Thus for n elements of mass within S, since our conclusion is independent of the position of dm_i within S, we write

$$\iint_S \left(\sum_{i=1}^{n} \mathbf{g}_i \right) \cdot d\mathbf{S} = -4\pi k^2 \sum_{i=1}^{n} dm_i \tag{1.1.2}$$

Now the only known ways to distribute mass within S are in volume, surface, line, and point distributions. In the last case, Equation (1.1.1) will hold. In the first three cases, we can divide the extended mass into elements, each of which is essentially a point mass. Then, since vector superposition is an experimental fact, we may apply (1.1.2) to the cases where the total mass M is obtained by summation, or by integration, or by both. Contributions from outside the surface, such as from dm_j in Figure 1.1.1, do not affect the value of the integral as is clear from Gauss's theorem. Thus, in forming $\mathbf{g}$ for use in (1.1.2), we need consider only those mass elements interior to S. Hence if $\mathbf{g}$ is the gravity field at P due to sources interior to the closed surface S and M is the total mass within S, we may write

$$\iint_S \mathbf{g} \cdot d\mathbf{S} = -4\pi k^2 M \qquad \text{(Gauss's law)} \tag{1.1.3}$$

Consider a sphere of mass M centered at point O for which the mass density is a function only of the distance from O. Consider a concentric spherical surface of radius r that contains the sphere. By symmetry, $\mathbf{g}$ is constant in magnitude on the surface of the containing sphere and is everywhere directed toward O. Hence, from (1.1.3), since $\mathbf{g}$ and the normal to $d\mathbf{S}$ are oppositely directed, $|\,\mathbf{g}\,| = k^2 M/r^2$, and

$$\mathbf{g} = \frac{-k^2 M}{r^3}\,\mathbf{r} \tag{1.1.4}$$

where $\mathbf{r}$ is the position vector directed away from O. This establishes Theorem 1.1.1, for (1.1.4) is exactly the expression we would write for the gravity field of a mass M concentrated at O.

Since the planets and the Sun appear to be essentially spheres composed of uniformly dense shells, we will treat them as "mass-points." Certainly the main part of the external gravity field of a planet is given by such a model. This fact, together with the vast distances that are generally involved in studies relating to the solar system, provides ample justification for this treatment. Later on in our work with artificial close–Earth satellites, however, we will see that it is necessary to consider the effects of deviations from the simple spherical model.

Vector notation will be used in what follows where convenient. In other places, scalar notation will be employed even though the use of a vector notation may be possible. Where it seems desirable, we will carry, simultaneously, both vector and nonvector notations, but this practice will not often be necessary.

We will shortly begin a discussion of the fundamental dynamics of n particles. In such a discussion one usually considers the following three types of forces to be present in the system:

1. Forces between the particles, equal and opposite, and acting along the lines joining them.
2. Forces between the particles, not necessarily equal and opposite, or not necessarily along the lines joining them.
3. Forces arising outside of the particles themselves.

The first two types of forces are called internal forces, whereas the last type is called external. Since forces of type 2 do not arise from gravitation, we will, in what follows, consider them to be absent.

In our discussion of particle dynamics, we will admit the possibility of variable masses that do not arise from relativistic considerations. One must distinguish between variations in mass due to high particle velocities and variations due to secreted or ejected material. In either case, the entire system must be considered, but, in the latter case, the system must specifically include the extra material. It is only the latter type of variable mass that will be included in our considerations.

We will now introduce the notion of a fixed or inertial system of axes.

1.2. NEWTONIAN REFERENCE SYSTEMS

Consider the rectangular coordinate system $Oxyz$ shown in Figure 1.2.1. In this system, Newton's second law, applied to a particle of constant mass m at position $\mathbf{r}$, in a force field $\mathbf{F}$ that depends on position only, is assumed to hold in the form

$$m\ddot{\mathbf{r}} = \mathbf{F} \tag{1.2.1}$$

where, here and elsewhere, the dot notation indicates differentiation with respect to the time t.

Let $O'x'y'z'$ be a second reference system, in which the x'-, y'-, and z'-axes always remain parallel to the x-, y-, and z-axes respectively. The origin of the second reference system is assumed to move relative to the first at a nonrelativistic velocity. The position vector $\mathbf{r}'$ in this second system is related to $\mathbf{r}$ and the vector $\mathbf{a}$, which gives the position of O' with respect to O, through the nonrelativistic transformation

$$\mathbf{r} = \mathbf{a} + \mathbf{r}' \tag{1.2.2}$$

In terms of acceleration (1.2.2) may be written

$$m\ddot{\mathbf{r}}' = \mathbf{F} - m\ddot{\mathbf{a}} \tag{1.2.3}$$

The additional term, $m\ddot{\mathbf{a}}$, in (1.2.3) is zero if the point O' is either at rest with respect to O or moves with constant velocity with respect to O, and the form of Newton's law will be then the same in the two reference systems.

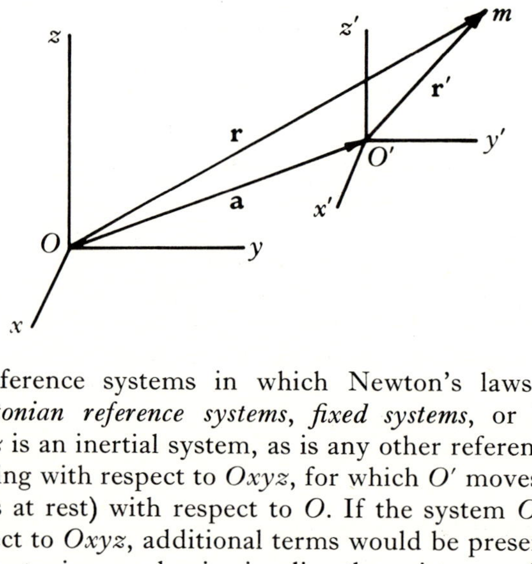

FIGURE 1.2.1

Reference systems in which Newton's laws hold are said to be *Newtonian reference systems*, *fixed systems*, or *inertial systems*. Thus $Oxyz$ is an inertial system, as is any other reference system $O'x'y'z'$, not rotating with respect to $Oxyz$, for which O' moves with constant velocity (or is at rest) with respect to O. If the system $O'x'y'z'$ is rotating with respect to $Oxyz$, additional terms would be present in (1.2.3).

Newtonian mechanics implies the existence of a preferred system in which $(d/dt)(m\dot{\mathbf{r}}) = \mathbf{F}$ provides an exact description of the motion of a particle. Newton assumed that there exists an absolute or immovable space and his laws hold only for reference frames that are at rest or fixed in absolute space or in uniform rectilinear motion with respect to it. From the point of view of relativity, such an assumption is meaningless;

nonetheless, a system of axes at rest with respect to the Earth approximates such an absolute situation for most practical purposes. A reference system fixed with respect to the Sun provides an even better approximation; or a still better reference system would be one that is at rest with respect to the average positions of the stars.

1.3. LINEAR MOMENTUM

The linear momentum $\mathbf{p}$ of a particle of mass m, with respect to a fixed or inertial origin, which we hereinafter designate by the symbol O, is defined by the equation

$$\mathbf{p} = m\mathbf{v} = m\dot{\mathbf{r}} \tag{1.3.1}$$

where the velocity $\mathbf{v} \equiv \dot{\mathbf{r}}$ is with respect to the fixed frame of reference.† Neither m nor $\mathbf{v}$ need be constant. In (1.3.1) $\mathbf{r}$ is the position vector of the particle referred to O (see Figure 1.3.1). We say nothing at this point about the type of coordinates (that is, spherical, cylindrical, Cartesian, etc.) that we will use in applications of the theory.

FIGURE 1.3.1

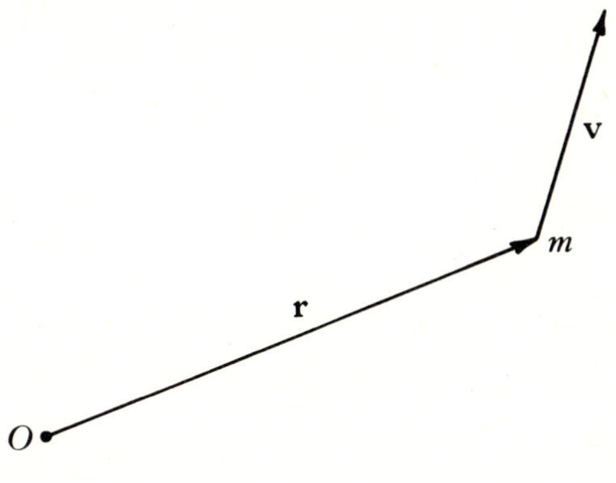

The total linear momentum $\mathbf{P}$, with respect to the origin O, of a system of n particles is by definition

$$\mathbf{P} = \sum_{i=1}^{n} \mathbf{p}_i = \sum_{i=1}^{n} m_i\mathbf{v}_i = \sum_{i=1}^{n} m_i\dot{\mathbf{r}}_i \tag{1.3.2}$$

where m_i designates the mass of the ith particle and $\mathbf{r}_i$ is the position vector of the ith particle referred to O (see Figure 1.3.2).

† For brevity, when we speak about a quantity (e.g., $\mathbf{p}$) as being referenced to a certain type of origin, an appropriate reference system at the origin is inferred.

Let $\mathbf{r}_i'$ be the position vector of the ith particle with respect to an origin O', fixed with respect to the origin O, and let $\mathbf{a}$ be the constant vector drawn from O to O'. Then $\mathbf{p}_i' = m_i\dot{\mathbf{r}}_i'$ (called the linear momentum of m_i with respect to O') is equal to $\mathbf{p}_i$, for

$$\mathbf{r}_i = \mathbf{a} + \mathbf{r}_i' \tag{1.3.3}$$

and

$$\dot{\mathbf{r}}_i = \dot{\mathbf{r}}_i'$$

so that

$$\mathbf{P} = \sum_{i=1}^{n} \mathbf{p}_i = \sum_{i=1}^{n} m_i\dot{\mathbf{r}}_i = \sum_{i=1}^{n} m_i\dot{\mathbf{r}}_i' = \mathbf{P}' \tag{1.3.4}$$

where $\mathbf{P}' \equiv \sum_{i=1}^{n} m_i\dot{\mathbf{r}}_i'$ is called the total linear momentum of the system of n particles with respect to O'. Thus we have:

Theorem 1.3.1. The total linear momentum of a system of n particles with respect to any point O', fixed relative to the origin O, is equal to the total linear momentum with respect to the origin O.

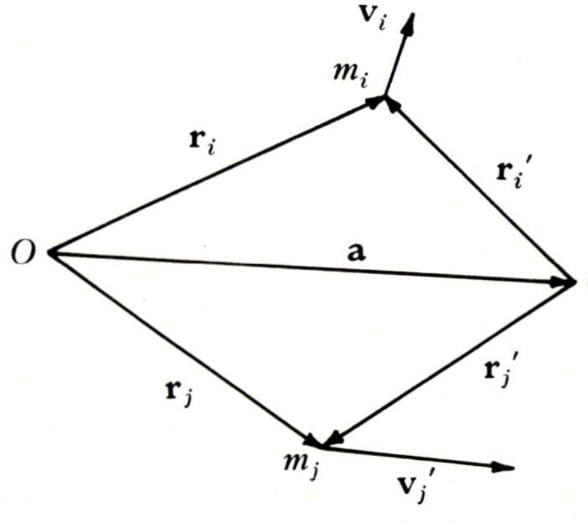

FIGURE 1.3.2

Now permit the point O' (see Figure 1.3.2) to move, in any manner, with respect to the origin O. Then, from (1.3.3) we have

$$\dot{\mathbf{r}}_i = \dot{\mathbf{a}} + \dot{\mathbf{r}}_i'$$

Replacing $\dot{\mathbf{r}}_i$ in (1.3.2), we obtain

$$\mathbf{P} = \sum_{i=1}^{n} m_i\dot{\mathbf{r}}_i = \sum_{i=1}^{n} m_i(\dot{\mathbf{a}} + \dot{\mathbf{r}}_i') = M\dot{\mathbf{a}} + \sum_{i=1}^{n} m_i\dot{\mathbf{r}}_i'$$

$$= M\dot{\mathbf{a}} + \mathbf{P}' \qquad \left(\mathbf{P}' \equiv \sum_{i=1}^{n} m_i\dot{\mathbf{r}}_i'\right) \tag{1.3.5}$$

where $M \equiv \sum_{i=1}^{n} m_i$ is the total mass of the system of n particles, and $\mathbf{P}'$ is called the total linear momentum of the system of particles with respect to the moving point O'. The quantity $M\dot{\mathbf{a}}$ is called *the linear momentum of transport at* O'.

Theorem 1.3.2. The total linear momentum of a system of n particles with respect to the origin O is equal to the sum of the total linear momentum with respect to the moving point O' and the linear momentum of transport at O'.

The *center of mass* of a system of n particles with respect to the origin O is defined as the terminus of the vector

$$\mathbf{R} = \sum_{i=1}^{n} m_i \mathbf{r}_i \Big/ \sum_{i=1}^{n} m_i \tag{1.3.6}$$

where $\mathbf{R}$, the position vector of the center of mass, is with respect to the origin O. Equation (1.3.6) is customarily written

$$M\mathbf{R} = \sum_{i=1}^{n} m_i \mathbf{r}_i \tag{1.3.7}$$

Note that neither the individual masses nor the total mass $M = \sum_{i=1}^{n} m_i$ need be constant.

Introduce now an origin O' at a position $\mathbf{a}$ with respect to the origin O, moving in any manner relative to O. Define a vector $\mathbf{R}'$ by the equation (see Figure 1.3.3)

$$\mathbf{R} = \mathbf{a} + \mathbf{R}'$$

where $\mathbf{R}'$ is referenced to the origin O'.

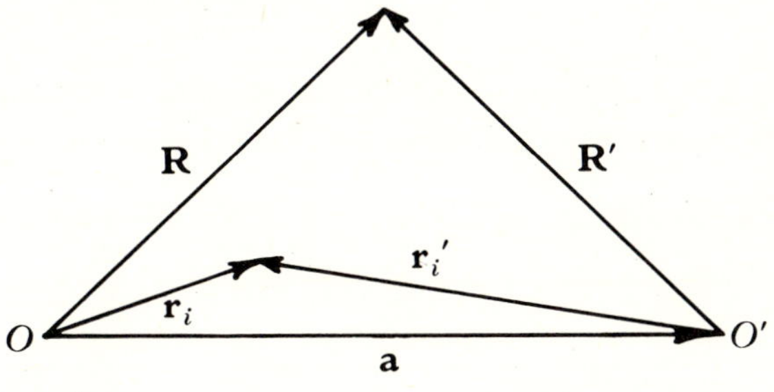

FIGURE 1.3.3

As before, $\mathbf{r}_i = \mathbf{a} + \mathbf{r}_i'$ so that

$$M\mathbf{R} = M(\mathbf{a} + \mathbf{R}') = M\mathbf{a} + M\mathbf{R}' = \sum_{i=1}^{n} m_i \mathbf{r}_i = \sum_{i=1}^{n} m_i(\mathbf{a} + \mathbf{r}_i')$$

or

$$M\mathbf{a} + M\mathbf{R}' = M\mathbf{a} + \sum_{i=1}^{n} m_i \mathbf{r}_i'$$

and so we write

$$M\mathbf{R}' = \sum_{i=1}^{n} m_i \mathbf{r}_i' \tag{1.3.8}$$

Here $\mathbf{R}'$ is called the position vector of the center of mass of the system of n particles with respect to the moving point O'.

Theorem 1.3.3. The form of the expression for the center of mass of a system of n particles is unchanged upon introduction of a new origin moving in any manner relative to the origin O.

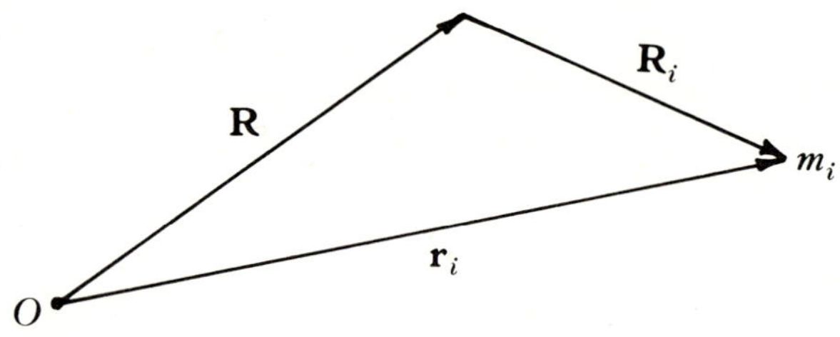

FIGURE 1.3.4

Let $\mathbf{R}_i$ denote the position vector of the ith mass with respect to the center of mass of the system of n particles. We can (see Figure 1.3.4) multiply each side of the equation

$$\mathbf{r}_i = \mathbf{R} + \mathbf{R}_i$$

by m_i and sum over the n particles to obtain

$$\sum_{i=1}^{n} m_i \mathbf{r}_i = \sum_{i=1}^{n} m_i \mathbf{R} + \sum_{i=1}^{n} m_i \mathbf{R}_i$$

Since

$$\sum_{i=1}^{n} m_i \mathbf{R} = M\mathbf{R} = \sum_{i=1}^{n} m_i \mathbf{r}_i$$

it follows that both

$$\sum_{i=1}^{n} m_i \mathbf{R}_i = \mathbf{0} \qquad \text{and} \qquad \frac{d}{dt}\left(\sum_{i=1}^{n} m_i \mathbf{R}_i\right) = \mathbf{0}$$

Designate the total linear momentum of the system of n particles with respect to their center of mass by $\mathbf{P}_c$. We define $\mathbf{P}_c$ by the equation

$$\mathbf{P}_c = \sum_{i=1}^{n} m_i \dot{\mathbf{R}}_i$$

Since

$$\frac{d}{dt}\left(\sum_{i=1}^{n} m_i \mathbf{R}_i\right) = \sum m_i \dot{\mathbf{R}}_i + \sum_{i=1}^{n} \dot{m}_i \mathbf{R}_i = \mathbf{0}$$

we note that

$$\mathbf{P}_c = \sum_{i=1}^{n} m_i \dot{\mathbf{R}}_i = -\sum_{i=1}^{n} \dot{m}_i \mathbf{R}_i \tag{1.3.9}$$

In particular, if the masses are constant, the total linear momentum of a system of n particles with respect to their center of mass is the zero vector; that is,

$$\mathbf{P}_c = \mathbf{0}$$

On the other hand, we can write

$$\mathbf{P} = \sum_{i=1}^{n} \mathbf{p}_i = \sum_{i=1}^{n} m_i \dot{\mathbf{r}}_i = \sum_{i=1}^{n} m_i \dot{\mathbf{R}} + \sum_{i=1}^{n} m_i \dot{\mathbf{R}}_i$$

or

$$\mathbf{P} = M\dot{\mathbf{R}} + \mathbf{P}_c \tag{1.3.10}$$

Theorem 1.3.4. The total linear momentum of a system of n particles with respect to O is equal to the linear momentum of transport at the center of mass of the n particles, plus the total linear momentum with respect to their center of mass. In particular, if the masses are constant, $\mathbf{P}_c = \mathbf{0}$ and the total linear momentum of the system with respect to O is equal to the linear momentum of transport at the center of mass.

We accept as a postulate Newton's second law of motion, which relates the time rate of change of momentum to the resultant force producing this change. Thus the $3n$, second-order differential equations governing the motions of a system of n particles can be written

$$\dot{\mathbf{p}}_i = \sum_{j=1}^{n} \mathbf{f}_{ij} + \mathbf{F}_i^E \qquad (\mathbf{f}_{ii} = \mathbf{0}) \quad (i = 1, 2, \ldots, n) \tag{1.3.11}$$

where we assume that only forces of types 1 and 3 are present in the system and that $\mathbf{f}_{ij}$ is the force on the ith particle by particle j and $\mathbf{F}_i^E$ is the resultant external force on the ith particle.

The complete solution of Equation (1.3.11) requires the determination of $6n$ integrals involving $6n$ arbitrary constants. The word integral as used here means a relationship between the coordinates and/or velocities and the time, which does not involve the accelerations. Again, it may be well to note that in (1.3.11) we have a symbolic representation of the equations of motion, without reference to the particular type of coordinates in which a specific problem will eventually be cast.

If we sum the equations in (1.3.11) over n, we have

$$\dot{\mathbf{P}} = \sum_{i=1}^{n} \dot{\mathbf{p}}_i = \sum_{i=1}^{n} \sum_{j=1}^{n} \mathbf{f}_{ij} + \sum_{i=1}^{n} \mathbf{F}_i^E \qquad (\mathbf{f}_{ii} = \mathbf{0})$$

The double summation above vanishes, for the forces of type 1 cancel in pairs; that is, $\mathbf{f}_{ij} = -\mathbf{f}_{ji}$. Thus

$$\dot{\mathbf{P}} = \sum_{i=1}^{n} \mathbf{F}_i^E \tag{1.3.12}$$

and we see that $\dot{\mathbf{P}}$ depends only on the external forces. If now, as is often the case in celestial mechanics, we can neglect the time variations of the n masses, we have from (1.3.10)

$$\dot{\mathbf{P}} = M\ddot{\mathbf{R}} = \sum_{i=1}^{n} \mathbf{F}_i^E \tag{1.3.13}$$

When the internal forces are of type 1 and the masses are considered to be constant, we see that the center of mass of a system of n particles moves under the influence of the total external force alone. The motion is as if the entire mass were concentrated at the center of mass. Note, however, that the total external force is, in general, the sum of the actual forces experienced by the individual particles in their respective actual positions. If we now add the additional restriction that the total external force vanishes—as we can meaningfully, for example, if the n particles comprise our solar system—we have

$$\ddot{\mathbf{R}} = \mathbf{0}, \qquad \dot{\mathbf{R}} = \mathbf{C}_1, \qquad \mathbf{R} = \mathbf{C}_1 t + \mathbf{C}_0 \tag{1.3.14}$$

where t is the time and $\mathbf{C}_1$ and $\mathbf{C}_0$ are constant vectors. Equation (1.3.14) says that the center of mass of the system of n particles moves on a straight line with constant speed. This can be seen by reference to Figure (1.3.5), and the facts that $\mathbf{R}(0) = \mathbf{C}_0$ and $\mathbf{R}(1) = \mathbf{C}_1 + \mathbf{C}_0$.

Thus the parametric representation of the line through the termini of $\mathbf{C}_0$ and $\mathbf{C}_1 + \mathbf{C}_0$ is given by (1.3.14).

Whenever Equations (1.3.13) can be completely integrated, we obtain six integrals involving six arbitrary constants. Thus six of the required $6n$

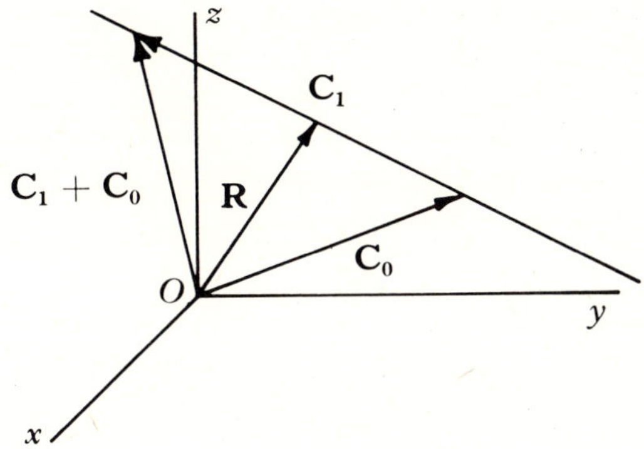

FIGURE 1.3.5

integrals have been obtained. Note also from (1.3.13) that when the total external force vanishes, $\dot{\mathbf{P}} = \mathbf{0}$ and $\mathbf{P}$ is a constant of the motion, say $\mathbf{P}_0$.

1.4. ANGULAR MOMENTUM

The angular momentum (moment of momentum) $\boldsymbol{l}$ of a particle of mass m, with respect to the origin O, is defined by the equation

$$\boldsymbol{l} = \mathbf{r} \times \mathbf{p} \tag{1.4.1}$$

where the linear momentum $\mathbf{p}$ and the position $\mathbf{r}$ are with respect to O (see Figure 1.4.1). We can write

$$\boldsymbol{l} = \mathbf{r} \times \mathbf{p} = l\mathbf{u}_A$$

where

$$l \equiv mrv \sin\theta \qquad (v = |\mathbf{v}|)$$

$\mathbf{u}_A \equiv$ a unit vector perpendicular to the plane determined by $\mathbf{r}$ and $\mathbf{p}$ such that $\mathbf{r}$, $\mathbf{p}$, and $\mathbf{u}_A$ form a right-handed system.

The angular momentum per unit mass $\mathbf{h}$ is given by the equation

$$\mathbf{h} = (1/m)\,\boldsymbol{l} = h\mathbf{u}_A \qquad (h \equiv l/m = |\mathbf{h}|) \tag{1.4.2}$$

The total angular momentum **L**, with respect to the origin O, of a system of n particles is by definition

$$\mathbf{L} = \sum_{i=1}^{n} \boldsymbol{l}_i = \sum_{i=1}^{n} \mathbf{r}_i \times \mathbf{p}_i \tag{1.4.3}$$

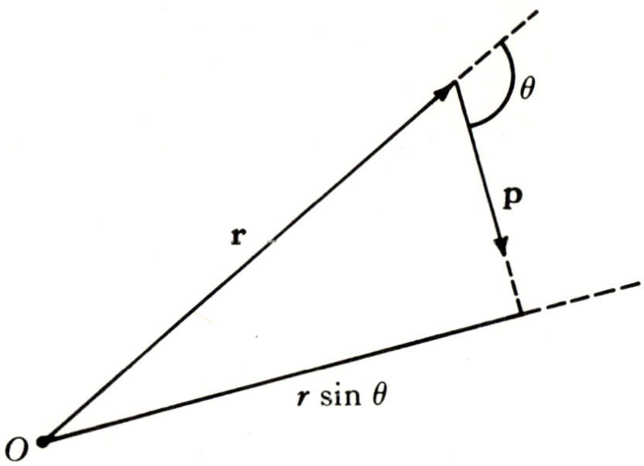

FIGURE 1.4.1

Let O' be a point fixed with respect to O. From (1.3.3) we have

$$\mathbf{r}_i = \mathbf{a} + \mathbf{r}_i', \qquad \dot{\mathbf{r}}_i = \dot{\mathbf{r}}_i', \qquad \mathbf{p}_i = \mathbf{p}_i' \tag{1.4.4}$$

so that

$$\begin{aligned}\mathbf{L} &= \sum_{i=1}^{n} \boldsymbol{l}_i = \sum_{i=1}^{n} \mathbf{r}_i \times \mathbf{p}_i = \sum_{i=1}^{n} (\mathbf{a} + \mathbf{r}_i') \times \mathbf{p}_i \\ &= \sum_{i=1}^{n} \mathbf{a} \times \mathbf{p}_i + \sum_{i=1}^{n} \mathbf{r}_i' \times \mathbf{p}_i = \mathbf{a} \times \mathbf{P} + \sum_{i=1}^{n} \mathbf{r}_i' \times \mathbf{p}_i' \\ &= \mathbf{a} \times \mathbf{P} + \mathbf{L}' \end{aligned} \tag{1.4.5}$$

where $\mathbf{L}' \equiv \sum_{i=1}^{n} \mathbf{r}_i' \times \mathbf{p}_i'$ is called the total angular momentum of the system of n particles with respect to O'.

Theorem 1.4.1. The total angular momentum with respect to O is equal to the sum of the total angular momentum with respect to O' plus the moment about O of the total linear momentum considered to reside at the new origin O'.

The result just stated holds for variable masses. We note that whereas the total linear momentum does not depend on the choice of the point O'

with respect to which it is calculated, the total angular momentum does depend on the choice of O'.

From (1.3.9) and (1.3.10) we have

$$\mathbf{P} = M\dot{\mathbf{R}} - \sum_{i=1}^{n} \dot{m}_i \mathbf{R}_i \tag{1.4.6}$$

Substitute (1.4.6) into (1.4.5) to obtain

$$\mathbf{L} = \mathbf{a} \times M\dot{\mathbf{R}} - \mathbf{a} \times \sum_{i=1}^{n} \dot{m}_i \mathbf{R}_i + \mathbf{L}' \tag{1.4.7}$$

and, for constant masses,

$$\mathbf{L} = \mathbf{a} \times M\dot{\mathbf{R}} + \mathbf{L}' \tag{1.4.8}$$

Theorem 1.4.2. If the masses remain constant, the total angular momentum with respect to O is equal to the sum of the angular momentum about O of a particle, whose mass is the entire mass of the system, situated at O' and moving with the velocity with respect to O of the center of mass with respect to O, and the total angular momentum of the system of particles about O'.

Now permit the point O' to move, in any manner, with respect to O. We have

$$\begin{aligned} \boldsymbol{l}_i &= \mathbf{r}_i \times \mathbf{p}_i = (\mathbf{a} + \mathbf{r}_i') \times \mathbf{p}_i = \mathbf{a} \times \mathbf{p}_i + \mathbf{r}_i' \times \mathbf{p}_i \\ &= \mathbf{a} \times \mathbf{p}_i + \mathbf{r}_i' \times m_i \dot{\mathbf{a}} + \mathbf{r}_i' \times m_i \dot{\mathbf{r}}_i' \end{aligned}$$

and

$$\mathbf{L} = \sum_{i=1}^{n} \boldsymbol{l}_i = \mathbf{a} \times \mathbf{P} + M\mathbf{R}' \times \dot{\mathbf{a}} + \mathbf{L}' \tag{1.4.9}$$

Substitute for **P** from (1.4.6) to obtain

$$\mathbf{L} = \mathbf{L}' + \mathbf{a} \times M\dot{\mathbf{R}} - \mathbf{a} \times \sum_{i=1}^{n} \dot{m}_i \mathbf{R}_i + M\mathbf{R}' \times \dot{\mathbf{a}} \tag{1.4.10}$$

Referring to (1.4.5), we see that the difference between **L** and the total moment of momentum **L**$'$ with respect to a fixed point O' is simply the moment of the total linear momentum $\mathbf{a} \times \mathbf{P}$. If $\mathbf{a} \times \mathbf{P}$ is constant, **L** and **L**$'$ differ by a constant vector. When, however, we permit the point O' to move in an arbitrary manner, we note from (1.4.9) that, generally, the difference $\mathbf{L} - \mathbf{L}'$ is not constant even if $\mathbf{a} \times \mathbf{P}$ is constant. If, however, the moving point O' is chosen to be the center of mass of the system ($\mathbf{a} = \mathbf{R}, \mathbf{R}' = \mathbf{0}$), the difference $\mathbf{L} - \mathbf{L}_c$ becomes constant when $\mathbf{R} \times \mathbf{P}$ is constant. For this choice,

$$\mathbf{L} = \mathbf{L}_c + \mathbf{R} \times \mathbf{P} \qquad (\mathbf{L}_c = \mathbf{L}') \tag{[1.4.9]}$$

a result that holds for variable masses. For constant masses, $\mathbf{P}_c = \mathbf{0}$ and

$$\mathbf{L} = \mathbf{L}_c + \mathbf{R} \times M\dot{\mathbf{R}} \tag{[1.4.10]}$$

Theorem 1.4.3. For constant masses, the total angular momentum about a fixed point O is equal to the sum of the angular momentum about O of a particle, whose mass is the entire mass of the system of n particles, situated at and moving with the center of mass, and the total angular momentum of the system of n particles about the center of mass.

The difference $\mathbf{L} - \mathbf{L}_c$ in [1.4.10] is constant if the center of mass moves with constant velocity, for then

$$\frac{d}{dt}(\mathbf{R} \times M\dot{\mathbf{R}}) = \mathbf{R} \times M\ddot{\mathbf{R}} + \dot{\mathbf{R}} \times M\dot{\mathbf{R}} = \mathbf{0}$$

and

$$\mathbf{R} \times M\dot{\mathbf{R}} = \mathbf{K} \qquad (\mathbf{K} = \text{constant vector})$$

Thus if the moving point O' is the center of mass of the system of n particles of constant mass and if the center of mass moves with constant velocity,

$$\mathbf{L} - \mathbf{L}_c = \mathbf{K} \tag{1.4.11}$$

The moment $\mathbf{n}$ of a force $\mathbf{F}$ (the torque) about the point O is by definition

$$\mathbf{n} = \mathbf{r} \times \mathbf{F} \tag{1.4.12}$$

where $\mathbf{r}$ is the position vector from O to the point of application of the force. The total torque $\mathbf{N}$ about O, resulting from forces $\mathbf{F}_i$ applied to a system of n particles, is by definition

$$\mathbf{N} = \sum_{i=1}^{n} \mathbf{n}_i = \sum_{i=1}^{n} \mathbf{r}_i \times \mathbf{F}_i \tag{1.4.13}$$

The cross product of (1.3.11) with $\mathbf{r}_i$ is

$$\mathbf{r}_i \times \dot{\mathbf{p}}_i = \sum_{j=1}^{n} \mathbf{r}_i \times \mathbf{f}_{ij} + \mathbf{r}_i \times \mathbf{F}_i^E$$

Since $\dot{\mathbf{r}}_i \times \mathbf{p}_i = \mathbf{0}$, then $\mathbf{r}_i \times \dot{\mathbf{p}}_i = \dot{\boldsymbol{l}}_i$; hence summing over i, we can write

$$\dot{\mathbf{L}} = \sum_{i=1}^{n} \dot{\boldsymbol{l}}_i = \mathbf{S} + \sum_{i=1}^{n} \mathbf{r}_i \times \mathbf{F}_i^E$$

where

$$\mathbf{S} = \sum_{i=1}^{n} \sum_{j=1}^{n} \mathbf{r}_i \times \mathbf{f}_{ij}$$

If the summation indices are interchanged, we can write

$$\mathbf{S} = \sum_{j=1}^{n} \sum_{i=1}^{n} \mathbf{r}_j \times \mathbf{f}_{ji} = \sum_{i=1}^{n} \sum_{j=1}^{n} \mathbf{r}_j \times \mathbf{f}_{ji}$$

and so

$$\mathbf{S} = \sum_{i=1}^{n} \sum_{j=1}^{n} \mathbf{r}_i \times \mathbf{f}_{ij} = \frac{1}{2} \sum_{i=1}^{n} \sum_{j=1}^{n} (\mathbf{r}_i \times \mathbf{f}_{ij} + \mathbf{r}_j \times \mathbf{f}_{ji})$$

If the internal forces are restricted to be of type 1 so that $\mathbf{f}_{ij} = -\mathbf{f}_{ji}$,

$$\mathbf{r}_i \times \mathbf{f}_{ij} + \mathbf{r}_j \times \mathbf{f}_{ji} = (\mathbf{r}_i - \mathbf{r}_j) \times \mathbf{f}_{ij} = \mathbf{0}$$

since $\mathbf{r}_i - \mathbf{r}_j$ is in the direction of $\mathbf{f}_{ij}$. Thus

$$\mathbf{S} = \mathbf{0} \tag{1.4.14}$$

and

$$\dot{\mathbf{L}} = \sum_{i=1}^{n} \dot{\boldsymbol{l}}_i = \sum_{i=1}^{n} \mathbf{r}_i \times \mathbf{F}_i^E$$

or

$$\dot{\mathbf{L}} = \mathbf{N}^E \tag{1.4.15}$$

where

$$\mathbf{N}^E = \sum_{i=1}^{n} \mathbf{r}_i \times \mathbf{F}_i^E$$

is the total torque of all the external forces about O.

Theorem 1.4.4. The time rate of change of the total angular momentum about the origin O depends only on the external forces and is equal to the total torque of these forces about O.

Equation (1.4.15) holds for variable masses. If, in particular, the masses are constant, we have from Equations (1.3.13) and [1.4.10]

$$M\ddot{\mathbf{R}} = \sum_{i=1}^{n} \mathbf{F}_i^E$$

$$\dot{\mathbf{L}} = \dot{\mathbf{L}}_c + \mathbf{R} \times M\ddot{\mathbf{R}}$$

If the first of these two expressions is substituted into the second, it follows that

$$\dot{\mathbf{L}} = \dot{\mathbf{L}}_c + \sum_{i=1}^{n} \mathbf{R} \times \mathbf{F}_i^E$$

Noting that $\mathbf{R} = \mathbf{r}_i - \mathbf{R}_i$, we have

$$\dot{\mathbf{L}} = \dot{\mathbf{L}}_c + \sum_{i=1}^{n} \mathbf{r}_i \times \mathbf{F}_i^E - \sum_{i=1}^{n} \mathbf{R}_i \times \mathbf{F}_i^E$$

$$= \dot{\mathbf{L}}_c + \mathbf{N}^E - \mathbf{N}_c^E \qquad \left(\mathbf{N}_c^E \equiv \sum_{i=1}^{n} \mathbf{R}_i \times \mathbf{F}_i^E\right)$$

But $\dot{\mathbf{L}} = \mathbf{N}^E$ from (1.4.15), so

$$\dot{\mathbf{L}}_c = \mathbf{N}_c^E \tag{1.4.16}$$

Theorem 1.4.5. For constant masses, the time rate of change of angular momentum for motion with respect to the center of mass is equal to the total torque $\mathbf{N}_c^E$ about the center of mass. Furthermore, this is true whether the center of mass is at rest or in translational motion, relative to a fixed primary coordinate system.

From (1.4.15) we observe that if the external forces are zero,

$$\dot{\mathbf{L}} = \mathbf{0}, \qquad \mathbf{L} = \text{constant} \tag{1.4.17}$$

and the total angular momentum is a constant of the motion. Note that this is true even if the masses are not constant. Note further that when any component of the external torque is zero, the corresponding component of the total angular momentum is constant even though the other components may not be constant. This, of course, follows directly from (1.4.15). For systems such as the solar system for which Equation (1.4.17) holds, three additional integrals containing three independent arbitrary scalar constants are obtained. In this sense, we say that a total of nine out of the $6n$ integrals we seek have been found.

From (1.3.12) and (1.4.15) it is clear that when the internal forces are of type 1 and the external forces are zero, both **P** and **L**, referred to the fixed origin O, are constants of the motion. These constants can be labeled $\mathbf{P}_0$ and $\mathbf{L}_0$. Consider the situation for some other origin O', either fixed or moving with respect to the origin O. Rearranging (1.3.5) and (1.4.9), we write

$$\mathbf{P}' = \mathbf{P}_0 - M\dot{\mathbf{a}}$$

$$\mathbf{L}' = \mathbf{L}_0 - \mathbf{a} \times \mathbf{P}_0 - M\mathbf{R}' \times \dot{\mathbf{a}}$$

If O' is free to move in any manner, it is clear that, in general, neither $\mathbf{P}'$ nor $\mathbf{L}'$ is constant. If, however, O' is fixed with respect to O, $\dot{\mathbf{a}} = \mathbf{0}$ and

$$\mathbf{P}' = \mathbf{P}_0 = \mathbf{P}_0', \qquad \mathbf{L}' = \mathbf{L}_0 - \mathbf{a} \times \mathbf{P}_0 = \mathbf{L}_0'$$

Thus, if the external forces are zero, the total linear momentum and total angular momentum, both with respect to O', are constants of the motion if O' is fixed.

If O' is taken at the center of mass, $\mathbf{R}' = \mathbf{0}$, $\dot{\mathbf{a}} = \dot{\mathbf{R}}$, and

$$\mathbf{P}_c = \mathbf{P}_0 - M\dot{\mathbf{R}}$$

$$\begin{aligned}\mathbf{L}_c &= \mathbf{L}_0 - \mathbf{R} \times \mathbf{P}_0 \\ &= \mathbf{L}_0 - \mathbf{R} \times \mathbf{P}_c + \mathbf{R} \times M\dot{\mathbf{R}}\end{aligned}$$

Thus, even if the external forces are zero, it is clear that $\mathbf{P}_c$ and $\mathbf{L}_c$ are not constants of the motion, as a rule. In addition, however, if the masses are constant,

$$\mathbf{P}_c = \mathbf{0}$$

and

$$\frac{d}{dt}(\mathbf{R} \times M\dot{\mathbf{R}}) = \mathbf{R} \times M\ddot{\mathbf{R}} = \mathbf{R} \times \sum_{i=1}^{n} \mathbf{F}_i^E = \mathbf{0}$$

so that $\mathbf{R} \times M\dot{\mathbf{R}}$ is a constant vector. Thus, for this case, both $\mathbf{P}_c$ and $\mathbf{L}_c$ are constants of the motion, say $\mathbf{P}_{c0}$ and $\mathbf{L}_{c0}$. The constant vector $\mathbf{L}_{c0}$ determines a plane through the center of mass, which remains forever parallel to itself. In the case of the solar system, it is called "the invariable plane of the solar system." Because the orbital angular momentum of Jupiter and Saturn comprise nearly 87 percent of the total angular momentum of the solar system, and because the orbital planes of these two major planets nearly coincide with the plane of the Earth's motion about the Sun (i.e., the ecliptic plane), the vector $\mathbf{L}_{c0}$ is directed only about

1°35′ from the pole of the ecliptic. The line along which $\mathbf{L}_{c0}$ is directed is called the invariable line. The inclinations of the orbits of Jupiter and Saturn to the ecliptic are given by the formulas

$$1°18'31.4'' - 20.5''T \quad \text{(Jupiter)}, \qquad 2°29'33.1'' - 14.0''T \quad \text{(Saturn)}$$

where T is measured in tropical centuries.† The invariable plane lies between these two orbital planes.

1.5. KINETIC AND POTENTIAL ENERGY

The kinetic energy T_i of a particle of mass m_i, with respect to O, is defined by the equation

$$T_i = \frac{1}{2m_i} \mathbf{p}_i \cdot \mathbf{p}_i \tag{1.5.1}$$

where $\mathbf{p}_i$ is with respect to O.

The total kinetic energy T, with respect to O, of a system of n particles is by definition

$$T = \sum_{i=1}^{n} T_i = \sum_{i=1}^{n} \frac{1}{2m_i} \mathbf{p}_i \cdot \mathbf{p}_i \tag{1.5.2}$$

Let O' be a point fixed with respect to O. We write

$$T = \sum_{i=1}^{n} T_i = \sum_{i=1}^{n} \frac{1}{2m_i} \mathbf{p}_i \cdot \mathbf{p}_i = \sum_{i=1}^{n} \frac{1}{2m_i} \mathbf{p}_i' \cdot \mathbf{p}_i' = T' \tag{1.5.3}$$

where

$$T' = \sum_{i=1}^{n} \frac{1}{2m_i} \mathbf{p}_i' \cdot \mathbf{p}_i'$$

is called the total kinetic energy of the system of n particles with respect to the fixed point O'.

Theorem 1.5.1. The total kinetic energy with respect to any point O' fixed relative to O is invariant in form.

Now permit the point O' to move, in any manner, with respect to O. We have from (1.3.3)

$$\mathbf{p}_i = m_i\dot{\mathbf{r}}_i = m_i(\dot{\mathbf{a}} + \dot{\mathbf{r}}_i')$$

† See pp. 103 and 108.

Thus

$$2T = \sum_{i=1}^{n} \frac{1}{m_i} \mathbf{p}_i \cdot \mathbf{p}_i = \sum_{i=1}^{n} m_i(\dot{\mathbf{a}} + \dot{\mathbf{r}}_i')^2$$

$$= \sum_{i=1}^{n} m_i \dot{\mathbf{a}} \cdot \dot{\mathbf{a}} + 2 \sum_{i=1}^{n} m_i \dot{\mathbf{a}} \cdot \dot{\mathbf{r}}_i' + \sum_{i=1}^{n} \frac{1}{m_i} \mathbf{p}_i' \cdot \mathbf{p}_i'$$

and we can write

$$T = \tfrac{1}{2} M \dot{\mathbf{a}}^2 + \dot{\mathbf{a}} \cdot \mathbf{P}' + T' \tag{1.5.4}$$

In particular, if O' is chosen to be the center of mass and the masses are constant, then $\mathbf{P}' = \mathbf{P}_c = \mathbf{0}$, and we have

$$T = \tfrac{1}{2} M \dot{\mathbf{R}}^2 + T_c, \qquad T_c = T' \tag{1.5.5}$$

Theorem 1.5.2. For constant masses, the total kinetic energy of a system of particles may be written as the sum of the kinetic energy of translation of the center of mass, plus the kinetic energy of the motion with respect to the center of mass.

If a particle experiences a force at each point in a region in space, there is said to be a force field $\mathbf{F}$ in the region. Suppose that the particle is at some reference point $\mathbf{r}(A)$ in a region in which a force field $\mathbf{F}$ exists and it is then transported along some continuous path Γ to some other point $\mathbf{r}$. We define a quantity W, which we call the total work done on the particle *by* the force field $\mathbf{F}$ when the particle passes from $\mathbf{r}(A)$ to $\mathbf{r}$ along the chosen path, by the line integral

$$W = \int_{\mathbf{r}(A)}^{\mathbf{r}} \mathbf{F} \cdot d\mathbf{r} \tag{1.5.6}$$

where the path of integration is understood to be Γ and the vector limits on the integral are to be interpreted as determining the two endpoints of this path. In general, the line integral in (1.5.6) depends on the path Γ. Of special interest to us, will be the class of force fields in which the value of the line integral between any two points in the field is independent of the particular path chosen. Such as force field is called a conservative force field and, for the present, we restrict our considerations to such force fields.

We also define a quantity V, called the total work done on the particle *against* the force field $\mathbf{F}$ when the particle is transported from $\mathbf{r}(A)$ to $\mathbf{r}$, by the equation

$$V = -W = -\int_{\mathbf{r}(A)}^{\mathbf{r}} \mathbf{F} \cdot d\mathbf{r} \tag{1.5.7}$$

We will now invoke, but not prove, a theorem from vector analysis which states that the line integral in (1.5.6) is independent of the path Γ connecting $\mathbf{r}(A)$ and $\mathbf{r}$ if and only if there exists a scalar function V^* of the coordinates such that $\mathbf{F} = -\nabla V^*$, where ∇ is the gradient operator. If we designate by (u_1, u_2, u_3) a set of orthogonal curvilinear coordinates and the corresponding set of mutually orthogonal unit vectors tangent to the coordinate curves by $\mathbf{u}_1$, $\mathbf{u}_2$, $\mathbf{u}_3$, the quantities referred to in the theorem just cited have the form

$$\mathbf{F} = \sum_{i=1}^{3} F_i(u_1, u_2, u_3)\,\mathbf{u}_i \tag{1.5.8}$$

$$V^* = V^*(u_1, u_2, u_3) \tag{1.5.9}$$

$$\nabla = \sum_{i=1}^{3} \frac{1}{h_i}\frac{\partial}{\partial u_i}\,\mathbf{u}_i$$

where F_1, F_2, and F_3 are the components of the force $\mathbf{F}$ along the coordinate axes. The unit vectors $\mathbf{u}_1$, $\mathbf{u}_2$, and $\mathbf{u}_3$ can be expressed in terms of the unit Cartesian basis vectors $\mathbf{i}$, $\mathbf{j}$, and $\mathbf{k}$ through the transformation equations

$$\mathbf{u}_i = \frac{1}{h_i}\frac{\partial \mathbf{r}}{\partial u_i}, \qquad h_i = \left|\frac{\partial \mathbf{r}}{\partial u_i}\right| \qquad (i = 1, 2, 3) \tag{1.5.10}$$

where

$$\mathbf{r} = x(u_1, u_2, u_3)\,\mathbf{i} + y(u_1, u_2, u_3)\,\mathbf{j} + z(u_1, u_2, u_3)\,\mathbf{k}$$

Operating on $V^*(u_1, u_2, u_3)$, we have

$$\mathbf{F} = \sum_{i=1}^{3} F_i\mathbf{u}_i = -\nabla V^* = -\sum_{i=1}^{3} \frac{1}{h_i}\frac{\partial V^*}{\partial u_i}\,\mathbf{u}_i \tag{1.5.11}$$

so that

$$F_i = -\frac{1}{h_i}\frac{\partial V^*}{\partial u_i} \qquad (i = 1, 2, 3) \tag{1.5.12}$$

Thus the components of the force at any point in a conservative field are the negatives of the corresponding components of the gradient of the scalar function from which the field is derived.

From the position vector $\mathbf{r} = \mathbf{r}(u_1, u_2, u_3)$ we obtain, by differentiation, the relations

$$d\mathbf{r} = \sum_{i=1}^{3} \frac{\partial \mathbf{r}}{\partial u_i}\,du_i = \sum_{i=1}^{3} h_i\,du_i\mathbf{u}_i \qquad \text{and} \qquad \dot{\mathbf{r}} = \sum_{i=1}^{3} h_i\dot{u}_i\mathbf{u}_i \tag{1.5.13}$$

Combining (1.5.8), (1.5.11), and (1.5.13), we can write

$$\mathbf{F}\cdot d\mathbf{r} = F_1h_1\,du_1 + F_2h_2\,du_2 + F_3h_3\,du_3$$
$$= -\sum_{i=1}^{3}\frac{\partial V^*}{\partial u_i}\,du_i = -dV^*$$

so that

$$V = -W = -\int_{\mathbf{r}(A)}^{\mathbf{r}} \mathbf{F}\cdot d\mathbf{r} = \int_{V^*[\mathbf{r}(A)]}^{V^*(\mathbf{r})} dV^* = V^*(\mathbf{r}) - V^*[\mathbf{r}(A)]$$

where $V^*[\mathbf{r}(A)]$ is a constant. Since

$$\frac{1}{h_i}\frac{\partial V}{\partial u_i} = \frac{1}{h_i}\frac{\partial V^*}{\partial u_i} \qquad (i = 1, 2, 3)$$

it follows immediately that

$$\nabla V = \sum_{i=1}^{3}\frac{1}{h_i}\frac{\partial V}{\partial u_i}\mathbf{u}_i = \sum_{i=1}^{3}\frac{1}{h_i}\frac{\partial V^*}{\partial u_i}\mathbf{u}_i = \nabla V^* = -\mathbf{F}$$

so that $\mathbf{F}$ may be obtained from V by differentiation.

For a more general mechanical system, the limits $\mathbf{r}(A)$ and $\mathbf{r}$ in (1.5.7) are to be replaced by the letters A and B respectively. Then (1.5.7) defines the total work done on the system *against* the force field $\mathbf{F}$ when the system is transported from an initial configuration A to a final configuration B. The quantity V is referred to as "the potential energy of the system at B with respect to the reference configuration A associated with the force field." When no ambiguity will arise, we speak of V more briefly as either "the potential energy of the system" or "the potential energy." As we have seen, V^* represents, to within an additive constant, $V^*[\mathbf{r}(A)]$, the work done on the particle against the force field in moving the particle from $\mathbf{r}(A)$ to $\mathbf{r}$. In celestial mechanics, one is generally interested in finite mass distributions for which the gravity forces fall off rapidly with distance from the source. For such fields, it is convenient to choose the reference point $\mathbf{r}(A)$ at infinity and $V^*[\mathbf{r}(\infty)] = 0$. Unless stated otherwise, we will choose the reference point $\mathbf{r}(A) = \mathbf{r}(\infty)$ so that $V = V^* = -W$.

Example 1.5.1. Assume a gravity field $\mathbf{F} = (-k^2mM/r^3)\,\mathbf{r}$ produced by a source of constant mass M. We consider moving a particle of constant mass m from infinity [i.e., $\mathbf{r}(\infty)$] to $\mathbf{r}$ in this field. We have

$$V = -\int_{\mathbf{r}(\infty)}^{\mathbf{r}} \mathbf{F}\cdot d\mathbf{r} = -\int_{\mathbf{r}(\infty)}^{\mathbf{r}} |\,\mathbf{F}\,|\,|\,d\mathbf{r}\,|\cos 0$$
$$= -\int_{\infty}^{r}\frac{k^2mM}{r^2}(-dr) = \frac{-k^2mM}{r}$$

Note that, by definition, $|\, d\mathbf{r} \,| = -dr$ if $dr < 0$. We can check our result by using V to deduce the force field $\mathbf{F}$. For this purpose, a formula from vector analysis, namely

$$\nabla f(r) = \frac{df(r)}{dr} \frac{\mathbf{r}}{r}$$

for determining the gradient of a function of r only, is useful. Applying the formula in the present case,we have

$$\nabla V = \nabla \left(\frac{-k^2 mM}{r} \right) = -k^2 mM \left(\frac{-1}{r^2} \right) \frac{\mathbf{r}}{r} = \frac{k^2 mM}{r^3} \mathbf{r} = -\mathbf{F}$$

We now define a quantity $\phi(u_1, u_2, u_3)$, called the potential, by the equation

$$\phi(u_1, u_2, u_3) = \frac{V}{m} \tag{1.5.14}$$

The function ϕ is the potential energy per unit mass at a point in the field $\mathbf{F}$ which is derivable from the potential energy V. We also define a vector quantity $\mathbf{E}$, called the field intensity of the force field $\mathbf{F}$ at $\mathbf{r}$, by the equation

$$\mathbf{E} = \frac{\mathbf{F}}{m} \tag{1.5.15}$$

The quantity $\mathbf{E}$ is the force experienced by unit mass when placed at $\mathbf{r}$ in the force field $\mathbf{F}$. It is understood that the field $\mathbf{F}$ is not altered by the introduction of m. If $\mathbf{F}$ is derived from V, we have

$$\mathbf{E} = \frac{\mathbf{F}}{m} = \frac{-1}{m} \nabla V = -\nabla \left(\frac{V}{m} \right) = -\nabla \phi \tag{1.5.16}$$

For the preceding example,

$$\phi = \frac{-k^2 M}{r}.$$

Assume that the external force field intensity $\mathbf{E}^E$ is derivable from a scalar potential $\phi^E(u_1, u_2, u_3)$. Then

$$\mathbf{E}^E = -\nabla \phi^E = \sum_{i=1}^{3} \frac{1}{h_i} \frac{\partial \phi^E}{\partial u_i} \mathbf{u}_i$$

A mass m_i placed at the point (u_{1i}, u_{2i}, u_{3i}) in the external field $\mathbf{F}^E$ will experience a force $\mathbf{F}_i^E$ given by

$$\begin{aligned} \mathbf{F}_i^E &= m_i \mathbf{E}_i^E(u_{1i}, u_{2i}, u_{3i}) = -\nabla_i m_i \phi_i^E(u_{1i}, u_{2i}, u_{3i}) \\ &= -\nabla_i V_i^E(u_{1i}, u_{2i}, u_{3i}) \end{aligned} \tag{1.5.17}$$

Here $\mathbf{E}_i^E$ and ϕ_i^E represent the field intensity and the potential at (u_{1i}, u_{2i}, u_{3i}), while V_i^E is the potential energy of the ith particle. All quantities are associated with $\mathbf{F}^E$. The subscript on the gradient operator means that the gradient is to be evaluated at the coordinates of the ith particle.

We will shortly deduce another integral, the tenth, of the equations of motion for a system of n particles. In this connection, we will want to express the summations

$$S_1 = \sum_{i=1}^{n} \dot{\mathbf{r}}_i \cdot \mathbf{F}_i^E \tag{1.5.18}$$

$$S_2 = \sum_{i=1}^{n} \sum_{j=1}^{n} \dot{\mathbf{r}}_i \cdot \mathbf{f}_{ij} \tag{1.5.19}$$

in the form of time derivatives of potential energy functions. To this end, we now consider the sum S_1.

Using (1.5.13) and (1.5.17), we write

$$\begin{aligned} \dot{\mathbf{r}}_i \cdot \mathbf{F}_i^E &= \Big(\sum_{k=1}^{3} h_{ki}\dot{u}_{ki}\mathbf{u}_{ki}\Big) \cdot \Big(-\sum_{k=1}^{3} \frac{1}{h_{ki}} \frac{\partial V_i^E}{\partial u_{ki}} \mathbf{u}_{ki}\Big) \\ &= -\sum_{k=1}^{3} \frac{\partial V_i^E}{\partial u_{ki}} \dot{u}_{ki} \end{aligned}$$

On the other hand, since V_i^E is a function of the coordinates, and the time through m_i, we have

$$\dot{V}_i^E = \sum_{k=1}^{3} \frac{\partial V_i^E}{\partial u_{ki}} \dot{u}_{ki} + \frac{\partial V_i^E}{\partial t} = \sum_{k=1}^{3} \frac{\partial V_i^E}{\partial u_{ki}} \dot{u}_{ki} + \phi_i^E \dot{m}_i$$

It follows immediately that

$$\dot{\mathbf{r}}_i \cdot \mathbf{F}_i^E = -\dot{V}_i^E + \phi_i^E \dot{m}_i$$

Returning to (1.5.18), we now write

$$S_1 = \sum_{i=1}^{n} \dot{\mathbf{r}}_i \cdot \mathbf{F}_i^E = -\dot{V}^E + \sum_{i=1}^{n} \phi_i^E \dot{m}_i \tag{1.5.20}$$

where $V^E = \sum_{i=1}^{n} V_i^E$ is the potential energy of the system of n particles due to the external force field. For constant masses, $\dot{m}_i = 0$ for every i, hence S_1 reduces to

$$S_1 = \sum_{i=1}^{n} \dot{\mathbf{r}}_i \cdot \mathbf{F}_i^E = -\dot{V}^E \tag{1.5.21}$$

For the case of internal forces, the situation is quite different because the particles themselves are the sources of the force field. The internal force field governing the motion of the ith particle is produced *by* the other $n - 1$ particles in the system, while the ith particle contributes *to* the internal force field experienced *by* each of the other $n - 1$ particles. We can, however, proceed in a simple fashion when the internal forces are conservative and of type 1. Then $\mathbf{f}_{ij} = -\mathbf{f}_{ji}$, where both $\mathbf{f}_{ij}$ and $\mathbf{f}_{ji}$ are derivable from a scalar function.

Let $\mathbf{g}_{ij}$ and ϕ_{ij} represent the field intensity and potential at the ith particle produced by the jth particle and let $\mathbf{g}_{ji}$ and ϕ_{ji} represent the field intensity and potential at the jth particle produced by the ith particle. Since the forces are conservative,

$$\mathbf{g}_{ij} = -\nabla_i \phi_{ij}, \qquad \mathbf{g}_{ji} = -\nabla_j \phi_{ji}$$

where

$$\nabla_i = \sum_{k=1}^{3} \frac{1}{h_{ki}} \frac{\partial}{\partial u_{ki}} \mathbf{u}_{ki}$$

$$\nabla_j = \sum_{k=1}^{3} \frac{1}{h_{kj}} \frac{\partial}{\partial u_{kj}} \mathbf{u}_{kj}$$

and u_{ki} and u_{kj}, $k = 1, 2, 3$, refer to the coordinates of the ith and jth particles respectively. Since $\mathbf{f}_{ij} = -\mathbf{f}_{ji}$ and

$$\mathbf{f}_{ij} = m_i \mathbf{g}_{ij} = -m_i \nabla_i \phi_{ij} = -\nabla_i (m_i \phi_{ij}) = -\nabla_i V_{ij}$$

$$\mathbf{f}_{ji} = m_j \mathbf{g}_{ji} = -m_j \nabla_j \phi_{ji} = -\nabla_j (m_j \phi_{ji}) = -\nabla_j V_{ji}$$

it follows that

$$\nabla_i V_{ij} = -\nabla_j V_{ji}$$

We now assume that the potential energies V_{ij} and V_{ji} depend only on the distance $|\mathbf{r}_j - \mathbf{r}_i|$ between the particles and on the product of their masses $(m_i m_j)$. Then $V_{ij} = V_{ji}$ and $\nabla_j V_{ji} = \nabla_j V_{ij}$.

We can write S_2 in the form

$$S_2 = \frac{1}{2} \sum_{i=1}^{n} \sum_{j=1}^{n} (\dot{\mathbf{r}}_i \cdot \mathbf{f}_{ij} + \dot{\mathbf{r}}_j \cdot \mathbf{f}_{ji}) \tag{1.5.22}$$

Hence we are led to consider terms of the form appearing in the double summation in (1.5.22). We have

$$\dot{\mathbf{r}}_i \cdot \mathbf{f}_{ij} + \dot{\mathbf{r}}_j \cdot \mathbf{f}_{ji} = -\dot{\mathbf{r}}_i \cdot \nabla_i V_{ij} - \dot{\mathbf{r}}_j \nabla_j V_{ji} = -\dot{\mathbf{r}}_i \nabla_i V_{ij} - \dot{\mathbf{r}}_j \cdot \nabla_j V_{ij}$$

$$= -\Big(\sum_{k=1}^{3} h_{ki}\dot{u}_{ki}\mathbf{u}_{ki}\Big) \cdot \Big(\sum_{k=1}^{3} \frac{1}{h_{ki}} \frac{\partial V_{ij}}{\partial u_{ki}} \mathbf{u}_{ki}\Big)$$

$$- \Big(\sum_{k=1}^{3} h_{kj}\dot{u}_{kj}\mathbf{u}_{kj}\Big) \cdot \Big(\sum_{k=1}^{3} \frac{1}{h_{kj}} \frac{\partial V_{ij}}{\partial u_{kj}} \mathbf{u}_{kj}\Big)$$

$$= -\sum_{k=1}^{3} \Big(\frac{\partial V_{ij}}{\partial u_{ki}} \dot{u}_{ki} + \frac{\partial V_{ij}}{\partial u_{kj}} \dot{u}_{kj}\Big) = -\dot{V}_{ij} + \frac{\partial V_{ij}}{\partial t}$$

We introduce the definitions

$$V_i^I = \sum_{j=1}^{n} V_{ij} \qquad (V_{ii} = 0)$$

$$V^I = \sum_{i=1}^{n} V_i^I$$

The quantity V_i^I is called the total potential energy of the ith particle due to the internal force field, while V^I is called the total potential energy due to the internal forces. Then we have

$$S_2 = \frac{1}{2}\sum_{i=1}^{n}\sum_{j=1}^{n} (\dot{\mathbf{r}}_i \cdot \mathbf{f}_{ij} + \dot{\mathbf{r}}_j \cdot \mathbf{f}_{ji}) = \frac{1}{2}\sum_{i=1}^{n}\sum_{j=1}^{n}\Big(-\dot{V}_{ij} + \frac{\partial V_{ij}}{\partial t}\Big)$$

$$= \frac{1}{2}\sum_{i=1}^{n}\Big(-\dot{V}_i^I + \frac{\partial V_i^I}{\partial t}\Big) = \frac{1}{2}\Big(-\dot{V}^I + \frac{\partial V^I}{\partial t}\Big) \tag{1.5.23a}$$

For constant masses, $\partial V^I/\partial t = 0$ and (1.5.23a) reduces to

$$S_2 = \frac{-\dot{V}^I}{2} \tag{1.5.23b}$$

If (1.5.2) is differentiated with respect to time, we obtain

$$\dot{T} = \sum_{i=1}^{n} \frac{1}{m_i} \mathbf{p}_i \cdot \dot{\mathbf{p}}_i - \sum_{i=1}^{n} \frac{\dot{m}_i}{2m_i^2} \mathbf{p}_i \cdot \mathbf{p}_i \tag{1.5.24a}$$

and, if the masses are constant,

$$\dot{T} = \sum_{i=1}^{n} \frac{1}{m_i} \mathbf{p}_i \cdot \dot{\mathbf{p}}_i \tag{1.5.24b}$$

Introducing the equations of motion into (1.5.24b) and using (1.3.11), we find that

$$\dot{T} = \sum_{i=1}^{n} \frac{1}{m_i} \mathbf{p}_i \cdot \left(\sum_{j=1}^{n} \mathbf{f}_{ij} + \mathbf{F}_i^E \right)$$

$$= \sum_{i=1}^{n} \sum_{j=1}^{n} \dot{\mathbf{r}}_i \cdot \mathbf{f}_{ij} + \sum_{i=1}^{n} \dot{\mathbf{r}}_i \cdot \mathbf{F}_i^E = S_1 + S_2$$

Using (1.5.21) and (1.5.23b) we can write

$$\dot{T} = -\dot{V}^E - \frac{\dot{V}^I}{2} \tag{1.5.25}$$

It follows from (1.5.25) that

$$E = T + V = \text{constant}, \qquad V \equiv V^E + \tfrac{1}{2}V^I \tag{1.5.26}$$

The quantity V is called the total potential energy of the system. The quantity $E = T + V$ is called the total energy of the system. We have thus found the tenth of the required $6n$ integrals. No more are known, in general.

Exercises

1. Assume Newton's law of gravitational attraction and the ideas of elementary integral calculus to show that a sphere, composed of uniformly dense shells, attracts other bodies as though the entire mass of the sphere were concentrated at its center.
2. The polar components of velocity parallel to and perpendicular to the radius vector are $2a\theta$ and br, where a and b are constants. Show that the path is $br = a\theta^2 + A$, where A is a constant.†
3. A heavy particle projected horizontally on a smooth table with speed v is reduced to rest by the resistance of the air after describing a distance s. Supposing the resistance of the air to be a uniform force, prove that when the particle is projected upward with any velocity, the squares of the times of ascent and descent to the point of projection are in the ratio $2gs - v^2$ to $2gs + v^2$. Here g is the magnitude of the gravitational acceleration, assumed constant.

† Exercises 2–5, 14, and 15 are reprinted from E. J. Routh, "A Treatise on Dynamics of a Particle," pp. 6, 9, 71, and 14. Copyright by Cambridge Univ. Press, London and New York. Used with permission of Cambridge Univ. Press.

4. Two points P and Q move on a straight line AB. The point P starts from A in the direction AB with speed u and uniform acceleration f, and at the same time Q starts from B in the direction BA with speed u' and uniform acceleration f'. If they pass one another at the midpoint of AB and arrive at the other ends of AB with equal speeds, prove that

$$(u + u')(f - f') = 8(fu' - f'u)$$

5. Two particles P, P' starting simultaneously from the points A, A' with initial speeds u, u' move in the straight line AA' with uniform accelerations f, f'. If v, v' are their speeds when the distance PP' exceeds the initial distance AA' by s, show that

$$(v' - v)^2 = (u' - u)^2 + 2(f' - f)s$$

6. Two masses m_1, m_2 are connected by a light inextensible string of length b. The masses are whirled around by hand and subsequently released. At the instant of release, m_1 is momentarily at rest in the hand and m_2 is traveling with linear speed v_0 with respect to m_1. Show that the tension S in the string after release is

$$m_1 m_2 v_0^2 / b(m_1 + m_2)^\dagger$$

7. A box filled with sand is sliding down a smooth hill of inclination α. Assume that sand is being ejected at a constant rate b in such a manner as always to reduce the sand to rest. Assume that the box starts from rest and show that the speed v of the box, as a function of time, is given by

$$v = \frac{(m_0 - \frac{1}{2}bt)\, tg \sin\alpha}{m_0 - bt}$$

where m_0 is the initial mass of the box and g is the magnitude of the gravitational acceleration.

8. A raindrop falls with constant acceleration through a uniform fog, sweeping out mist (i.e., collecting mass) from the region it traverses. If it starts from rest with zero radius, and remains at all times spherical, deduce that the acceleration with which it falls is $g/7$.

† Exercises 6–8 are reprinted from R. A. Becker, "Introduction to Theoretical Mechanics," pp. 167, 186, and 189. Copyright 1954 by McGraw-Hill, New York. Used with permission of McGraw-Hill Book Company.

9. A raindrop falls through a uniform fog, sweeping out mist (i.e., collecting mass) from the region it traverses. If it starts from rest with zero radius, remains at all times spherical, and falls with constant acceleration $g/7$, deduce that the radius σ of the raindrop satisfies the differential equation

$$\sigma\ddot{\sigma} - \tfrac{1}{2}\dot{\sigma}^2 = 0$$

By solving this differential equation, subject to the stated boundary conditions, deduce that the radius σ is proportional to the square of the time of fall.

10. Deduce the acceleration of the raindrop in Exercise 8 under the assumption that the sum of its potential and kinetic energies remains constant. Does the value of the acceleration obtained here agree with the value given in Exercise 8? Comment.

11. Represent the vector

$$\mathbf{g} = \text{grad}\, u + \tfrac{1}{2}\,\text{grad}(w^2 r^2 \cos^2\theta)$$

where

$$u = \frac{\mu}{r} + \frac{\mu a^2 J}{r^3}\left(\frac{1}{3} - \sin^2\theta\right)$$

and w, μ, a, J are constants, in terms of the unit base vectors $\mathbf{e}_r$, $\mathbf{e}_\theta$ and $\mathbf{e}_\phi$ of the spherical coordinate system depicted in Figure A.

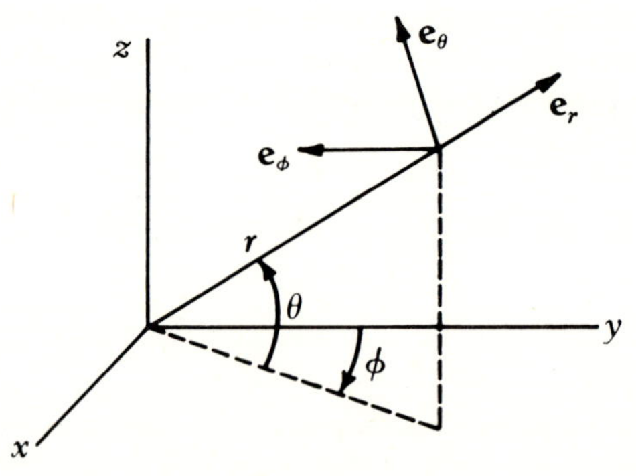

FIGURE A

12. Consider a satellite of unit mass (constant) in motion about an oblate Earth subject only to conservative gravitational forces that are derivable from the scalar function

$$V = \frac{\mu}{r} + \frac{\mu a^2 J}{r^3}\left(\frac{1}{3} - \cos^2\theta\right)$$

Write the three second-order scalar differential equations of motion in spherical coordinates (r, θ, ϕ) (see Figure B).

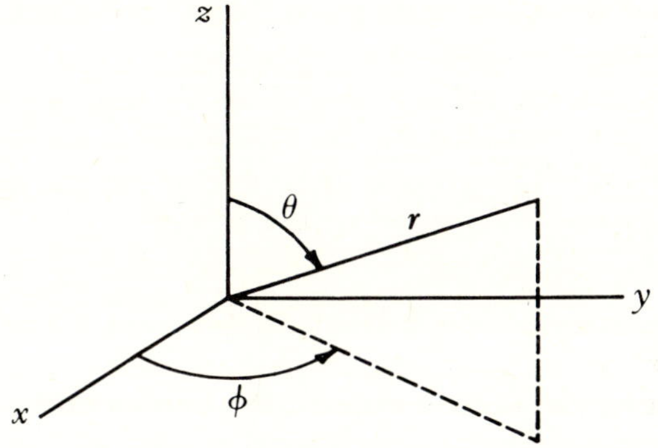

FIGURE B

13. A particle falls to the Earth from a height h. Assume a spherical Earth with uniform mass density. Prove that the time of falling is

$$\left(1 + \frac{5h}{6R}\right)\left(\frac{2h}{g}\right)^{1/2}$$

where R is the Earth's radius, g is the acceleration due to gravity at its surface, and the square of h/R is neglected.

14. A body of mass M is moving in a straight line with speed u and is followed at a distance r by a smaller body of mass m moving in the same line with speed v, $v < u$. The bodies attract each other according to the law of gravitation. Prove that the smaller body will overtake the other after a time

$$\frac{1}{(k^2M')^{1/2}}\left(\frac{r}{1+w}\right)^{3/2}[\pi + (1 - w^2)^{1/2} + \cos^{-1} w]$$

where

$$1 - w = \frac{r(v-u)^2}{k^2M'}, \qquad M' = m + M$$

15. A point P describes a circle of 1 foot radius in 1 hour, and a point Q describes a concentric circle of 4 feet radius in 14 hours, both points move in the counterclockwise direction. Show that the line joining them rotates in the counterclockwise direction for a period of $43\frac{1}{13}$ minutes followed by a period of $21\frac{7}{13}$ minutes in the clockwise direction.

16. A column of troops 3 miles in depth marches along a road. An officer rides at a uniform rate from the rear of the column to the head and back again at once, reaching the rear of the column just as an advance of 4 miles has been made. How far did the officer ride?

17. A baseball outfielder has to judge a catch which he can secure with equal ease at any height above the ground between k_1 and k_2, $k_2 > k_1$. Assuming that he plans to catch the ball after it passes through the greatest height h which it attains, show that he must estimate his position within a length

$$R\left[\left(1-\frac{k_1}{h}\right)^{1/2}-\left(1-\frac{k_2}{h}\right)^{1/2}\right]$$

where $2R$ is the range on the horizontal.†

18. Consider a rocket with initial mass m_0, composed entirely of fuel, launched vertically from the surface of the Earth in a gravity field $\mathbf{g}$ assumed constant. Let m and X be its instantaneous mass and distance above the surface of the Earth respectively. Let $w = -d(mg)/dt$ be the rate of burning of the fuel and u the constant speed of efflux. Develop an equation of motion for the rocket and then show that the height X_b at burn out is given by

$$X_b = \frac{m_0 g}{w}u - \frac{g}{2}\left(\frac{m_0 g}{w}\right)^2$$

† From A. E. H. Love, "Theoretical Mechanics," p. 72. Copyright by Cambridge Univ. Press, London and New York. Used with permission of Cambridge Univ. Press.

Chapter 2

The One-Body and Two-Body Problems

2.1. EQUATIONS OF MOTION OF A SYSTEM OF n PARTICLES WITH RESPECT TO AN UNACCELERATING ORIGIN

Equations (1.3.11) give the $3n$ equations of motion, with respect to O, of a system of n particles when only forces of types 1 and 3 are considered. If

$$\mathbf{F}_i^T = \sum_{j=1}^{n} \mathbf{f}_{ij} + \mathbf{F}_i^E$$

these equations have the form

$$\dot{\mathbf{p}}_i = \mathbf{F}_i^T = m_i\ddot{\mathbf{r}}_i + \dot{m}_i\dot{\mathbf{r}}_i \qquad (i = 1, 2,..., n) \tag{2.1.1}$$

where $\mathbf{F}_i^T$ is the total force on the ith particle due to both internal and external sources. In terms of a moving origin O', we have [see (1.3.3)]

$$\mathbf{p}_i = m_i\dot{\mathbf{r}}_i = m_i(\dot{\mathbf{a}} + \dot{\mathbf{r}}_i')$$

It follows that

$$\dot{\mathbf{p}}_i' = \mathbf{F}_i^T - m_i\ddot{\mathbf{a}} - \dot{m}_i\dot{\mathbf{a}} \tag{2.1.2}$$

Comparing (2.1.1) and (2.1.2), we see that if

$$m_i\ddot{\mathbf{a}} + \dot{m}_i\dot{\mathbf{a}} = \mathbf{0} \tag{2.1.3}$$

the form of the equations of motion is the same when referenced to either O or O'. Thus if O' is fixed with respect to O so that $\dot{\mathbf{a}} = \ddot{\mathbf{a}} = \mathbf{0}$ or if the masses are constant ($\dot{m}_i = 0$) while O' is unaccelerated ($\ddot{\mathbf{a}} = \mathbf{0}$), the equations of motion have the same form when referenced to O' as they do when they are referenced to O. In particular, if for constant masses the unaccelerated origin is at the center of mass, we have

$$\dot{\mathbf{p}}_{ic} = m_i\ddot{\mathbf{R}}_i = \mathbf{F}_i^T \tag{2.1.4}$$

It turns out that in many cases, including those we will examine, the use of the center of mass as an origin provides little or no help in obtaining the solution of the equations of motion. A far more useful notion for our purpose will be to reference the equations of motion to an origin at one of the particles. We now consider this notion.

2.2. EQUATIONS OF MOTION OF A SYSTEM OF n PARTICLES WITH RESPECT TO AN ORIGIN AT ONE OF THE PARTICLES

When one of the n particles has a mass much larger than any of the others, which is the case of the Sun in the solar system for example, the choice of an origin at that massive particle assists in determining the relative motions of the particles. Since this is likely to be the main problem of interest, we single out one of the particles and designate it by the subscript 0. The remaining particles in the system will be designated by the integer subscripts 1 through $n - 1$. Since unprimed notation will be reserved for motion with respect to an origin at m_0, we write first the equations of motion with respect to an origin O' fixed with respect to O. For particles 1 through $n - 1$, we have

$$\dot{\mathbf{p}}_i' = \sum_{j=0}^{n-1} \mathbf{f}_{ij} + \mathbf{F}_i^E \qquad (i = 1,..., n - 1) \tag{2.2.1}$$

If the equations

$$\dot{\mathbf{p}}_i' = m_i\ddot{\mathbf{r}}_i' + \dot{m}_i\dot{\mathbf{r}}_i' \tag{2.2.2}$$

$$\dot{\mathbf{p}}_0' = m_0\ddot{\mathbf{r}}_0' + \dot{m}_0\dot{\mathbf{r}}_0' \tag{2.2.3}$$

are substituted into (2.2.1), we can write

$$\ddot{\mathbf{r}}_i' = \frac{1}{m_i}\sum_{j=0}^{n-1} \mathbf{f}_{ij} + \frac{1}{m_i}\mathbf{F}_i^E - \frac{\dot{m}_i}{m_i}\dot{\mathbf{r}}_i' \tag{2.2.4}$$

$$\ddot{\mathbf{r}}_0' = \frac{1}{m_0}\sum_{j=1}^{n-1} \mathbf{f}_{0j} + \frac{1}{m_0}\mathbf{F}_0^E - \frac{\dot{m}_0}{m_0}\dot{\mathbf{r}}_0' \tag{2.2.5}$$

Let $\mathbf{r}_i$ represent the position vector of the ith particle with respect to an origin at m_0 and write (see Figure 2.2.1)

$$\mathbf{r}_i' = \mathbf{r}_0' + \mathbf{r}_i \tag{2.2.6}$$

$$\ddot{\mathbf{r}}_i' = \ddot{\mathbf{r}}_0' + \ddot{\mathbf{r}}_i \tag{2.2.7}$$

FIGURE 2.2.1

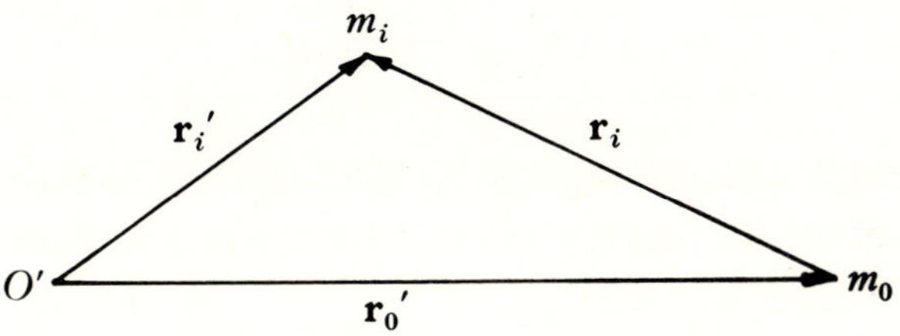

If (2.2.5) is subtracted from (2.2.4),and (2.2.7) is used, it follows that

$$\ddot{\mathbf{r}}_i = \frac{1}{m_i}\sum_{j=0}^{n-1}\mathbf{f}_{ij} - \frac{1}{m_0}\sum_{j=1}^{n-1}\mathbf{f}_{0j} + \frac{1}{m_i}\mathbf{F}_i^E - \frac{1}{m_0}\mathbf{F}_0^E - \frac{\dot{m}_i\dot{\mathbf{r}}_i'}{m_i} + \frac{\dot{m}_0}{m_0}\dot{\mathbf{r}}_0' \tag{2.2.8}$$

For m_0, the equation of motion is

$$\dot{\mathbf{p}}_0' = \sum_{j=1}^{n-1}\mathbf{f}_{0j} + \mathbf{F}_0^E \tag{2.2.9}$$

Equation (2.2.8) represents the motion of the particles, 1 to $n - 1$, with respect to an origin at m_0. Equation (2.2.9) represents the motion of m_0 with respect to a fixed origin O'. If the masses are constant (the situation in which we will be primarily interested), Equation (2.2.8) reduces to

$$\ddot{\mathbf{r}}_i = \frac{1}{m_i}\sum_{j=0}^{n-1}\mathbf{f}_{ij} - \frac{1}{m_0}\sum_{j=1}^{n-1}\mathbf{f}_{0j} + \frac{1}{m_i}\mathbf{F}_i^E - \frac{1}{m_0}\mathbf{F}_0^E \tag{2.2.10}$$

or, where $\mathbf{E}_i^E \equiv \mathbf{F}_i^E/m_i$,

$$\ddot{\mathbf{r}}_i = \frac{1}{m_i}\sum_{j=0}^{n-1}\mathbf{f}_{ij} - \frac{1}{m_0}\sum_{j=1}^{n-1}\mathbf{f}_{0j} + \mathbf{E}_i^E - \mathbf{E}_0^E \tag{2.2.11}$$

If the intensity of the external fields is constant, $\mathbf{E}_i^E - \mathbf{E}_0^E = \mathbf{0}$ and the relative motion of the n particles is unaffected by the presence of the external field. For such a case, the equations of relative motion, for constant masses, become

$$\ddot{\mathbf{r}}_i = \frac{1}{m_i} \sum_{j=0}^{n-1} \mathbf{f}_{ij} - \frac{1}{m_0} \sum_{j=1}^{n-1} \mathbf{f}_{0j} \qquad (i = 1, 2, \ldots, n-1) \tag{2.2.12}$$

The order of the system of equations (2.2.12) is $6n - 6$. The remaining six constants to complete the set of $6n$ constants must come from the solution of the independent equation (2.2.9).

2.3. EQUATIONS OF RELATIVE MOTION WHEN THE INTERNAL FORCES DEPEND ONLY UPON THE DISTANCES BETWEEN THE PARTICLES

It is now assumed that the internal force between each pair of particles depends only on a function $F(\rho_{ij})$ of the distance between the pair in the form

$$\mathbf{f}_{ij} = k^2 m_i m_j F(\rho_{ij})\, \mathbf{u}_{ij} \qquad (i = 1, 2, \ldots, n-1;\ \ j = 1, 2, \ldots, n-1) \tag{2.3.1}$$

where ρ_{ij} is the distance between particles m_i and m_j, $\mathbf{u}_{ij}$ is a unit vector directed from m_i towards m_j, and k^2 is a gravitational constant whose numerical value depends on the units employed. In particular, the symbol r_k will be used instead of either ρ_{0k} or ρ_{k0}, $k = 1, 2, \ldots, n-1$. Equation (2.2.11) becomes

$$\ddot{\mathbf{r}}_i = k^2 \sum_{j=0}^{n-1} m_j F(\rho_{ij})\, \mathbf{u}_{ij} - k^2 \sum_{j=1}^{n-1} m_j F(r_j)\, \mathbf{u}_{0j} + \mathbf{E}_i^E - \mathbf{E}_0^E \tag{2.3.2}$$

Equation (2.3.2) can be written

$$\ddot{\mathbf{r}}_i = k^2 m_0 F(r_i)\, \mathbf{u}_{i0} + k^2 \sum_{\substack{j=1 \\ j \neq i}}^{n-1} m_j F(\rho_{ij})\, \mathbf{u}_{ij} - k^2 m_i F(r_i)\, \mathbf{u}_{0i}$$

$$- k^2 \sum_{\substack{j=1 \\ j \neq i}}^{n-1} m_j F(r_j)\, \mathbf{u}_{0j} + \mathbf{E}_i^E - \mathbf{E}_0^E$$

or

$$\ddot{\mathbf{r}}_i = -k^2 (m_0 + m_i) F(r_i)\, \mathbf{u}_{0i}$$

$$+ k^2 \sum_{\substack{j=1 \\ j \neq i}}^{n-1} m_j [F(\rho_{ij})\, \mathbf{u}_{ij} - F(r_j)\, \mathbf{u}_{0j}] + \mathbf{E}_i^E - \mathbf{E}_0^E \tag{2.3.3}$$

Equations (2.3.3) can also be written in the more convenient form

$$\ddot{\mathbf{r}}_i + k^2 M_i F(r_i)\, \mathbf{u}_{0i} = \nabla_i R_i + \mathbf{E}_i^E - \mathbf{E}_0^E \tag{2.3.4}$$

where

$$\nabla_i = \frac{\partial}{\partial x_i}\mathbf{i} + \frac{\partial}{\partial y_i}\mathbf{j} + \frac{\partial}{\partial z_i}\mathbf{k}$$

$$M_i = m_0 + m_i$$

$$R_i \equiv -k^2 \sum_{\substack{j=1\\ j\neq i}}^{n-1} m_j \left[P_{ij} + \frac{F(r_j)}{r_j}(x_i x_j + y_i y_j + z_i z_j)\right]$$

and P_{ij} is any differentiable function of ρ_{ij} such that

$$\frac{dP_{ij}}{d\rho_{ij}} = F(\rho_{ij})$$

To show that (2.3.4) is equivalent to (2.3.3), we first observe that

$$\mathbf{u}_{ij} = \frac{x_j - x_i}{\rho_{ij}}\mathbf{i} + \frac{y_j - y_i}{\rho_{ij}}\mathbf{j} + \frac{z_j - z_i}{\rho_{ij}}\mathbf{k}$$

$$\mathbf{u}_{0j} = \frac{x_j}{r_j}\mathbf{i} + \frac{y_j}{r_j}\mathbf{j} + \frac{z_j}{r_j}\mathbf{k}$$

$$\rho_{ij}^2 = (x_j - x_i)^2 + (y_j - y_i)^2 + (z_j - z_i)^2$$

so that

$$\frac{\partial P_{ij}}{\partial x_i} = \frac{dP_{ij}}{d\rho_{ij}}\frac{\partial \rho_{ij}}{\partial x_i} = -F(\rho_{ij})\frac{x_j - x_i}{\rho_{ij}}$$

with similar expressions for $\partial P_{ij}/\partial y_i$ and $\partial P_{ij}/\partial z_i$. Thus adding the three expressions,

$$\frac{\partial R_i}{\partial x_i}\mathbf{i} = k^2 \sum_{\substack{j=1\\ j\neq i}}^{n-1} m_j \left[F(\rho_{ij})\frac{x_j - x_i}{\rho_{ij}}\mathbf{i} - F(r_j)\frac{x_j}{r_j}\mathbf{i}\right]$$

$$\frac{\partial R_i}{\partial y_i}\mathbf{j} = k^2 \sum_{\substack{j=1\\ j\neq i}}^{n-1} m_j \left[F(\rho_{ij})\frac{y_j - y_i}{\rho_{ij}}\mathbf{j} - F(r_j)\frac{y_j}{r_j}\mathbf{j}\right]$$

$$\frac{\partial R_i}{\partial z_i}\mathbf{k} = k^2 \sum_{\substack{j=1\\ j\neq i}}^{n-1} m_j \left[F(\rho_{ij})\frac{z_j - z_i}{\rho_{ij}}\mathbf{k} - F(r_j)\frac{z_j}{r_j}\mathbf{k}\right]$$

we have

$$\nabla_i R_i = k^2 \sum_{\substack{j=1\\ j\neq i}}^{n-1} m_j [F(\rho_{ij})\,\mathbf{u}_{ij} - F(r_j)\,\mathbf{u}_{0j}]$$

The function R_i is called the disturbing or perturbative function associated with the internal forces in the system of particles. In certain cases it will be useful to consider the external field intensity without regard to the perturbing function. For such cases, we write

$$\ddot{\mathbf{r}}_i + k^2M_iF(r_i)\,\mathbf{u}_{0i} = \mathbf{E}_i^E - \mathbf{E}_0^E \qquad (i = 1, 2, \ldots, n-1) \tag{2.3.5}$$

2.4. THE TWO-BODY PROBLEM AND THE EQUIVALENT ONE-BODY PROBLEM

The equation for the relative motion in a uniform external force field of two particles of masses m_0 and m, acting upon each other with a force that depends only on the distance r between them, is

$$\ddot{\mathbf{r}} + k^2M\frac{F(r)}{r}\,\mathbf{r} = \mathbf{0} \qquad (M = m_0 + m) \tag{2.4.1}$$

where $\mathbf{r}$ is the position vector of the mass m referred to the particle m_0 as origin. The force $\mathbf{F}$ acting on the particle of mass m is

$$\mathbf{F} = -k^2m_0mF(r)\,\mathbf{u}_{0m} \tag{2.4.2}$$

From (1.3.8) we know that the form of the expression for the center of mass of a system of n particles is unchanged when referenced to an origin moving in any manner with respect to an inertial origin O. Thus the vector $\mathbf{R}$ from m_0 to the center of mass of the two particles is given by the expression (see Figure 2.4.1)

$$M\mathbf{R} = \sum_i m_i\mathbf{r}_i = m_0\mathbf{0} + m\mathbf{r} = m\mathbf{r}$$

or

$$\mathbf{R} = \frac{m}{M}\mathbf{r}$$

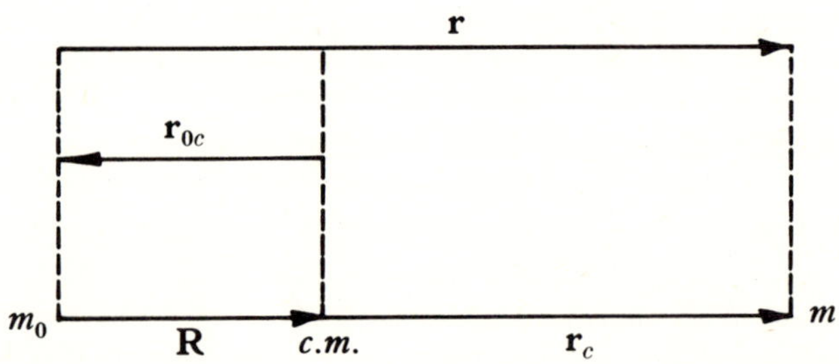

FIGURE 2.4.1

The vector $\mathbf{r}_c$ from the center of mass to m is related to $\mathbf{r}$ through the equation

$$\mathbf{r} - \mathbf{R} = \left(1 - \frac{m}{M}\right)\mathbf{r} = \mathbf{r}_c$$

or

$$\mathbf{r} = \frac{M}{m_0}\,\mathbf{r}_c \tag{2.4.3}$$

The equation

$$\ddot{\mathbf{r}}_c + k^2 m_0 \frac{F(r)}{r_c}\,\mathbf{r}_c = \mathbf{0} \qquad (r_c \equiv |\,\mathbf{r}_c\,|) \tag{2.4.4}$$

for the motion of particle m about the center of mass is obtained from (2.4.1) through the use of (2.4.3). To express the motion of m_0 about the center of mass, the position vector $\mathbf{r}_{0c}$, from the center of mass to the mass m_0, is introduced. Since $\mathbf{r}_{0c} = -\mathbf{R}$ we have

$$\mathbf{r} = \frac{M}{m}\,\mathbf{R} = -\frac{M}{m}\,\mathbf{r}_{0c} \tag{2.4.5}$$

Hence the equation

$$\ddot{\mathbf{r}}_{0c} + k^2 m \frac{F(r)}{r_{0c}}\,\mathbf{r}_{0c} = \mathbf{0} \qquad (r_{0c} = |\,\mathbf{r}_{0c}\,|) \tag{2.4.6}$$

which describes the motion of m_0 about the center of mass, is obtained from (2.4.1) through the use of (2.4.5).

In summary, the three equations for relative motion are

(a) $$\ddot{\mathbf{r}} = -k^2 M \frac{F(r)}{r}\,\mathbf{r} \qquad \text{(motion of } m \text{ about } m_0) \qquad [2.4.1]$$

(b) $$\ddot{\mathbf{r}}_c = -k^2 m_0 \frac{F(r)}{r_c}\,\mathbf{r}_c \qquad \text{(motion of } m \text{ about the center of mass)} \qquad [2.4.4]$$

(c) $$\ddot{\mathbf{r}}_{0c} = -k^2 m \frac{F(r)}{r_{0c}}\,\mathbf{r}_{0c} \qquad \text{(motion of } m_0 \text{ about the center of mass)} \qquad [2.4.6]$$

It is clear that if $F(r)$ is a homogeneous function of r, the orbits (a), (b), and (c), whatever they are, differ only in scale. Moreover, the particles are similarly placed in their respective orbits at any instant.

The equation of motion for a particle of mass m acted on by a force $\mathbf{F} = -k^2 mM[F(r)/r]\,\mathbf{r}$, always directed toward a fixed (inertial) origin, would be exactly the same as (2.4.1). Thus the two-body problem described by (2.4.1) is said to be equivalent to a one-body problem in which the particle is acted on by a force $\mathbf{F} = -k^2 mM[F(r)/r]\,\mathbf{r}$ directed toward a fixed origin. Such a force or, more generally, any force (not necessarily a function of r alone) that is always directed toward a fixed origin is called a central force.

Specifically, the differential equation of motion for the problem of a single particle of mass m in motion about a fixed center of force under the influence of a central force of the form

$$\mathbf{F} = -k^2 m(m_0 + m)\frac{F(r)}{r}\mathbf{r}$$

is

$$\ddot{\mathbf{r}} = -k^2(m_0 + m)\frac{F(r)}{r}\mathbf{r} \tag{2.4.7}$$

where $\mathbf{r}$ is the position vector from the fixed center to the mass m.

The operation of forming the cross product of $\mathbf{r}$ with (2.4.7) yields

$$\mathbf{r} \times \ddot{\mathbf{r}} = (\mathbf{r} \times \dot{\mathbf{r}})^{\cdot} = \dot{\mathbf{h}} = \mathbf{0} \qquad (2.4.8)^{\dagger}$$

where $\mathbf{h}$ is the angular momentum per unit mass of the mass m referenced to the fixed origin. An integral of (2.4.8) is [see (1.4.2)]:

$$\mathbf{h} = h\mathbf{u}_A = \mathbf{h}_0$$

where $\mathbf{h}_0$ is a constant vector that can, in some convenient nonrotating rectangular reference frame, be written in the form

$$\mathbf{h}_0 = c_1\mathbf{i} + c_2\mathbf{j} + c_3\mathbf{k} \tag{2.4.9}$$

where c_1, c_2, and c_3 are real constants and $|\,\mathbf{h}\,| = (c_1^2 + c_2^2 + c_3^2)^{1/2}$. The vector $\mathbf{h}$ is always perpendicular to $\mathbf{r} \times \dot{\mathbf{r}}$, and since $\mathbf{h}$ is a constant vector, the motion of m takes place in a plane perpendicular to $\mathbf{h}$, the equation of which may be written in the rectangular form

$$c_1\xi + c_2\eta + c_3\zeta = 0 \tag{2.4.10}$$

where $\mathbf{r} = \xi\mathbf{i} + \eta\mathbf{j} + \zeta\mathbf{k}$. Since (2.4.10) is the equation of a plane that contains the origin, it is clear that the motion of m takes place in a plane containing the fixed origin. This plane will be called the orbital plane.

The general solution of (2.4.7) must contain six independent constants of integration. The integration of $\dot{\mathbf{h}} = \mathbf{0}$ introduced the three constants of integration c_1, c_2, c_3 and reduced the problem to that of solving the first-order differential equation

$$\mathbf{h} = h\mathbf{u}_A = \mathbf{r} \times \dot{\mathbf{r}} = c_1\mathbf{i} + c_2\mathbf{j} + c_3\mathbf{k} \tag{2.4.11}$$

Hence three more constants of integration will be required to specify the general solution.

† The superscript dot on parentheses is used throughout the text to indicate time differentiation of the quantity enclosed therein.

2.5. SPATIAL ORIENTATION OF THE ORBITAL PLANE

In order to interpret, in three-dimensional space, the motion described by (2.4.7), the orientation of the orbital plane with respect to a rectangular reference frame $O\xi\eta\zeta$ with origin at m_0 will be given. The orientation with respect to space of the reference frame $O\xi\eta\zeta$ will not be specified, for the particular choice of orientation will depend on the problem to be studied. The following discussion will be referenced to Figure 2.5.1.

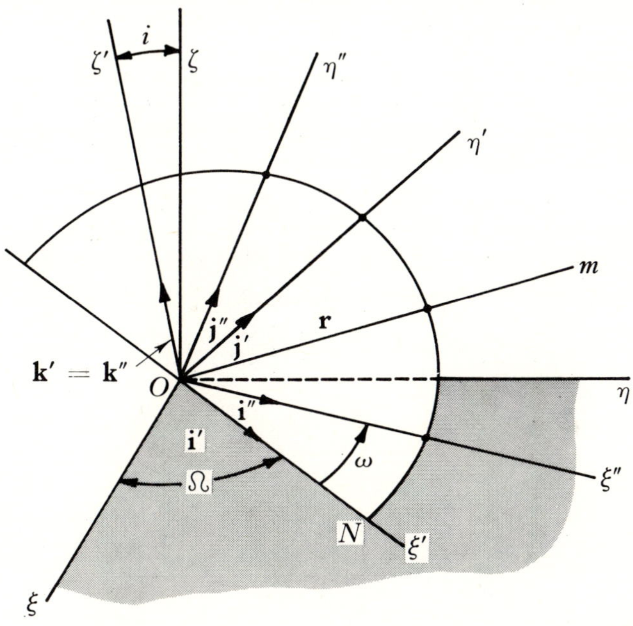

FIGURE 2.5.1

Let the orbital plane intersect the $\xi\eta$-plane in a line $N'ON$ called the line of nodes. This line is defined whenever the orbital plane and the $\xi\eta$-plane are not parallel. Let the angle Ω, $0° \leqslant \Omega \leqslant 360°$, be defined by rotating the $O\xi\eta\zeta$-frame about the ζ-axis, counterclockwise as viewed from positive ζ. Label the new reference frame obtained from this rotation $O\xi'\eta^*\zeta$. Let the angle i, $0° < i < 180°$, be defined by rotating the $O\xi'\eta^*\zeta$-axes about the ξ'-axis counterclockwise as viewed from positive ξ' to form the new reference frame $O\xi'\eta'\zeta'$. Let the angle ω be defined by rotating the $O\xi'\eta'\zeta'$-frame about the ζ'-axis, counterclockwise as viewed from positive ζ', to form the new reference frame $O\xi''\eta''\zeta''$. Let **i, j, k** represent an orthogonal basis for the reference frame

$O\xi\eta\zeta$ in the usual order. In a similar manner, $\mathbf{i}'$, $\mathbf{j}'$, $\mathbf{k}'$ and $\mathbf{i}''$, $\mathbf{j}''$, $\mathbf{k}''$ will designate orthogonal bases for the rectangular reference frames $O\xi'\eta'\zeta'$ and $O\xi''\eta''\zeta''$, respectively, in the usual order.

The equations of transformation that relate the several reference frames described in the preceding paragraph are readily deduced by considering the separate rotations. They are given below in matrix notation. The most frequently used rotation matrices and their transposes (superscript T) are

$$A = \begin{pmatrix} \cos\omega & \sin\omega & 0 \\ -\sin\omega & \cos\omega & 0 \\ 0 & 0 & 1 \end{pmatrix}, \qquad A^T = \begin{pmatrix} \cos\omega & -\sin\omega & 0 \\ \sin\omega & \cos\omega & 0 \\ 0 & 0 & 1 \end{pmatrix}$$

$$B = \begin{pmatrix} 1 & 0 & 0 \\ 0 & \cos i & \sin i \\ 0 & -\sin i & \cos i \end{pmatrix}, \qquad B^T = \begin{pmatrix} 1 & 0 & 0 \\ 0 & \cos i & -\sin i \\ 0 & \sin i & \cos i \end{pmatrix}$$

$$C = \begin{pmatrix} \cos\Omega & \sin\Omega & 0 \\ -\sin\Omega & \cos\Omega & 0 \\ 0 & 0 & 1 \end{pmatrix}, \qquad C^T = \begin{pmatrix} \cos\Omega & -\sin\Omega & 0 \\ \sin\Omega & \cos\Omega & 0 \\ 0 & 0 & 1 \end{pmatrix}$$

$$BC = \begin{pmatrix} \cos\Omega & \sin\Omega & 0 \\ -\sin\Omega\cos i & \cos\Omega\cos i & \sin i \\ \sin\Omega\sin i & -\cos\Omega\sin i & \cos i \end{pmatrix}$$

$$(BC)^T = \begin{pmatrix} \cos\Omega & -\sin\Omega\cos i & \sin\Omega\sin i \\ \sin\Omega & \cos\Omega\cos i & -\cos\Omega\sin i \\ 0 & \sin i & \cos i \end{pmatrix}$$

$$D = ABC = \begin{pmatrix} D_{11} & D_{12} & D_{13} \\ D_{21} & D_{22} & D_{23} \\ D_{31} & D_{32} & D_{33} \end{pmatrix}, \qquad D^T = \begin{pmatrix} D_{11} & D_{21} & D_{31} \\ D_{12} & D_{22} & D_{32} \\ D_{13} & D_{23} & D_{33} \end{pmatrix}$$

where

$$\begin{aligned} D_{11} &= \cos\omega\cos\Omega - \sin\omega\sin\Omega\cos i \\ D_{12} &= \cos\omega\sin\Omega + \sin\omega\cos\Omega\cos i \\ D_{13} &= \sin\omega\sin i \\ D_{21} &= -\sin\omega\cos\Omega - \cos\omega\sin\Omega\cos i \\ D_{22} &= -\sin\omega\sin\Omega + \cos\omega\cos\Omega\cos i \\ D_{23} &= \cos\omega\sin i \\ D_{31} &= \sin\Omega\sin i \\ D_{32} &= -\cos\Omega\sin i \\ D_{33} &= \cos i \end{aligned} \tag{2.5.1}$$

The transformations between the coordinates (and basis vectors) are explicitly:

$$\begin{pmatrix} \xi' \\ \eta^* \\ \zeta \end{pmatrix} = C \begin{pmatrix} \xi \\ \eta \\ \zeta \end{pmatrix}, \qquad \begin{pmatrix} \xi \\ \eta \\ \zeta \end{pmatrix} = C^T \begin{pmatrix} \xi' \\ \eta^* \\ \zeta \end{pmatrix} \tag{2.5.2}$$

$$\begin{pmatrix} \xi' \\ \eta' \\ \zeta' \end{pmatrix} = B \begin{pmatrix} \xi' \\ \eta^* \\ \zeta \end{pmatrix}, \qquad \begin{pmatrix} \xi' \\ \eta^* \\ \zeta \end{pmatrix} = B^T \begin{pmatrix} \xi' \\ \eta' \\ \zeta' \end{pmatrix} \tag{2.5.3}$$

$$\begin{pmatrix} \xi'' \\ \eta'' \\ \zeta'' \end{pmatrix} = A \begin{pmatrix} \xi' \\ \eta' \\ \zeta' \end{pmatrix}, \qquad \begin{pmatrix} \xi' \\ \eta' \\ \zeta' \end{pmatrix} = A^T \begin{pmatrix} \xi'' \\ \eta'' \\ \zeta'' \end{pmatrix} \tag{2.5.4}$$

$$\begin{pmatrix} \xi' \\ \eta' \\ \zeta' \end{pmatrix} = BC \begin{pmatrix} \xi \\ \eta \\ \zeta \end{pmatrix}, \qquad \begin{pmatrix} \xi \\ \eta \\ \zeta \end{pmatrix} = (BC)^T \begin{pmatrix} \xi' \\ \eta' \\ \zeta' \end{pmatrix} \tag{2.5.5}$$

$$\begin{pmatrix} \xi'' \\ \eta'' \\ \zeta'' \end{pmatrix} = D \begin{pmatrix} \xi \\ \eta \\ \zeta \end{pmatrix}, \qquad \begin{pmatrix} \xi \\ \eta \\ \zeta \end{pmatrix} = D^T \begin{pmatrix} \xi'' \\ \eta'' \\ \zeta'' \end{pmatrix} \tag{2.5.6}$$

The constants c_1, c_2, and c_3 in (2.4.9) are determined when the initial conditions on $\mathbf{r}$ and $\dot{\mathbf{r}}$ are prescribed. From (2.5.5) it follows that

$$c_1\mathbf{i} + c_2\mathbf{j} + c_3\mathbf{k} = h \sin \Omega \sin i\mathbf{i} - h \cos \Omega \sin i\mathbf{j} + h \cos i\mathbf{k}$$

whence

$$\sin \Omega \sin i = \frac{c_1}{h} = \frac{c_1}{(c_1^2 + c_2^2 + c_3^2)^{1/2}} \tag{2.5.7a}$$

$$\cos \Omega \sin i = -\frac{c_2}{h} = -\frac{c_2}{(c_1^2 + c_2^2 + c_3^2)^{1/2}} \tag{2.5.7b}$$

$$\cos i = \frac{c_3}{h} = \frac{c_3}{(c_1^2 + c_2^2 + c_3^2)^{1/2}} \tag{2.5.7c}$$

The angle i is uniquely determined by (2.5.7c) because it must lie in the first or second quadrant, depending on whether $\cos i$ is greater than or less than zero. The value of i, so determined, is used to enter (2.5.7a) and (2.5.7b). The signs in the number pair ($\sin \Omega$, $\cos \Omega$) serve to determine the quadrant of Ω uniquely. Thus the constants c_1, c_2, c_3 determine Ω and i uniquely and hence orient the orbital plane in space with respect to the rectangular coordinate system $O\xi\eta\zeta$.

It is important to notice the geometrical significance of the vector **h**. The rate at which the radius vector joining an origin O (either moving or fixed) to a moving point P sweeps out a surface is called the areal velocity relative to the origin O (see Figure 2.5.2). Let $\Delta\mathbf{A}$ represent vectorially the area

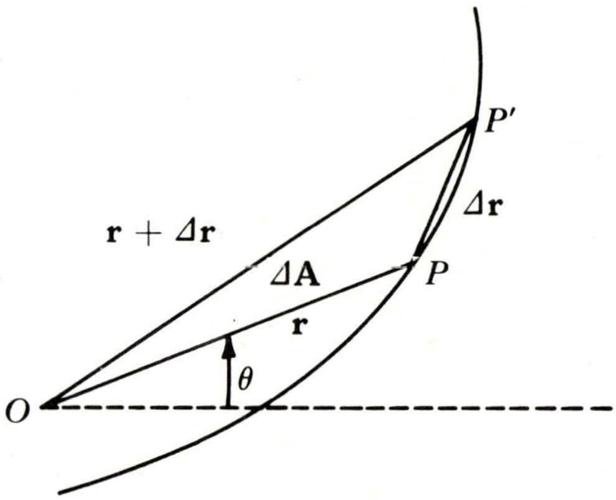

FIGURE 2.5.2

of the small triangle OPP' swept out by the radius vector in time Δt. The sense of $\Delta\mathbf{A}$ is determined by the equation

$$\Delta\mathbf{A} = \tfrac{1}{2}[\mathbf{r} \times (\mathbf{r} + \Delta\mathbf{r})] = \tfrac{1}{2}\mathbf{r} \times \Delta\mathbf{r} \tag{2.5.8}$$

Since the areal velocity $\dot{\mathbf{A}}$ at P is, by definition, $\lim_{\Delta t \to 0}(\Delta\mathbf{A}/\Delta t)$, we have

$$\dot{\mathbf{A}} = \tfrac{1}{2}\mathbf{r} \times \dot{\mathbf{r}} = \tfrac{1}{2}\mathbf{h} \tag{2.5.9a}$$

Thus the angular momentum per unit mass as viewed at O is equal to twice the rate as viewed at O at which area is being swept out by the radius vector **r**. We can also write

$$\dot{A} = \tfrac{1}{2}h \qquad (\dot{A} = |\,\dot{\mathbf{A}}\,|) \tag{2.5.9b}$$

whence

$$A = \tfrac{1}{2}ht + c \qquad (A = |\,\mathbf{A}\,|) \tag{2.5.10a}$$

where c is the fourth integration constant to appear in the development thus far. Taking $A = 0$, at some initial time t_0, we have

$$c = -\tfrac{1}{2}ht_0 \tag{2.5.10b}$$

2.6. GENERAL SOLUTION OF THE EQUIVALENT ONE-BODY PROBLEM IN PRINCIPLE

Although we are interested in obtaining the general solution of (2.4.1), it is clear that if the general solution of the equivalent one-body problem is obtained, in which the force $\mathbf{F}$ on particle m is taken to be

$$\mathbf{F} = -\mu m \frac{F(r)\,\mathbf{r}}{r}, \qquad \mu \equiv k^2(m_0 + m)$$

directed toward a fixed center, the results can be interpreted in terms of the relative motion of m about m_0. Thus we continue our study of the problem of a single particle of mass m, moving about a fixed center of force (taken to be the origin), under the influence of a central force of the form

$$\mathbf{F} = -\mu m \frac{F(r)}{r}\,\mathbf{r} \tag{2.6.1}$$

Since the curl of a vector field of the form given in (2.6.1) vanishes identically, the field is conservative and we can associate a potential energy $V(r)$ with it such that $\mathbf{F} = -\nabla V(r)$. In the problem to be studied, all the conditions necessary for the application of (1.5.26) are met; hence

$$T + V(r) = \tfrac{1}{2}mv^2 + V(r) = E \tag{2.6.2}$$

where E is the total energy of the system.

Since the motion in a central field is confined to a plane, a polar coordinate system (r, θ) will be chosen in this plane with the pole at the center of force. For the moment, the orientation of the reference line from which θ is to be measured will not be further specified.

The velocity $\dot{\mathbf{r}}$ and the acceleration $\ddot{\mathbf{r}}$ expressed in terms of these polar coordinates have the forms (see Appendix 1)

$$\dot{\mathbf{r}} = \dot{r}\mathbf{u}_r + r\dot{\theta}\mathbf{u}_\theta \tag{2.6.3}$$

$$\ddot{\mathbf{r}} = (\ddot{r} - r\dot{\theta}^2)\,\mathbf{u}_r + (2\dot{r}\dot{\theta} + r\ddot{\theta})\,\mathbf{u}_\theta \tag{2.6.4}$$

where $\mathbf{u}_r = \mathbf{r}/r$ and $\mathbf{u}_\theta$ is a unit vector in the plane, perpendicular to $\mathbf{u}_r$, in the direction of increasing θ. Thus (2.6.2) can be written

$$\tfrac{1}{2}m(\dot{r}^2 + r^2\dot{\theta}^2) + V(r) = E \tag{2.6.5}$$

The angular momentum per unit mass $\mathbf{h}$ has already been found to be a constant of the motion. It then follows directly from this fact and from (2.6.3) that

$$\mathbf{h} = \mathbf{r} \times \dot{\mathbf{r}} = r\mathbf{u}_r \times (\dot{r}\mathbf{u}_r + r\dot{\theta}\mathbf{u}_\theta) = r^2\dot{\theta}\mathbf{u}_A = h\mathbf{u}_A \tag{2.6.6}$$

whence

$$r^2\dot{\theta} = h \tag{2.6.7}$$

where h is a constant.† The quantity $\dot{\theta}^2$ in (2.6.5) can now be replaced by h^2/r^4 and the resulting equation solved for $\dot{r}$ to obtain

$$\dot{r} = \pm\left[\frac{2}{m}\left(E - V - \frac{mh^2}{2r^2}\right)\right]^{1/2} \tag{2.6.8}$$

or

$$dt = \pm \frac{dr}{\{(2/m)[E - V - (mh^2/2r^2)]\}^{1/2}} \tag{2.6.9}$$

At time $t = 0$, let $r = r_0$. Integrating (2.6.9), we have

$$t = \pm\int_{r_0}^{r} \frac{dr}{\{(2/m)[E - V - (mh^2/2r^2)]\}^{1/2}} \tag{2.6.10}$$

Equation (2.6.10) gives t as a function of r, E, r_0, h. In principle, it might be solved to give r as a function of t, E, r_0, h in the form $r = r(E, r_0, h, t)$. Then (2.6.7) can be used to write

$$d\theta = \frac{h}{r^2}\,dt \tag{2.6.11}$$

Let $\theta = \theta_0$ when $t = 0$. The integral of (2.6.11) is

$$\theta(\theta_0, E, r_0, h, t) = \theta_0 + \int_0^t \frac{h}{r^2(t)}\,dt \tag{2.6.12}$$

Formally, the problem has been reduced to two quadratures, with the four constants of integration E, h, r_0, θ_0. We have already obtained the four constants of integration c_1, c_2, c_3, and c from our previous integration of Equations (2.4.7) and (2.5.9b).

At first it might appear that we have obtained eight independent constants of integration for a vector differential equation (2.4.7) which requires but six independent constants of integration for its general solution. Recalling, however, that $h = (c_1^2 + c_2^2 + c_3^2)^{1/2}$ and noting

† Since $h = |\mathbf{h}|$ is a positive constant, θ must be measured so that $\dot{\theta}$ is positive.

that the constant c arose in integrating an alternative form of (2.6.7), we see that θ_0 and c are not independent. Thus we have but six independent constants, say E, c_1, c_2, c_3, r_0, θ_0.

Although the equivalent one-body problem has been solved formally, practically speaking, the integrals (2.6.10) and (2.6.12) are often analytically quite intractable, and it is usually more convenient to perform the integration in some other fashion.

Thus far we have been discussing the solution to the equivalent one-body problem as though it meant finding r and θ as functions of the time with E, h, etc. as constants of integration. Actually, we are more often interested in the dependence of r on θ, which we can determine by eliminating the parameter t. Because of the intrinsic importance of the case with $F(r) = 1/r^2$, we now take up the detailed treatment of this problem from the point of view of attempting to obtain first a relationship between r, θ and the constants of integration.

2.7. THE ONE-BODY PROBLEM FOR AN INVERSE SQUARE LAW OF FORCE

Consider the case for which $F(r) = r^{-2}$ so that (2.6.1) becomes

$$\mathbf{F} = -\mu m r^{-3}\mathbf{r} \tag{2.7.1}$$

To find V we note that $\nabla V(r) = V'(r)\, r^{-1}\mathbf{r} = -\mathbf{F}$ so that

$$V'(r) = \frac{dV}{dr} = \frac{\mu m}{r^2}$$

and

$$V(r) = -\mu m r^{-1}$$

where the constant of integration has been taken to be zero. The energy equation (2.6.2) can be written

$$\frac{v^2}{2} - \frac{\mu}{r} = \alpha \qquad \left(\alpha = \frac{E}{m}\right) \tag{2.7.2}$$

It is clear that from (2.6.7)

$$\frac{d}{dt} = \frac{d}{d\theta}\frac{d\theta}{dt} = \frac{h}{r^2}\frac{d}{d\theta} \tag{2.7.3}$$

Eliminating t from (2.6.9) through the use of (2.7.3), we write

$$d\theta = \pm \frac{h\,dr}{r^2\{(2/m)[E + (\mu m/r) - (mh^2/2r^2)]\}^{1/2}}$$

$$= \pm \frac{r^{-2}\,dr}{[(2E/mh^2) + (2\mu/h^2r) - (1/r^2)]^{1/2}} \tag{2.7.4}$$

Let $u = r^{-1}$, so that $du = -r^{-2}\,dr$. If (2.7.4) is integrated indefinitely, we can write

$$\theta = \omega^* \pm \frac{\pi}{2} \mp \int \frac{du}{[(2E/h^2m) + (2\mu/h^2)\,u - u^2]^{1/2}} \tag{2.7.5}$$

The quantities $\omega^* \pm (\pi/2)$ are constants of integration determined by the initial conditions. Neither of these constants will necessarily be the same as the initial angle θ_0 at time $t = 0$. The terms $\pm\pi/2$ are added because it will be more convenient to express the orbit in terms of a cosine rather than a sine function of θ (see Appendix 2).

The indefinite integral in (2.7.5) is readily evaluated. It is found that

$$\int \frac{du}{[(2E/mh^2) + (2\mu u/h^2) - u^2]^{1/2}} = \sin^{-1} \frac{2u - B}{(B^2 + 4C)^{1/2}} \tag{2.7.6}$$

where

$$B = \frac{2\mu}{h^2}, \qquad C = \frac{2E}{mh^2}$$

If the result given in (2.7.6) is substituted into (2.7.5), we have

$$\mp\sin^{-1} \frac{2u - B}{(B^2 + 4C)^{1/2}} = \theta - \omega^* \mp \frac{\pi}{2}$$

so that

$$\frac{h^2u\mu^{-1} - 1}{[1 + (2Eh^2/m\mu^2)]^{1/2}} = \sin\left[\mp\left(\theta - \omega^* \mp \frac{\pi}{2}\right)\right]$$

$$= \sin\left[\frac{\pi}{2} \mp (\theta - \omega^*)\right] = \cos(\theta - \omega^*)$$

Solving for $h^2u\mu^{-1}$, we have

$$h^2u\mu^{-1} = 1 + \left(1 + \frac{2Eh^2}{m\mu^2}\right)^{1/2} \cos(\theta - \omega^*)$$

or, in terms of r,

$$r = \frac{h^2\mu^{-1}}{1 + [1 + (2h^2E/m\mu^2)]^{1/2} \cos w} \tag{2.7.7}$$

where the angle w has been written for $\theta - \omega^*$.†

Equation (2.7.7) is the equation of a conic in which $2ep = h^2/\mu$. Thus $p = h^2/2\mu e$ is positive. The eccentricity of the conic is given by

$$e = \left(1 + \frac{2h^2E}{m\mu^2}\right)^{1/2} \tag{2.7.8}$$

The point on a conic orbit closest to the occupied focus (in this case, closest to the origin O) is called *the pericenter*. Thus the polar coordinates of pericenter are $[h^2\mu^{-1}/(1 + e), 0]$.

Equation (2.7.8) can be rewritten in the form

$$e^2 = 1 + \frac{2h^2\alpha}{\mu^2} \tag{2.7.9}$$

For an ellipse, $0 < e < 1$ and $2ep = a(1 - e^2)$, where a is the semimajor axis of the ellipse. For a hyperbola, $e > 1$, and $2ep = a(e^2 - 1)$, where a $(a > 0)$ is the semitransverse axis. For a parabola, $e = 1$, and $2p = 2q$, where q is called the pericentric distance (see Appendix 2). Using these facts, plus the fact that $h^2\mu^{-1} = 2ep$, we can write

$$h^2 = \mu a(1 - e^2) \qquad \text{(ellipse)} \tag{2.7.10a}$$

$$h^2 = \mu a(e^2 - 1) \qquad \text{(hyperbola)} \tag{2.7.10b}$$

$$h^2 = 2\mu q \qquad \text{(parabola)} \tag{2.7.10c}$$

If (2.7.9) is solved for α, we can use (2.7.2) and write

$$\alpha = \frac{\mu^2}{2h^2}(e^2 - 1) = \frac{v^2}{2} - \frac{\mu}{r} \tag{2.7.11}$$

For elliptic orbits $(e^2 < 1)$ α is negative, for parabolic orbits $(e^2 = 1)$ α is zero, and for hyperbolic orbits $(e^2 > 1)$ α is positive. If the expressions (2.7.10a) and (2.7.10b) are substituted into (2.7.9) and the resulting equations are solved for α, it is readily seen that

$$\alpha = -\frac{\mu}{2a} \qquad \text{(ellipse of semimajor axis } a) \tag{2.7.12a}$$

$$\alpha = \frac{\mu}{2a} \qquad \text{(hyperbola of semitransverse axis } a) \tag{2.7.12b}$$

$$\alpha = 0 \qquad \text{(parabola)} \tag{2.7.12c}$$

† The angle w is called the *true anomaly* and it must be measured so that $\dot{w}$ is positive (see Appendix 2).

If the expressions (2.7.12a), (2.7.12b), and (2.7.12c) are substituted into (2.7.11) and the resulting equations are solved for v^2, it is found that

$$v^2 = \mu\left(\frac{2}{r} - \frac{1}{a}\right) \qquad \text{(ellipse)} \qquad (2.7.13a)$$

$$v^2 = \mu\left(\frac{2}{r} + \frac{1}{a}\right) \qquad \text{(hyperbola)} \qquad (2.7.13b)$$

$$v^2 = \frac{2\mu}{r} \qquad \text{(parabola)} \qquad (2.7.13c)$$

In (2.7.13c) v is referred to as the escape velocity (speed) at a distance r.

Let t_0 represent a certain instant of time and T represent the *time of pericenter passage* in a conic orbit ($t_0 < T$). If the motion is in an ellipse, it is to be understood that T is referenced to a specific revolution. Let A_T represent the area described in the time interval $T - t_0$. Using (2.5.10a) and (2.5.10b), we can write

$$A_T = \tfrac{1}{2}h(T - t_0) \qquad (2.7.14)$$

Alternatively, we have

$$dA = \frac{1}{2}h\,dt = \frac{1}{2}r^2\,dw = \frac{h^4}{2\mu^2}\frac{dw}{(1 + e\cos w)^2}$$

so that

$$\frac{1}{2}h(T - t_0) = \frac{h^4}{2\mu^2}\int_{w_0}^{w_T}(1 + e\cos w)^{-2}\,dw = A_T(h, e, w_0) \qquad (2.7.15)$$

where w_T is some nonnegative multiple of 2π. The definite integral appearing in (2.7.15) can readily be integrated for any of the three conic orbits (see Chapter 3), so that the time of pericentric passage can be determined from (2.7.15) if h, e, t_0, and w_0 are known. Clearly, a similar analysis can be made if $t_0 > T$.

The angle ω^* is the angle between the line joining the focus and pericenter (the pericentric line) and the polar axis $\theta = 0$. If the polar axis is chosen to coincide with the line of nodes, as will be done here, ω^* is called the *argument of pericenter* and it is replaced by the symbol ω. The symbol θ is then replaced by $u = w + \omega$, and u is called the *argument of latitude*.

The parameters a (or q), e, i, ω, Ω, T are called orbital elements, and they form a set of six independent constants for the one-body problem with an inverse square law of force. They are related to the six independent constants of integration α, ω, c, c_1, c_2, and c_3 through the equations that have already been derived. Thus i and Ω are determined by c_1, c_2, c_3, as is h, through (2.5.7a), (2.5.7b), and (2.5.7c). This latter set of constants is determined when the initial position $\mathbf{r}_0$ and the initial velocity $\dot{\mathbf{r}}_0$ are given at some instant t_0. To see this, note that the constants c_1, c_2, c_3, $h = (c_1^2 + c_2^2 + c_3^2)^{1/2}$ and $c = -\frac{1}{2}ht_0$ are determined by the relationship $\mathbf{r}_0 \times \dot{\mathbf{r}}_0 = c_1\mathbf{i} + c_2\mathbf{j} + c_3\mathbf{k}$. If $\mathbf{r}_0$ and $\dot{\mathbf{r}}_0$ are prescribed, r_0 and v_0 are known, and hence α is determined by (2.7.11) As we shall see in Chapter 6, the argument of latitude, u, is related to the coordinate ζ, $\dot{\zeta}$ and the inclination i through the equation $\zeta = r \sin i \sin u$. The angles i and Ω are determined by (2.5.7a), (2.5.7b), and (2.5.7c) and ζ_0, $\dot{\zeta}_0$, r_0, and v_0 are prescribed; hence the initial value u_0 is known uniquely. The initial value w_0 may be calculated uniquely from (2.7.7) and (2.7.7)$^{\cdot}$. Thus ω is known. This completes the determination of the integration constants α, ω, c, c_1, c_2, c_3, from a prescription of the values of $\mathbf{r}_0$ and $\dot{\mathbf{r}}_0$ at t_0.

In the process of determining the integration constants α, ω, c, c_1, c_2, c_3 from the initial conditions, we obtained the values of the three orbital elements ω, i, and Ω. The values of a (or q) and e are readily obtained from (2.7.9), (2.7.10a), (2.7.10b), and (2.7.10c) when h and α are known. The remaining orbital element T can, in principle, be calculated from (2.7.15), for it is the only unknown remaining in the equation. In practice, the determination of T is more easily carried out through the use of auxiliary variables to be introduced in Chapter 3.

An expression for the period of revolution P for elliptic motion about a fixed origin may readily be derived. The area swept out by the radius vector in time $t - t_0$, $t > t_0$, as given by (2.5.10a), is

$$A = \tfrac{1}{2}h(t - t_0)$$

When $t = t + P$, $P > 0$, the area swept out is

$$A + \pi ab = \tfrac{1}{2}h(t + P - t_0) \qquad [b = a(1 - e^2)^{1/2}]$$

Subtracting the former of the last two equations from the latter, we obtain

$$P = 2\pi abh^{-1} = \frac{2\pi\sqrt{a^3}}{\sqrt{\mu}} \tag{2.7.16}$$

The main daily motion n of a mass m in an elliptic orbit about the force center is defined by the equation

$$n = \frac{2\pi}{P} \tag{2.7.17}$$

The parameter n is the angle described in unit time by an imaginary point moving with constant speed in a circle around the force center and accomplishing a full revolution in P units of time. If (2.7.16) is substituted into (2.7.17), we may write

$$n^2a^3 = \mu \tag{2.7.18}$$

2.8. THE TWO-BODY PROBLEM FOR AN INVERSE SQUARE LAW OF FORCE

In carrying out the development of the one-body problem in the preceding two sections, the equation

$$\frac{v^2}{2} - \frac{\mu}{r} = \alpha \qquad \left(\alpha \equiv \frac{E}{m}\right) \qquad [2.7.2]$$

was introduced as a special case of the energy equation

$$T + V(r) = E \qquad [2.6.2]$$

where the quantities T and E in (2.6.2) were defined in terms of a fixed origin. There is a counterpart energy equation, similar in form to (2.7.2), for the relative motion of m and m_0. To see this, return to the equation

$$\ddot{\mathbf{r}} + \mu \frac{F(r)}{r} \mathbf{r} = \mathbf{0}, \qquad \mu \equiv k^2(m_0 + m) \qquad [2.4.1]$$

for the relative motion of m about m_0 and form the dot product of each side of (2.4.1) with $\dot{\mathbf{r}}$. Since

$$\dot{\mathbf{r}} \cdot \ddot{\mathbf{r}} = \frac{1}{2}(\dot{\mathbf{r}} \cdot \dot{\mathbf{r}})^{\cdot} = \frac{d}{dt}\left(\frac{v^2}{2}\right)$$

$$\dot{\mathbf{r}} \cdot \mathbf{r} = (\dot{r}\mathbf{u}_r + r\dot{\theta}\mathbf{u}_\theta) \cdot (r\mathbf{u}_r) = r\dot{r}$$

where v is the speed of m with respect to an origin at m_0, we can write

$$\frac{d}{dt}\left(\frac{v^2}{2}\right) + \mu F(r)\,\dot{r} = 0 \tag{2.8.1}$$

Integrating (2.8.1), we find

$$\tfrac{1}{2}v^2 + \mu \int F(r)\, dr = \text{constant} \tag{2.8.2}$$

Introducing $F(r) = 1/r^2$ into (2.8.2), we obtain

$$\frac{1}{2}v^2 - \frac{\mu}{r} = \alpha^* \tag{2.8.3}$$

Equation (2.8.3) is similar in form to (2.7.2), but the constant α^* is not to be interpreted as the total energy per unit mass; and the speed v that appears in (2.8.3) is to be interpreted as the speed of m in a nonrotating system of axes whose origin moves with m_0.

With this understanding of the meaning of (2.8.3), it is clear that the dynamical problem of the motion of m relative to m_0 is the same as the dynamical problem of the motion of m about a fixed force center under the influence of a central force given by

$$\mathbf{F} = -\mu m \frac{F(r)}{r} \mathbf{r}, \qquad \mu \equiv k^2(m_0 + m)$$

All the equations already derived for the fixed force center problem hold equally well for the two-body problem.

For two-body motion, we rewrite (2.7.16) in the form

$$P = \frac{2\pi(a^3)^{1/2}}{k(m_0 + m)^{1/2}} \tag{2.8.4}$$

Equation (2.8.4) applies to every planet and periodic comet in the solar system. Thus if m_0 refers to the Sun and m and m_1 refer to the masses of any two bodies revolving about the Sun in elliptical orbits, we have

$$\frac{P^2(m_0 + m)}{P_1^2(m_0 + m_1)} = \frac{a^3}{a_1^3} \tag{2.8.5}$$

Equation (2.8.5) is usually referred to as Kepler's third law in its generalized form. When the masses of the planets are neglected in comparison to the mass of the Sun, we obtain the more frequently encountered form of Kepler's third law, namely,

$$\frac{P^2}{P_1^2} = \frac{a^3}{a_1^3} \tag{2.8.6}$$

2.9. NUMERICAL VALUES OF THE GRAVITATIONAL CONSTANT

The numerical value of the gravitational constant k depends on the choice of units of mass, length, and time that are to be used in (2.7.16). The particular choice of units of mass, length, and time, however, is determined by the problem of interest. For example, if we study the the motion of a planet about the Sun, it is not convenient to choose the same units of mass, length, and time we would use to study the motion of an artificial satellite about the Earth. In the present work, we will be concerned primarily with orbital motion about either the Sun or the Earth; hence we will restrict our considerations to the choices of units for these two cases.

Case 1. The Orbit of a Planet about the Sun

To determine the constant of gravitation, it is clearly sufficient to know the mass of any one of the planets, its period of revolution and semimajor axis, and the mass of the Sun. For convenience, in most applications of celestial mechanics to the motion of bodies in the solar system, the mass of the Sun is chosen as the unit of mass. This unit is called one solar mass (sm). The unit of time is taken to be the mean solar day (actually, the ephemeris day†), where the mean solar day (msd) is defined to be the imagined interval between transits of a fictitious "mean" Sun having a prescribed uniform sidereal (with respect to the stars) motion in the plane of the Earth's equator.‡ Finally, because of the difficulty of measuring distances in the solar system, an arbitrary numerical value

$$k_n = 0.017\,202\,098\,95 \tag{2.9.1}$$

is adopted for k. This value is then used in defining a unit of distance called an astronomical unit (au).

Definition of astronomical unit. An astronomical unit is the length, predicted by (2.8.4), of the semimajor axis of the orbit of a hypothetical planet of mass 1/354 710 sm and period 365.256 383 5 msd in elliptic motion about the Sun; that is,

$$a = \left[\frac{0.017\,202\,098\,95\left(1 + \dfrac{1}{354\,710}\right)^{1/2} 365.256\,383\,5}{2\pi}\right]^{2/3} = 1 \text{ au} \tag{2.9.2}$$

The semimajor axis of a planet of mass m, in orbit around the Sun, is expressed in astronomical units when in the expression

$$a = \left[\frac{k(1+m)^{1/2} P}{2\pi}\right]^{2/3} \tag{2.9.3}$$

† The variable t in Newton's second law is referred to as ephemeris time. It differs slightly from mean solar time due to slight irregularities in the Earth's rate of rotation.

‡ The constants for this "mean" Sun are specified by Newcomb [1].

k has the numerical value given in (2.9.1), P is expressed in mean solar days and $(1 + m)^{1/2}$ is in solar mass.

The quantity k is called the Gaussian gravitational constant. Its numerical value was arrived at by Gauss in the following way. For the Earth–Moon mass m_1, the period P_1 of the Earth's revolution about the Sun, and the semimajor axis a_1 of its orbit, he took the values

$$m_1 = \frac{1}{354\,710} = 0.000\,002\,819\,20 \text{ sm}$$

$$P_1 = 365.256\,383\,5 \text{ msd} \qquad \text{(1 sidereal year)}$$

$$a_1 = 1 \text{ au}$$

$$(1 + m_1)^{1/2} = 1.000\,001\,409\,60 \sqrt{\text{sm}}$$

and calculated k from (2.8.4). He found that

$$k = \frac{2\pi}{P_1(1 + m_1)^{1/2}} = 0.017\,202\,098\,95 \frac{\sqrt{\text{au}^3}}{\text{msd} \sqrt{\text{sm}}} \tag{2.9.4}$$

The data used by Gauss have been revised. Better values are

$$m_1 = \frac{1}{328\,466} \text{ sm}$$

$$P_1 = 365.256\,365\,56 - 0.000\,000\,11\, T \text{ msd}$$

where T is measured in Julian centuries of 36 525 msd from 1900. On the basis of such revised data, the semimajor axis a of the Earth's orbit calculated from (2.9.3) is given as

$$a = 1.000\,000\,230 \text{ au} \tag{2.9.5}$$

The semimajor axis of a planetary orbit when calculated from Kepler's third law is determined without taking into account perturbations due to the disturbing influences of other planets in the solar system. Actually, an ellipse with a semimajor axis of 1.000 000 20 au is a better approximation to the Earth's actual orbit than is the the one with a given by (2.9.5).

For many purposes, it is convenient to use a system of units in which $k_n = 1$. Returning to Kepler's third law, we write

$$k = \frac{\sqrt{a^3}}{(P/2\pi)(1 + m)^{1/2}} \tag{2.9.6}$$

Define new units of mass and time by the equations[†]

$$1 \text{ augmented solar mass (asm)} \equiv (1 + m) \quad \text{solar mass} \tag{2.9.7}$$

$$\begin{aligned} 1 \text{ long day (ld)} &\equiv \frac{365.256\,383\,5}{2\pi} \quad \text{msd} \\ &= 58.132\,358\,9 \quad \text{msd} \end{aligned} \tag{2.9.8}$$

If, in Kepler's third law, time is now expressed in long days, mass in augmented solar mass, and distance in astronomical units, then k is numerically equal to 1. Including units, we have

$$k = 1 \, \frac{\sqrt{\text{au}^3}}{\text{ld} \sqrt{\text{asm}}} \tag{2.9.9}$$

Since the requirements at launch for interplanetary vehicles must be stated in terms of laboratory units, the conversion factor from the astronomical unit to a laboratory unit of distance (for example, kilometer) is needed. The measurement of the astronomical unit is a very difficult problem and it is not, at present, known with sufficient accuracy to take care of the precision required for interplanetary navigation.

One method of relating the astronomical unit to laboratory units used the orbit of the minor planet Eros, which in one of its close approaches came within 14 000 000 miles of the Earth. Eros has a bright, starlike image and it is fairly easy to observe. Its distance in laboratory units is measurable using principles of triangulation. The value of the astronomical unit can then be found by equating the observed distance to the distance predicted from theory.

A second method utilizes the perturbations of the orbit of Eros during a close approach to Earth. The orbit of Eros about the Sun can be predicted, taking into account the effect of important planetary perturbations. This orbit is then compared with the observed motion, and the mass of the Sun plus the Earth–Moon system (that is, $1 + m_1$) is deduced from the observed differences. Then the astronomical unit is connected with the Earth–Moon mass by Kepler's third law.

Still another method was introduced in February 1958 when Price *et al.* of Lincoln Laboratories of MIT sent radar beams (440 MHz) to Venus and received the echo. If the known velocity of light, the

[†] The astronomer does not give these new units of mass and time labels to distinguish them from the units of mass and time given earlier in this section. Nonetheless, it is helpful to nonspecialists in astronomy to make such a distinction. For this reason, the unconventional labels asm and ld are introduced here and the label lm is introduced later in (2.9.18).

accurately known distance between the centers of Venus and Earth in astronomical units and the observed travel time of the radio beam is used, the value of the astronomical unit in kilometers can be determined.

If the distance, in kilometers, between any two members of the solar system is known at some instant, then all its measurements are known and any interplanetary distance can be expressed in terms of kilometers. The value of

$$1 \text{ au} = 1.494\,67 \times 10^8 \text{ km} \tag{2.9.10}$$

based on radar measurements, is adopted here and will be used for purposes of calculation. Then the unit of velocity that we will employ is

$$1 \frac{\text{au}}{\text{msd}} = 1729.94 \frac{\text{km}}{\text{sec}} \tag{2.9.11}$$

Case 2. The Orbit of a Close Satellite of the Earth

In problems involving close artificial satellites of the Earth, it is not convenient to use units designed for the study of planets in orbit about the Sun. However, the same point of view can be adopted here as was adopted for heliocentric orbits.

The mass of the Earth m_e will be chosen as the unit of mass. Since the mass m_s of an artificial satellite is entirely negligible compared to the Earth's mass, we take $1 + m_s$ equal to one Earth mass (em) in all cases. Thus $k_n = \sqrt{\mu}$ numerically. The minute is chosen as a convenient unit of time, for the periods of artificial Earth satellites are customarily of the order of 100 minutes or so. The equatorial radius of the Earth, a_e, as it is now best known, is adopted as a unit of distance and called a g-radius (gr). Specifically, we take

$$1 \text{ g-radius} = 6378.270 \text{ km} = 3963.272\,646 \text{ statute miles} \tag{2.9.12}$$

An arbitrary value of

$$k = 0.074\,366\,16 \quad \frac{\sqrt{\text{gr}^3}}{\text{min}\sqrt{\text{em}}} \tag{2.9.13}$$

is also adopted, based on calculations of k from the formula

$$g_e = \frac{k^2 m_e{}^*}{a_e{}^2} \tag{2.9.14}$$

In the determination of k, the best measurements currently available on (1) the acceleration g_e due to gravity at the Earth's equator and (2) the effective mass $m_e{}^*$ of the Earth, taking into account the effects of its atmosphere and rotation, have been used. As better data become available, we will have new preferred values for the Earth's equatorial radius, which we can call an e-radius (er). The e-radius will vary as better measurements are obtained, but the value of k and the g-radius are fixed for all time by (2.9.12) and (2.9.13).

The quantity $\sqrt{\mu}$ is given by

$$\sqrt{\mu} = 0.074\,366\,16 \;\frac{\sqrt{\mathrm{gr}^3}}{\mathrm{min}} \tag{2.9.15}$$

The semimajor axis a of a close-Earth satellite orbit will be expressed in g-radii, when, in the expression

$$a = \left(\frac{\sqrt{\mu}\,P}{2\pi}\right)^{2/3} \tag{2.9.16}$$

$\sqrt{\mu}$ has the value given in (2.9.18) and P is expressed in minutes. If $a = 1$ gr, then P is equal to 84.489 844 6 min.

For many purposes it is convenient to use a system of units in which $\sqrt{\mu} = 1$ numerically. We may write

$$\sqrt{\mu} = \frac{\sqrt{a^3}}{P/2\pi} \tag{2.9.17}$$

Define a new unit of time by the equation†

$$\begin{aligned} 1 \text{ long minute (lm)} &= \frac{84.489\,844\,6}{2\pi}\,\mathrm{min} = 13.446\,976\,4 \text{ min} \\ &= 806^{s}.818\,584 \\ (1 \text{ min} &= 0.074\,366\,16 \text{ lm}) \end{aligned} \tag{2.9.18}$$

Then we have

$$\sqrt{\mu} = 1\,\frac{\sqrt{\mathrm{gr}^3}}{\mathrm{lm}} \tag{2.9.19}$$

When P is expressed in long minutes in (2.9.16), we can take $\mu = 1$ numerically.

For the unit of velocity for geocentric orbits, we have

$$1\,\frac{\mathrm{gr}}{\mathrm{lm}} = 7.905\,457\,\frac{\mathrm{km}}{\mathrm{sec}} = 4.912\,220\,\frac{\mathrm{mi}}{\mathrm{sec}} \tag{2.9.20}$$

† See footnote for Equation (2.9.7).

Exercises

1. A particle is moving in a parabola, and at a distance r from the focus its speed is v. Show that its acceleration is composed of

$$\frac{1}{4r}\frac{d}{dr}(v^2 r)$$

parallel to the axis and

$$\frac{r}{4}\frac{d}{dr}\left(\frac{v^2}{r}\right)$$

along the radius vector outward.†

2. A particle moves in the axis x with acceleration μ/x^2 toward the origin, where μ is a constant. If the particle starts from rest at $x = a$, show that the time of arriving at a distance x is

$$\left(\frac{a^3}{2\mu}\right)^{1/2}\left[\cos^{-1}\left(\frac{x}{a}\right)^{1/2} + \left(\frac{x}{a} - \frac{x^2}{a^2}\right)^{1/2}\right]$$

3. A particle is projected vertically upward with speed v; after a time t, a second particle is projected vertically with speed v' ($<v$). If the particles meet as soon as possible, show that

$$t = \frac{v - v' + (v^2 - v'^2)^{1/2}}{g}$$

4. A particle is projected upward with speed V in a medium resisting with a force per unit mass, $\mathbf{F}$. Assume that $|\mathbf{F}|$ is proportional to the cube of the speed v, that is, $|\mathbf{F}| = Kv^3$, where K is a constant. If g is the magnitude of the acceleration of gravity and L is the *terminal* speed, given by the equation $KL^3 = g$, show that if $V = L$, the whole time T and the height of ascent s are connected by the equation

$$s + LT = \frac{2\pi}{3\sqrt{3}}\frac{L^2}{g}\ ‡$$

† Exercises 1, 2, 3, 10(a) and (b), and 13 are reprinted from A. E. H. Love, "Theoretical Mechanics," pp. 70, 69, 53, 55, and 79. Copyright by Cambridge Univ. Press, London and New York. Used with permission of Cambridge Univ. Press.

‡ Exercises 4 and 7 are reprinted from E. J. Routh, "A Treatise on Dynamics of a Particle," pp. 60 and 71. Copyright by Cambridge Univ. Press, London and New York. Used with permission of Cambridge Univ. Press.

5. A particle of mass m is subject to a repulsive force $Kr^{-6}\mathbf{r}$, where K is a constant. Initially it is at a very large distance from the force center O and has a velocity of magnitude such that if it were directed along a path through the center of force, the distance of closest approach would be a. Actually, it is projected, with the same velocity, along a path that would pass the force center at a distance b if it were not deflected. Find the minimum linear speed (in terms of k, a, b, m) that the particle experiences as it follows its trajectory.†
6. A particle of mass m is initially at rest at a distance a from a fixed force center. The law of force is $-Kr^{-2}$. Show that the time of fall is

$$\left(\frac{2m}{K}\right)^{1/2} \frac{\pi a^{3/2}}{4}$$

and that the ratio of the time average of the speed over the first half of the path to that over the second half is $(\pi - 2)/(\pi + 2)$.
7. Two particles A, B, mutually attracting each other according to the Newtonian law, are placed at rest at a given distance a apart. The particle B is now constrained to move away from A along a straight line joining them with a uniform speed u. Show that A will catch up if $u^2 < 2\mu/a$, where $\mu = k^2 m_B$ and m_B is the mass of B. Show that the time will be

$$\frac{1}{2}(\pi + 2\beta + \sin 2\beta)\left(\frac{b^3}{2\mu}\right)^{1/2}$$

where

$$\cos^2 \beta = \frac{a}{b} \qquad \text{and} \qquad \frac{2\mu}{b} = \frac{2\mu}{a} - u^2$$

8. A particle is projected at a distance a from the center of force with speed $2\mu^{1/2}/a$ at an angle $\pi/4$ with the radial line, the attractive force per unit mass being

$$-\frac{\mu}{r^3}\left(3 + \frac{a^2}{r^2}\right)\frac{\mathbf{r}}{r}$$

Determine the orbit corresponding to a finite time to the center of force and show that this time is

$$\frac{a^2}{(2\mu)^{1/2}}\left(2 - \frac{\pi}{2}\right)^{\ddagger}$$

† Exercises 5 and 6 are reprinted from R. A. Becker, "Introduction to Theoretical Mechanics," p. 245. Copyright 1954 by McGraw-Hill, New York. Used with permission of McGraw-Hill Book Company.

‡ Exercises 8, 11, and 12 are reprinted from J. M. A. Danby, "Fundamentals of Celestial Mechanics," pp. 84, 141, and 140. Copyright 1962 by Macmillan, New York. Used with permission of Macmillan.

9. An apsis (plural: apsides) is a point of a central orbit at which the tangent is at right angles to the radius vector. The angle between consecutive apsides, in the order in which the moving point passes through them, is called an apsidal angle. If the central acceleration is a single-valued function of r alone, show that

   (a) The radius vector to any apsis divides the orbit symmetrically.

   (b) There are, at most, two different magnitudes of apsidal distances in any one orbit and but one magnitude of apsidal angle.

10. (a) Show that when the orbit is an ellipse described about the center of force, the acceleration is proportional to the radius vector.

    (b) Show that if any conic is described as a central orbit about a focus, the acceleration is $-\mu\mathbf{r}/r^3$ toward the focus where μ is a constant.

11. Consider the problem of finding the value to ascribe to a for the Earth, as a result of better values of m and P than those used by Gauss, while retaining his value of k. Let m_1 and P_1 be new values for the mass and period in question, and show that

$$a = \left(\frac{1 + m_1}{1 + m}\right)^{1/3} \left(\frac{P_1{}^3}{P^3}\right)^{1/2}$$

and thus, to terms of the first order,

$$a = 1 + \frac{1}{3}\left[m_1 - m + \frac{2(P_1 - P)}{P}\right]$$

Take Newcomb's values $1/m_1 = 329\,390$, $P_1 = 365.256\,360\,42$, and show that they give $a = 1.000\,000\,030$.

12. In order to find the Gaussian constant k, it is necessary to know the mass m_1 of the Earth–Moon system. To find m_1, first show that, to an accuracy greater than 1 part in 300 000,

$$\frac{1}{m_1} = \frac{P_1{}^2}{a_1{}^3}$$

where a_1 is the semimajor axis of the Moon's orbit about the Earth in astronomical units, P_1 is its sidereal period in sidereal years, and m_1 is expressed in units of solar mass. Hence show that

$$\frac{1}{m_1} = \left(\frac{\sin \pi_m}{\sin \pi_\odot}\right)^3 P_1{}^2$$

where π_m and $\pi_{\odot}$ are the parallaxes of the Moon and Sun (i.e., the angles subtended by the Earth's equatorial radius at the mean distance of these objects). The principal error in this method comes from the smallness of $\pi_{\odot}$.

13. If the perihelion distance of a comet is $(1/n)$th of the radius of the Earth's orbit, supposed circular, show that the comet will remain within the Earth's orbit for

$$\frac{\sqrt{2}}{3\pi}\left(1+\frac{2}{n}\right)\left(1-\frac{1}{n}\right)^{1/2} \quad \text{years}$$

the comet's orbit being parabolic.

14. The energy equation for the dynamical problem of the motion of mass m relative to mass M is given by (2.8.3). Show that the total energy of the two masses, in the system of the center of mass, is

$$\frac{\alpha^* mM}{m+M}$$

Chapter 3

The Determination of Position in the Orbital Plane

3.1. POSITIONS IN A CONIC ORBIT

A problem of considerable interest in celestial mechanics can be stated simply as follows: Given the orbital elements a or q, e, i, ω, Ω, T for a mass m, in relative motion about m_0 under their mutual gravitation, determine the coordinates r, w in the orbital plane at a given time t. We consider this problem next, deferring until later the more general problem of determining the coordinates with respect to the rectangular reference axes (for example, $O\xi\eta\zeta$ in Figure 2.5.1).

With reference to Figure 3.1.1, let C be a circle of radius a concentric with the orbital ellipse AQB, both curves touching at A and at B. The curve C is called the auxiliary circle for the ellipse AQB. Let F represent the occupied focus and QN, $Q'N$ represent line segments perpendicular to the major axis of the ellipse. The true anomaly w is shown and the eccentric anomaly E is defined by the figure.

To express r in terms of E, we note that for a given value of x, $-a \leqslant x \leqslant a$, the ordinate y_e of the ellipse is related to the ordinate y_c of the circle by the equation

$$\frac{y_e}{y_c} = \frac{b}{a} = \frac{QN}{Q'N} = \frac{a(1 - e^2)^{1/2}}{a} = (1 - e^2)^{1/2}$$

From this relationship and the geometry of the ellipse, it follows that

$$\begin{aligned} r &= [(FN)^2 + (QN)^2]^{1/2} = [(a \cos E - ae)^2 + (1 - e^2)\, a^2 \sin^2 E]^{1/2} \\ &= (a^2 - 2a^2 e \cos E + a^2 e^2 \cos^2 E)^{1/2} = a(1 - e \cos E) \end{aligned} \tag{3.1.1}$$

When $w = j\pi$, j an integer, $w = E$. For other values of w, the relationships

$$\cos w = \frac{FN}{r} = \frac{\cos E - e}{1 - e \cos E} \tag{3.1.2}$$

$$\sin w = \frac{QN}{r} = \frac{(1 - e^2)^{1/2} \sin E}{1 - e \cos E} \tag{3.1.3}$$

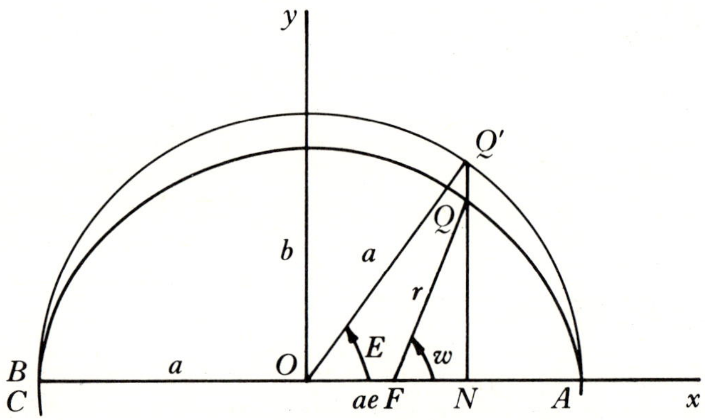

FIGURE 3.1.1

are readily deduced from Figure 3.1.1 and Equation (3.1.1). It follows directly that

$$\begin{aligned} \tan \frac{w}{2} &= \frac{\sin(w/2)}{\cos(w/2)} = \left(\frac{1 - \cos w}{1 + \cos w}\right)^{1/2} = \left(\frac{1 + e}{1 - e}\right)^{1/2} \left(\frac{1 - \cos E}{1 + \cos E}\right)^{1/2} \\ &= \frac{1}{\sqrt{\epsilon}} \tan \frac{E}{2} \end{aligned} \tag{3.1.4}$$

where

$$\epsilon \equiv \frac{1 - e}{1 + e} \tag{3.1.5}$$

In (3.1.4), $w/2$ and $E/2$ always lie in the same quadrant. When $0 < w < \pi$, they lie in the first quadrant; when $\pi < w < 2\pi$, they lie in the second quadrant. A similar argument holds for integer multiples of π.

The parameter ϵ is zero when $e = 1$ and positive when $e < 1$. For $e > 1$, $\epsilon < 0$ and $\sqrt{\epsilon}$ is a pure imaginary. We can, however, in a purely formal way define an auxiliary variable F for the hyperbola. With reference to Figure 3.1.1, we have, for an ellipse,

$$x = a\cos E = a\,\frac{e^{iE} + e^{-iE}}{2} = a\cosh F$$

$$y = b\sin E = b\,\frac{e^{iE} - e^{-iE}}{2i} = \frac{b\sinh F}{i} \qquad (i = \sqrt{-1})$$

where E is a real variable and $F = iE$ is a pure imaginary. The two equations just preceding suggest that, for the hyperbola, we define a new variable F by the equations $\cosh F = x/a$, $\sinh F = iy/b$. For the hyperbola, F is a real variable since $b = a(1 - e)^{1/2}$ is a pure imaginary. Since

$$\cosh^2 F - \sinh^2 F = \frac{x^2}{a^2} + \frac{y^2}{b^2} = 1$$

the form of the equation relating the coordinates x and y is the same as for the ellipse. Thus the equations that relate the polar coordinates r and w to the variable F can be obtained from the corresponding equations for the elliptic case by continuing the demand that r be positive and by replacing $\sin E$ by $\sinh F/i$ wherever it appears in these equations. This procedure leads to the relations

$$r^2 = a^2(e\cosh F - 1)^2 \tag{3.1.6a}$$

$$r = a(e\cosh F - 1) \qquad (r > 0) \tag{3.1.6b}$$

$$\cos w = \frac{e - \cosh F}{e\cosh F - 1} \tag{3.1.6c}$$

$$\sin w = \frac{(e^2 - 1)^{1/2}\sinh F}{e\cosh F - 1} \tag{3.1.6d}$$

$$\tan\frac{w}{2} = \frac{1}{\sqrt{-\epsilon}}\tanh\frac{F}{2} \tag{3.1.6e}$$

3.2. INTRODUCTION OF THE MEAN ANOMALIES FOR CONIC ORBITS

We now return to (2.7.7) and write

$$r = \frac{2ep}{1 + e\cos w} \tag{3.2.1}$$

where, from (2.7.10a), (2.7.10b), and (2.7.10c), $2ep = a(1 - e^2)$, $2q$, or $a(e^2 - 1)$, depending on whether $0 < e < 1$, $e = 1$, or $e > 1$ respectively. Since $w = \theta - \omega$, where ω is a constant, we can replace $\dot{\theta}$ by $\dot{w}$ in (2.6.7) and obtain the equation

$$\frac{h}{4e^2p^2}\, dt = \frac{dw}{(1 + e \cos w)^2} \tag{3.2.2}$$

Choosing as an epoch time, a time of pericenter passage T, we can integrate (3.2.2) and obtain

$$\frac{h}{4e^2p^2}(t - T) = I(w) \tag{3.2.3}$$

where

$$I(w) \equiv \int_0^w \frac{dw}{(1 + e \cos w)^2} \tag{3.2.4}$$

The factor $h/4e^2p^2$ appearing on the left-hand side of (3.2.3) can be written in each of the three forms

$$\frac{h}{4e^2p^2} = \frac{\sqrt{\mu}}{[a(1 - e^2)]^{3/2}} = \frac{n}{(1 - e^2)^{3/2}} \qquad \text{(ellipse)} \tag{3.2.5a}$$

$$\frac{h}{4e^2p^2} = \frac{\sqrt{\mu}}{[a(e^2 - 1)]^{3/2}} = \frac{\nu}{(e^2 - 1)^{3/2}} \qquad \text{(hyperbola)} \tag{3.2.5b}$$

$$\frac{h}{4e^2p^2} = \frac{\sqrt{\mu}}{(2q)^{3/2}} = \frac{\sigma}{2} \qquad \text{(parabola)} \tag{3.2.5c}$$

For the ellipse, we have from (2.7.18), $n = (\mu/a^3)^{1/2}$. By analogy, we define a parameter $\nu \equiv (\mu/a^3)^{1/2}$ for the hyperbola and a parameter $\sigma \equiv (\mu/2q^3)^{1/2}$ for the parabola. These quantities may be called mean daily motions for the respective conics, but they are of course in no way connected with the concept of a period of revolution for hyperbolic or parabolic orbits. We can now write

$$M \equiv n(t - T) = (1 - e^2)^{3/2}\, I(w) \qquad \text{(ellipse)} \tag{3.2.6a}$$

$$N \equiv \nu(t - T) = (e^2 - 1)^{3/2}\, I(w) \qquad \text{(hyperbola)} \tag{3.2.6b}$$

$$Q \equiv \sigma(t - T) = 2I(w) \qquad \text{(parabola)} \tag{3.2.6c}$$

where the parameters M, N, and Q defined by Equations (3.2.6) are called the *mean anomalies* for the respective conics. For the ellipse, the mean anomaly M is the angle that the direction of the planet would make with the direction to pericenter were the mass m to move with its mean daily motion n around m_0 in a circular orbit.

It is now clear that Equations (3.2.6) give the relationships between w and t which we seek. If w is known in terms of the time, the dependence of r on time may be obtained from (3.2.1). It remains to integrate $I(w)$, and to do this we introduce the standard change of variable

$$z = \tan \frac{w}{2} \tag{3.2.7}$$

The auxiliary equations

$$\cos w = \frac{1 - z^2}{1 + z^2} \tag{3.2.8a}$$

$$\sin w = \frac{2z^2}{1 + z^2} \tag{3.2.8b}$$

$$1 + e \cos w = \frac{(1 + e)(1 + \epsilon z^2)}{1 + z^2} \tag{3.2.8c}$$

$$r = \frac{2ep(1 + z^2)}{(1 + e)(1 + \epsilon z^2)} \tag{3.2.8d}$$

follow immediately from (3.2.7). When the transformation given in (3.2.7) is made, we obtain the equation

$$I^* \equiv \frac{(1 + e)^2}{2} I = \int_0^z \frac{1 + z^2}{(1 + \epsilon z^2)^2} dz \tag{3.2.9}$$

The integral I^* can be evaluated in terms of elementary functions, the forms of which depend on the parameter ϵ. We find that

$$I^* = \frac{(\epsilon - 1)\, z}{2\epsilon(1 + \epsilon z^2)} + \frac{\epsilon + 1}{2\epsilon \sqrt{\epsilon}} \tan^{-1} (\sqrt{\epsilon}\, z) \qquad (\epsilon > 0) \tag{3.2.10a}$$

$$I^* = \frac{(\epsilon - 1)\, z}{2\epsilon(1 + \epsilon z^2)} + \frac{\epsilon + 1}{2\epsilon \sqrt{-\epsilon}} \tanh^{-1} (\sqrt{-\epsilon}\, z) \qquad (\epsilon < 0) \tag{3.2.10b}$$

$$I^* = z + \frac{z^3}{3} \qquad (\epsilon = 0) \tag{3.2.10c}$$

It follows directly from the definitions of ϵ, E, F and from (3.2.7) that

$$\frac{\epsilon - 1}{2\epsilon} = \frac{-e}{1 - \epsilon}, \qquad \frac{\epsilon + 1}{2\epsilon} = \frac{1}{1 - e}$$

$$\sqrt{\epsilon}\, z = \tan \frac{E}{2}, \qquad \sqrt{-\epsilon}\, z = \tanh \frac{F}{2}$$

and if these results are substituted into (3.2.10), we find that

$$I^* = \frac{1}{2(1-e)\sqrt{\epsilon}}(E - e\sin E) \qquad (\epsilon > 0)$$

$$I^* = \frac{1}{2(1-e)\sqrt{-\epsilon}}(F - e\sinh F) \qquad (\epsilon < 0)$$

$$I^* = z + \frac{z^3}{3} \qquad (\epsilon = 0)$$

Equations (3.2.6) can now be written in the form

$$M - n(t - T) = E - e\sin E \qquad \text{(ellipse)} \tag{3.2.11}$$

$$N = \nu(t - T) = e\sinh F - F \qquad \text{(hyperbola)} \tag{3.2.12}$$

$$Q = \sigma(t - T) = \tan\frac{w}{2} + \frac{1}{3}\tan^3\frac{w}{3} \qquad \text{(parabola)} \tag{3.2.13}$$

Equation (3.2.11), which is known as Kepler's equation, is a fundamental equation in celestial mechanics. We will examine it in considerable detail in Section 3.4, as well as in Chapter 4.

Several important points should be noted in handling Equations (3.2.11), (3.2.12), and (3.2.13). Differentiating (3.2.11) with respect to E, (3.2.12) with respect to F, and (3.2.13) with respect to w, we note that

$$\frac{dM}{dE} = 1 - e\cos E > 0$$

$$\frac{dN}{dF} = e\cosh F - 1 > 0$$

$$\frac{dQ}{dw} = \frac{1}{4}\sec^4\frac{w}{2} > 0, \qquad -\pi < w < \pi$$

Thus M, N, and Q are strictly increasing functions of the variables E, F, and w respectively. Hence E, F, and w are uniquely determined for given values of M, N, and Q. We will defer discussion of just how E and F are to be determined for given values of M and N.

For the parabola, we have

$$Q(z) = z + \frac{z^3}{3} \qquad [Q = \sigma(t - T)] \tag{3.2.14}$$

Now

$$Q(0) = 0, \qquad \frac{dQ}{dz} > 0 \qquad \text{for all} \quad z$$

and the function Q is a strictly increasing function of z that takes on the value 0 when $z = 0$. This means that there is a unique real value of z for each value of Q. To find this unique value of z, we introduce a variable x by the equation

$$z = \sqrt[3]{x} - \frac{1}{\sqrt[3]{x}} \tag{3.2.15a}$$

so that

$$x - \frac{1}{x} = 3Q \tag{3.2.15b}$$

From its definition, if x is a real solution of (3.2.15b) for a given value of Q, then z calculated from (3.2.15a) is a real solution of (3.2.14) for the same Q. Since this solution for z is unique, it is the one we seek.

There is no ambiguity in the fact that Equation (3.2.15b) is quadratic in x. It can be verified readily that if x_1 is a solution of (3.2.15b), then the other solution is $-1/x_1$. Furthermore, either solution substituted into (3.2.15a) leads to the same value of z because

$$\left(-\frac{1}{x_1}\right)^{1/3} - \frac{1}{\left(-\dfrac{1}{x_1}\right)^{1/3}} = \sqrt[3]{x_1} - \frac{1}{\sqrt[3]{x_1}}$$

Since either solution of (3.2.15b) leads to the same value of z, choose the real solution

$$x = \frac{3Q + (9Q^2 + 4)^{1/2}}{2} \tag{3.2.15c}$$

Then, to solve (3.2.14) for z when $Q = \sigma(t - T)$ is prescribed, calculate x from (3.2.15c) and z from (3.2.15a). The true anomaly w is uniquely determined, for $w/2$ must lie in the open interval $(0, \pi/2)$ or $(-\pi/2, 0)$, depending on whether z is positive or negative.

3.3. DETERMINATION OF POSITION IN THE ORBITAL PLANE

We say that we have the complete solution to the two-body problem in the plane when we are able to determine the coordinates r and w at a prescribed time t. This requires that the appropriate elements (a or q, e and T) and the masses be known for the particular conic orbit of interest. Earlier we saw that if $\mathbf{r}$ and $\dot{\mathbf{r}}$ are prescribed at some time t, we can determine the six constants of integration and hence the six orbital elements a or q, e, i, ω, Ω, T. For the planar problem, we need determine only a or q, e and T from the initial values of $\mathbf{r}$ and $\dot{\mathbf{r}}$. Specific formulas for this purpose are tabulated. They are readily derived from the equations already developed in this and the preceding chapter.

Hyperbolic orbit:

$$a = \frac{r}{(rv^2/\mu) - 2} \tag{3.3.1a}$$

$$e = \left[1 + \frac{rv^2}{\mu}\left(\frac{rv^2}{\mu} - 2\right)\sin^2\theta\right]^{1/2} \tag{3.3.1b}$$

$$\cosh F = \frac{a + r}{ae} \tag{3.3.1c}$$

$$N = e \sinh F - F \tag{3.3.1d}$$

$$T = t - N\left(\frac{a^3}{\mu}\right)^{1/2} \tag{3.3.1e}$$

Elliptic orbit:

$$a = \frac{r}{2 - (rv^2/\mu)} \tag{3.3.2a}$$

$$e = \left[1 - \frac{rv^2}{\mu}\left(2 - \frac{rv^2}{\mu}\right)\sin^2\theta\right]^{1/2} \tag{3.3.2b}$$

$$\cos E = \frac{a - r}{ae} \tag{3.3.2c}$$

$$M = E - e \sin E \tag{3.3.2d}$$

$$T = t - M\left(\frac{a^3}{\mu}\right)^{1/2} \tag{3.3.2e}$$

Parabolic orbit:

$$q = \frac{r^2v^2 \sin^2\theta}{2\mu} \tag{3.3.3a}$$

$$e = 1 \tag{3.3.3b}$$

$$\cos w = \frac{2q - r}{r} \tag{3.3.3c}$$

$$Q = \tan\frac{w}{2} + \frac{1}{3}\tan^3\frac{w}{2} \tag{3.3.3d}$$

$$T = t - \sqrt{2}Q\left(\frac{q^3}{\mu}\right)^{1/2} \tag{3.3.3e}$$

Values for a or q, e and T can be obtained from (3.3.1), (3.3.2), or (3.3.3) when values of $\mathbf{r} = \xi\mathbf{i} + \eta\mathbf{j} + \zeta\mathbf{k}$ and $\dot{\mathbf{r}} = \dot{\xi}\mathbf{i} + \dot{\eta}\mathbf{j} + \dot{\zeta}\mathbf{k}$ are prescribed. The angle θ is the angle between $\mathbf{r}$ and $\dot{\mathbf{r}}$ and $\mu = k^2(m_0 + m)$. Since $\mathbf{r} \cdot \dot{\mathbf{r}} = rv \cos\theta = \xi\dot{\xi} + \eta\dot{\eta} + \zeta\dot{\zeta} = r\dot{r}$, it is clear that $\cos\theta$ is determined when $\mathbf{r}$ and $\dot{\mathbf{r}}$ are prescribed. Furthermore, θ must lie in the interval $[0, \pi]$; hence the quadrant of θ is determined by the sign of $\cos\theta$. We also note that $\dot{r} = v\cos\theta$; hence the sign of $\cos\theta$ tells us whether r is increasing or decreasing. This, in turn, permits us to determine a unique value of the appropriate anomaly. The following example will serve to fix the ideas.

Example 3.3.1. The following initial data are available for a close-Earth satellite at $t_0 = 1966$ February $1^{\mathrm{d}}.0625$ U.T.†

$$r_0 = 7822.5 \text{ km}, \qquad \mu = 40.02 \times 10^4 \frac{\text{km}^3}{\text{sec}^2}$$

$$v_0 = 7.1779 \frac{\text{km}}{\text{sec}}, \qquad \cos\theta_0 = -0.061746$$

It is required to find the orbital elements e, a, P and the time T, just preceding t_0, at which the satellite passed through pericenter.

We determine e, a, and P from the equations

$$e = \left[1 - \frac{r_0 v_0^2}{\mu}\left(2 - \frac{r_0 v_0^2}{\mu}\right)\sin^2\theta_0\right]^{1/2} = 0.062\,689$$

$$a = r_0\left(2 - \frac{r_0 v_0^2}{\mu}\right)^{-1} = 7878.4 \text{ km}, \qquad P = 2\pi\left(\frac{a^3}{\mu}\right)^{1/2} = 1.9292 \text{ hr}$$

The eccentric anomaly is obtained from the equation

$$\cos E_0 = \frac{a - r_0}{ae} = 0.11318$$

which has the two solutions $E_0 = 83°.490$, $E_0 = 276°.51$ on $[0, 2\pi]$. Since $\cos\theta_0 < 0$, $\dot{r}_0 < 0$ and the value $E_0 = 276°.51$ must be taken. The mean anomaly at t_0 is obtained from Kepler's equation

$$M_0 = E_0 - e\sin E_0 = 4.7638 \text{ rad}$$

The time of pericenter passage just preceding t_0 is given by

$$T = t_0 - M_0\left(\frac{a^3}{\mu}\right)^{1/2} = t_0 - 1.4628 \text{ hr}$$
$$= 1966 \text{ February } 1^{\mathrm{d}}.00155 \text{ U.T.}$$

† Universal time (U.T.) is defined in Section 5.3, Chapter 5.

Once the orbital elements (a or q, e, T) have been determined, the polar coordinates can be calculated at any instant t. For convenience, and in order to facilitate the subsequent discussion, we recapitulate the formulas for calculating r and w in the order in which they are applied when the time, masses, and orbital elements are given.

Hyperbolic orbit:

$$\mu = k^2(m_0 + m) \tag{3.3.4a}$$

$$\nu = \left(\frac{\mu}{a^3}\right)^{1/2} \tag{3.3.4b}$$

$$N = \nu(t - T) \tag{3.3.4c}$$

$$N = e \sinh F - F \tag{3.3.4d}$$

$$r = a(e \cosh F - 1) \tag{3.3.4e}$$

$$\tan \frac{w}{2} = \frac{1}{\sqrt{-\epsilon}} \tanh \frac{F}{2} \tag{3.3.4f}$$

Elliptic orbit:

$$\mu = k^2(m_0 + m) \tag{3.3.5a}$$

$$n = \left(\frac{\mu}{a^3}\right)^{1/2} \tag{3.3.5b}$$

$$M = n(t - T) \tag{3.3.5c}$$

$$M = E - e \sin E \tag{3.3.5d}$$

$$r = a(1 - e \cos E) \tag{3.3.5e}$$

$$\tan \frac{w}{2} = \frac{1}{\sqrt{\epsilon}} \tan \frac{E}{2} \tag{3.3.5f}$$

Parabolic orbit:

$$\mu = k^2(m_0 + m) \tag{3.3.6a}$$

$$\sigma = \left(\frac{\mu}{2q^3}\right)^{1/2} \tag{3.3.6b}$$

$$Q = \sigma(t - T) \tag{3.3.6c}$$

$$2x = 3Q + (9Q^2 + 4)^{1/2} \tag{3.3.6d}$$

$$z = \tan \frac{w}{2} = \sqrt[3]{x} - \frac{1}{\sqrt[3]{x}} \tag{3.3.6e}$$

$$r = \frac{2q}{1 + \cos w} = q \sec^2 \frac{w}{2} \tag{3.3.6f}$$

To illustrate, we consider the case of the elliptic orbit. Given the masses, we determine μ from (3.3.5a) and n from (3.3.5b) since a is given. We can now calculate M from (3.3.5c). Then E is given uniquely by (3.3.5d). The details of just how (3.3.5d) is to be solved for E will be described in the next section. If E is known, r can be obtained from (3.3.5e) since e is given. Finally, knowing E and e, we can determine w from (3.3.5f). Again we observe that $w/2$ and $E/2$ will always lie in the same quadrant.

3.4. SOLUTION OF KEPLER'S EQUATION

We return now to Kepler's equation

$$M = E - e \sin E$$

and examine the problem of solving the equation for E when M is known. It is to be noted that M and E always lie in the same semiellipse, as can readily be shown by the following argument.

We rewrite Kepler's equation in the form

$$\phi(E) \equiv E - e \sin E - M = 0 \tag{3.4.1}$$

When $M = n\pi$, where n is any integer, the value $E = n\pi$ uniquely satisfies (3.2.11). Assume that M has some given value between $n\pi$ and $(n + 1)\,\pi$. If ϕ is evaluated at $E = n\pi$ and $(n + 1)\,\pi$, respectively, we see that

$$\phi(n\pi) = n\pi - M < 0, \qquad \phi[(n + 1)\,\pi] = (n + 1)\,\pi - M > 0$$

so that the value of E for which $\phi(E)$ vanishes must lie within the interval $[n\pi, (n + 1)\,\pi]$. On the other hand, since

$$\frac{d\phi}{dE} = 1 - e \cos E > 0$$

the function $\phi(E)$ increases monotonically with E and hence can take on the value zero but once. Thus when M has some given value between $n\pi$ and $(n + 1)\,\pi$, there is but one real value of E satisfying Kepler's equation and it lies between $n\pi$ and $(n + 1)\,\pi$.

If M does not already lie in the closed interval $[0, 2\pi]$, by the addition of a suitable positive or negative multiple of 2π to both M and E we can reduce the problem to one in which the given value of M, and hence E, lies in $[0, 2\pi]$.

If Kepler's equation is written in the form

$$E = M + \psi, \qquad \psi = e \sin E = e \sin(M + \psi) \tag{3.4.2}$$

the following iteration scheme will yield a unique solution for E to any desired degree of accuracy.† Given M, take $\psi_0 = 0$ as a first estimate of ψ. The estimate ψ_0 is then used to determine a second estimate, ψ_1, of ψ from the relation

$$\psi_1 = e \sin(M + \psi_0) = e \sin M$$

Third-, fourth-, and higher-order estimates of ψ are obtained successively from

$$\begin{aligned} \psi_2 &= e \sin(M + \psi_1) \\ \psi_3 &= e \sin(M + \psi_2) \\ &\vdots \\ \psi_{k+1} &= e \sin(M + \psi_k) \end{aligned} \tag{3.4.3}$$

where k is a nonnegative integer. The iteration scheme just described yields a sequence of estimates of ψ. It is shown in what follows that (for a given value of M) this sequence of estimates converges uniformly to some unique value ψ^* which satisfies the equation $\psi = e \sin(M + \psi)$. In order to establish the convergence of the sequence, use will be made of the law of the mean, which states that if f is a function which is continuous at each point of a closed interval $a \leqslant x \leqslant b$ and has a derivative at each point of the open interval $a < x < b$, there exists a point $x = \xi$ in the open interval $a < \xi < b$ such that

$$f(b) - f(a) = (b - a) f'(\xi)$$

If $\psi_{k+1} = \psi_k$ at any stage of the iteration, the sequence converges uniformly to $\psi^* = \psi_k$ since

$$\psi_{k+1} = e \sin(M + \psi_k) = \psi_k$$

$$\psi_{k+2} = e \sin(M + \psi_{k+1}) = e \sin(M + \psi_k) = \psi_k, \quad \text{etc.}$$

Thus in the following material it is assumed that $\psi_{k+1} \neq \psi_k$.

† This iteration scheme can be found in the book by Kurth [2]. It is given here in detail, not only because of its intrinsic relevance to the subject matter of this section but also because it illustrates a rigorous approach to the iteration process which can be applied successfully to other problems (see Section 3.5 and Exercise 17 in Chapter 3).

For a given value of M, we consider the functions

$$f(\psi) = \sin(M + \psi) \tag{3.4.4}$$

$$f'(\psi) = \cos(M + \psi) \tag{3.4.5}$$

These two functions certainly satisfy the conditions of the law of the mean on any interval on the ψ-axis, since f is differentiable for all values of ψ. Let I_1 and I_2 represent the closed intervals

$$I_1 = [\psi_k, \psi_{k+1}], \qquad \psi_{k+1} > \psi_k$$

$$I_2 = [\psi_{k+1}, \psi_k], \qquad \psi_{k+1} < \psi_k$$

On either I_1 or I_2, the law of the mean can be applied to the functions $f(\psi)$ and $f'(\psi)$. In particular, if θ_k is a point lying between ψ_k and ψ_{k+1}, the expression

$$|\sin(M + \psi_{k+1}) - \sin(M + \psi_k)| = |\cos(M + \theta_k)| \, |\psi_{k+1} - \psi_k|$$

holds on either interval.

If the absolute value of the difference between two successive estimates

$$\psi_{k+2} = e \sin(M + \psi_{k+1}), \qquad \psi_{k+1} = e \sin(M + \psi_k)$$

is formed, we obtain

$$\begin{aligned} |\psi_{k+2} - \psi_{k+1}| &= e\,|\sin(M + \psi_{k+1}) - \sin(M + \psi_k)| \\ &= e\,|\cos(M + \theta_k)| \, |\psi_{k+1} - \psi_k| \\ &\leqslant e\,|\psi_{k+1} - \psi_k| \end{aligned} \tag{3.4.6}$$

whence, since $e < 1$, we have

$$|\psi_{k+2} - \psi_{k+1}| < |\psi_{k+1} - \psi_k| \tag{3.4.7}$$

It follows from (3.4.6) that

$$\begin{aligned} |\psi_2 - \psi_1| &\leqslant e\,|\psi_1 - 0| = e^2\,|\sin M| \\ |\psi_3 - \psi_2| &\leqslant e\,|\psi_2 - \psi_1| \leqslant e^3\,|\sin M| \\ &\vdots \\ |\psi_{k+1} - \psi_k| &\leqslant e^{k+1}\,|\sin M| \end{aligned} \tag{3.4.8}$$

The identity

$$\begin{aligned}\psi_{k+1} &= (\psi_1 - \psi_0) + (\psi_2 - \psi_1) + \cdots + (\psi_k - \psi_{k-1}) + (\psi_{k+1} - \psi_k) \\ &= \sum_{k=0}^{k} (\psi_{k+1} - \psi_k)\end{aligned}$$

is now considered. If ψ_{k+1} approaches a limit as k approaches infinity, designate this limit by S. Then

$$S = \lim_{k\to\infty} \psi_{k+1} = \lim_{k\to\infty} \sum_{k=0}^{k} (\psi_{k+1} - \psi_k) = \sum_{k=0}^{\infty} (\psi_{k+1} - \psi_k) \tag{3.4.9}$$

The geometric series

$$\sum_{k=0}^{\infty} e^k = 1 + e + e^2 + e^3 + \cdots$$

converges at each point e, $0 < e < 1$, and, from (3.4.8), since $e^{k+1} < e^k$, it follows that the inequality

$$| \psi_{k+1} - \psi_k | \leqslant e^k$$

is satisfied for all k. Thus the series S converges uniformly to some value ψ^*, by the Weierstrass test†; that is,

$$\lim_{k\to\infty} \psi_{k+1} = \psi^* \tag{3.4.10}$$

It must be shown that ψ^* satisfies the equation

$$\psi = e \sin(M + \psi)$$

Since $-\psi_{k+1} + e \sin(M + \psi_k) = 0$, for every k, we write

$$\begin{aligned}| \psi^* - e \sin(M + \psi^*)| &= | \psi^* - \psi_{k+1} + e \sin(M + \psi_k) - e \sin(M + \psi^*)| \\ &= | \psi^* - \psi_{k+1} - e[\sin(M + \psi^*) - \sin(M + \psi_k)]| \\ &\leqslant | \psi^* - \psi_{k+1} | + e \, | \cos(M + \theta^*)| \, | \psi^* - \psi_k | \\ &\leqslant | \psi^* - \psi_{k+1} | + e \, | \psi^* - \psi_k |\end{aligned}$$

† *Weierstrass M-test for uniform convergence*: Suppose that $\sum_{n=0}^{\infty} A_n$ is a convergent sequence of nonnegative numbers and that $\{u_1, u_2, \ldots\}$ is a sequence of functions for which $| u_n(z)| \leqslant A_n$ for each positive integer n and each z in a set D. Then $\sum_{n=0}^{\infty} u_n$ converges absolutely at each point of D and converges uniformly on D.

where θ^* lies between ψ^* and ψ_k. As $k \to \infty$, $\psi_k \to \psi_{k+1} \to \psi^*$ since $|\psi_{k+1} - \psi_k| \leqslant e^{k+1} |\sin M|$. Thus the right-hand side of the inequality just preceding must vanish and ψ^* is a solution of $\psi = e\sin(M + \psi)$. That ψ^* is a unique solution can be argued as follows. Suppose that there were two solutions ψ^* and $\bar{\psi}$ so that

$$\bar{\psi} = e\sin(M + \bar{\psi}), \qquad \psi^* = e\sin(M + \psi^*)$$

and

$$|\bar{\psi} - \psi^*| = e\,|\sin(M + \bar{\psi}) - \sin(M + \psi^*)|$$

Applying the law of the mean for θ' between $\bar{\psi}$ and ψ^*, we can write

$$|\bar{\psi} - \psi^*| = e\,|\cos(M + \theta')|\,|\bar{\psi} - \psi^*| \leqslant e\,|\bar{\psi} - \psi^*|$$

This is not possible for $e < 1$ unless $\bar{\psi} = \psi^*$. Thus the solution is unique.

Finally, an estimate of the truncation error will be obtained. To this end, we write

$$\begin{aligned}\psi^* = \lim_{k\to\infty} \psi_{k+1} = \sum_{k=0}^{\infty} (\psi_{k+1} - \psi_k) &= (\psi_1 - \psi_0) + (\psi_2 - \psi_1) + \cdots + (\psi_k - \psi_{k-1}) \\ &\quad + (\psi_{k+1} - \psi_k) + (\psi_{k+2} - \psi_{k+1}) + \cdots + \cdots \\ &= \psi_k + (\psi_{k+1} - \psi_k) + (\psi_{k+2} - \psi_{k+1}) + (\psi_{k+3} - \psi_{k+2}) + \cdots\end{aligned}$$

or, in terms of absolute values,

$$\begin{aligned}|\psi^* - \psi_k| &= |(\psi_{k+1} - \psi_k) + (\psi_{k+2} - \psi_{k+1}) + (\psi_{k+3} - \psi_{k+2}) + \cdots| \\ &\leqslant |\psi_{k+1} - \psi_k| + |\psi_{k+2} - \psi_{k+1}| + \cdots \\ &\leqslant e^{k+1}|\sin M| + e^{k+2}|\sin M| + \cdots \\ &\leqslant \frac{e^{k+1}}{1-e}|\sin M| \qquad (3.4.11)\end{aligned}$$

3.5. SOLUTION OF $N = e \sinh F - F$

The iteration scheme discussed in Section 3.4 can be applied to solve the equation

$$N = e \sinh F - F$$

for F when N is prescribed. Given N, take F_0 to be a first estimate of F and form the sequence of estimates[†]

$$\begin{aligned} F_1 &= \operatorname{arcsinh}\left(\frac{N+F_0}{e}\right) \\ F_2 &= \operatorname{arcsinh}\left(\frac{N+F_1}{e}\right) \\ &\vdots \\ F_{k+1} &= \operatorname{arcsinh}\left(\frac{N+F_k}{e}\right) \end{aligned} \tag{3.5.1}$$

This sequence of estimates can be shown to converge uniformly to a unique value F^* which satisfies the equation

$$F = \operatorname{arcsinh}\left(\frac{N+F}{e}\right)$$

An estimate of the truncation error is given by the formula

$$|F^* - F_k| = \frac{\epsilon^k}{1-\epsilon}\left|\operatorname{arcsinh}\left(\frac{N+F_0}{e}\right) - F_0\right| \qquad \left(\frac{1}{e} = \epsilon\right)$$

The details of the development are left as an exercise for the reader.

Exercises

1. (a) Express the ratio of the orbital speed at pericenter to that at apocenter as a function of the parameter $\epsilon = (1-e)/(1+e)$.
   (b) Express in terms of e the ratios of the orbital speeds at pericenter and apocenter to the speed in a circular orbit having a radius equal to the pericentric distance.
   (c) A body moves in an elliptic orbit of eccentricity e under the action of a central force directed toward one focus. When the body is at pericenter, the center of force is transferred to the other focus. Show that the eccentricity of the new orbit is $e(3+e)/(1-e)$.[‡]

[†] The idea of iterating on the equation $F = \operatorname{arcsinh}[(N+F)/e]$ was suggested to the author by B. Tully, Department of Physics, Auburn University, Auburn, Alabama.

[‡] Exercises 1(c), 12, and 13 are reprinted from E. J. Routh, "A Treatise on Dynamics of a Particle," pp. 242, 257, and 70. Copyright by Cambridge Univ. Press, London and New York. Used with permission of Cambridge Univ. Press.

2. An artificial Earth satellite is injected into an elliptic orbit at a height h above the Earth's surface and parallel to it with velocity v. Give an analytical criterion for deciding whether the injection point is at the perigee or the apogee of the orbit.

3. Show that as r approaches infinity $\dot{r}$ is asymptotically proportional to $1/\sqrt{r}$ for parabolic motion, whereas $\dot{r}$ approaches a constant value for hyperbolic motion. Also show that as $|t|$ approaches infinity, r is asymptotically proportional to $\sqrt[3]{t^2}$ for parabolic motion and asymptotically proportional to $|t|$ for hyperbolic motion.

4. A body of mass km describes an ellipse of eccentricity e and semi-major axis a under the action of a fixed gravitating body of mass m. Prove that if m is let go when the distance between the bodies is R, the eccentricity e' of the subsequent relative orbit is given by the equation

$$e'^2 - e^2 = \frac{k(1 - e^2)}{(1 + k)^2} [k + 2(1 - aR^{-1})]^\dagger$$

5. Prove that the least velocity with which a body could be projected from the North Pole so as to meet the Earth's surface at the equator is nearly 4.5 mps, and that the angle of elevation is 22.5 deg.

6. A particle is projected from the Earth's surface so as to describe a portion of an ellipse whose major axis is $\frac{3}{2}$ of the Earth's radius. Prove that if the direction of projection makes an angle 30 deg with the vertical, the time of flight is

$$\left(\frac{3}{4}\right)\left(\frac{3a}{g}\right)^{1/2} \left(\tan^{-1} \sqrt{6} + \sqrt{\frac{2}{3}}\right)$$

where a is the Earth's radius and g is the value of gravity at its surface.

7. When a particle is at the nearer apsis of an ellipse of eccentricity e described about the focus, the force on unit mass at unit distance is increased by the small fraction $1/n$ of itself. When the particle is at the further apsis, the force becomes less than its original amount by the same value. Show that the time taken in this revolution is less than the original period by the fraction $6e/n(1 - e^2)$ of itself.

8. Two stars of masses m_1 and m_2 move under their mutual gravitation.

† Exercises 4–8, 10, 11, 15, 16 are reprinted from A. E. H. Love, "Theoretical Mechanics," pp. 157, 162, 168, 167, 163, 167, 83, and 164. Copyright by Cambridge Univ. Press, London and New York. Used with permission of Cambridge Univ. Press.

Prove that if, when m_1 is at rest, the velocity of m_2 is at right angles to the line joining the two masses and equal in magnitude to

$$\left(\frac{\mu}{d}\right)^{1/2}, \qquad \mu = k^2(m_1 + m_2)$$

where d is the distance of the two bodies, then the path of m_1 in space is a succession of cycloids, the body coming to rest at a cusp at equal intervals of time.

9. When

$$F(r) = \frac{1}{r^2} + \frac{\nu}{\mu r^3}$$

show that if $\nu < h^2$, the general equation of the orbit described has the form

$$r = \frac{a}{1 + e\cos(k\theta)}$$

where a, e, and k are the constants depending upon the initial conditions and μ and ν. Observe that this orbit may be regarded as being a conic section whose major axis revolves around the focus with the mean angular speed

$$n = (1 - k)\left(\frac{2\pi}{T}\right)$$

where T is the period of revolution.†

10. A particle moves under a conservative central force and is projected with speed v_0 from a point at distance r_0 in a direction making an angle a with the radial line. Prove that the apsidal distances are the real roots of the equation for r

$$\frac{Wr^2}{r_0^2 \sin^2 a - r^2} = \frac{1}{2} v_0^2$$

where W is the work done per unit mass by the central attractive force as the particle moves from the point of projection to any point r, θ.

11. In a system of two gravitating bodies of masses M and m, the relative orbit is an ellipse of semiaxes a and b. Prove that if the mass of the second body could be suddenly doubled, the eccentricity of the new orbit would be

$$\frac{1}{M + 2m}\left[(M + 2m)^2 - \frac{b^2}{a^2}(M + m)(M + 2m) - \frac{b^2}{a}\frac{m}{\gamma}v^2\right]^{1/2}$$

† Reprinted from F. R. Moulton, "An Introduction to Celestial Mechanics," 2nd rev. ed., p. 96. Copyright 1914 by Macmillan, New York. Used with permission of Macmillan.

where v is the relative speed at the instant of the change and $\gamma = k^2$ is the gravitational constant.

12. The constituents of a double star are observed to describe circles relative to each other with a period T of relative motion. If they are suddenly deprived of velocity, show that they will collide after a time

$$\frac{T}{4\sqrt{2}}$$

13. The Earth and Moon being held at rest, show that the least speed V with which a particle must be projected from the Moon to reach the Earth is $V = 1.5$ mps.
14. A particle describes a central orbit about the origin with acceleration $\mu u^3(n^2 + 1 - 2n^2a^2u^2)(\mathbf{r}/r)$ starting from an apsis at distance a with a velocity equal in magnitude to that which it would have acquired if it had dropped from infinity. Prove that it describes the curve

$$r = a \cosh n\theta \qquad \left(r = \frac{1}{u}\right)$$

15. If the central acceleration is $2\mu(u^3 - a^2u^5)(\mathbf{r}/r)$ and a particle is projected from an apsis at distance a with speed $\sqrt{\mu}/a$, show that the time until the distance r is

$$\frac{1}{2\sqrt{\mu}}\left\{a^2 \log\left[\frac{r + (r^2 - a^2)^{1/2}}{a}\right] + r(r^2 - a^2)^{1/2}\right\}$$

16. A particle moves under a central acceleration

$$\frac{\mu(1 + 8\kappa \cos 2\theta)}{r^2}\frac{\mathbf{r}}{r}$$

being projected from an apsis on the initial line at distance c with speed $\sqrt{(\mu/c)}$; show that the next apsidal distance is $c/(1 + 3\kappa)$.
17. Show that the iteration scheme described in Section 3.5 converges to a unique value F^* of the equation

$$F = \operatorname{arcsinh}\left(\frac{N + F}{e}\right)$$

Deduce the formula given there for the truncation error.
18. Estimate E from the data (1) $e = 0.2$, $M = 765°$ and (2) $e = 0.2$, $M = -765°$. Determine the truncation error $|\psi^* - \psi_4|$.

Chapter 4

Expansions in Elliptic Motion

4.1. INTRODUCTORY COMMENTS

As we have seen in Chapter 3, the equations tabulated in (3.3.4), (3.3.5), and (3.3.6) provide a means of calculating the polar coordinates r and w at any given instant of time, t, if initial values of position and velocity are prescribed at some time t_0.

For many purposes, notably in connection with their use in the expansion of the disturbing function R_i, it is necessary to express the coordinates as periodic functions in the form of trigonometric series in the mean anomaly M. In obtaining solutions to Kepler's equation, it is also desirable to have an expression for E in the form of a series expansion in M. In particular, since $M = n(t - T)$, a series expansion in M is, in effect, a series in t. Thus such series representations of the coordinates provide an alternative method of calculating the position in a conic orbit.

We begin our discussion of series expansions in elliptic motion by introducing an expansion theorem due to Lagrange.

4.2. STATEMENT AND APPLICATION OF THE LAGRANGE EXPANSION THEOREM

In this section, Lagrange's expansion theorem will be stated and an illustrative example given in which the theorem is used to obtain a series inversion of Kepler's equation. A proof of the theorem, which includes a convergence criterion, is given in Appendix 3.

When applied to functions of real variables, the theorem says: If there is a relation between y and x given by

$$y = x + \alpha h(y) \tag{4.2.1}$$

in which $h(y)$ is a function of y, with derivatives of all orders on some interval, and α is a small parameter, then any function $F(y)$ that has derivatives of all orders on the same interval can be developed in a power series in α given by

$$F(y) = F(x) + \sum_{n=1}^{\infty} \frac{\alpha^n}{n!} \frac{d^{n-1}}{dx^{n-1}} \left\{ \frac{dF(x)}{dx} [h(x)]^n \right\} \tag{4.2.2}$$

Clearly E and $\sin E$ have derivatives of all orders on any interval on the E-axis so that the theorem can be applied to the equation $E = M + e \sin E$ by setting $F(E) = E$, $x = M$, $\alpha = e$, and $h(E) = \sin E$. We can then write

$$E = M + \sum_{n=1}^{\infty} \frac{e^n}{n!} \frac{d^{n-1}}{dM^{n-1}} (\sin^n M)$$

The calculation of the derivatives appearing in the series for E is tedious but straightforward. The development of the series through the sixth order in e is given below:

$$\begin{aligned} E = M &+ e \sin M + \frac{e^2}{2} \sin 2M \\ &+ \frac{e^3}{3!\, 2^2} (3^2 \sin 3M - 3 \sin M) \\ &+ \frac{e^4}{4!\, 2^3} (4^3 \sin 4M - 4 \cdot 2^3 \sin 2M) \\ &+ \frac{e^5}{5!\, 2^4} (5^4 \sin 5M - 5 \cdot 3^4 \sin 3M + 10 \sin M) \\ &+ \frac{e^6}{6!\, 2^5} (6^5 \sin 6M - 6 \cdot 4^5 \sin 4M + 15 \cdot 2^5 \sin 2M) \end{aligned} \tag{4.2.3}$$

4.3. BESSEL COEFFICIENTS

It is important to be able to express the coordinates as periodic functions of the time through the use of Fourier series in the mean anomaly M and the related anomalies w and E. These Fourier series expansions are treated in Section 4.5. In developing these series representations, it will be seen that Bessel functions of the first kind of integral order appear as the coefficients of the trigonometric functions. For this reason, those aspects of the theory of Bessel functions that will be drawn upon in the subsequent development are treated briefly in Section 4.4.

4.4. BESSEL FUNCTIONS OF THE FIRST KIND OF INTEGRAL ORDER

To introduce the discussion, we consider the function

$$u \equiv \exp\left[\frac{x}{2}(z - z^{-1})\right] = \exp\left(\frac{x}{2}z\right)\exp\left(-\frac{xz^{-1}}{2}\right) \tag{4.4.1}$$

Since the series expansion of e^{ξ} about $\xi = 0$ is given by

$$e^{\xi} = \sum_{n=0}^{\infty} \frac{\xi^n}{n!}$$

Equation (4.4.1) can be written

$$u = \left[\sum_{j=0}^{\infty} \frac{1}{j!}\left(\frac{x}{2}\right)^j z^j\right]\left[\sum_{j=0}^{\infty} \frac{(-1)^j}{j!}\left(\frac{x}{2}\right)^j z^{-j}\right] \tag{4.4.2}$$

When these two series are multiplied together, the only time a term containing z^p, p an integer, is obtained is when a term from the first series containing z^{p+j} multiplies a term from the second series containing z^{-j}. To find the total coefficient of the term containing z^p, we have to sum over j. This coefficient, $J_p(x)$, is given by

$$J_p(x) = \sum_{j=0}^{\infty}\left[\frac{1}{(p+j)!}\left(\frac{x}{2}\right)^{p+j}\right]\left[\frac{(-1)^j}{j!}\left(\frac{x}{2}\right)^j\right]$$

$$= \sum_{j=0}^{\infty} \frac{(-1)^j}{j!\,(p+j)!}\left(\frac{x}{2}\right)^{p+2j} \tag{4.4.3}$$

Collect the coefficient of the term z^{-p} in the product series. A term containing z^{-p} is obtained only if a term in the second series containing z^{-p-j} multiplies a term in the first series containing z^j. We designate by $J_{-p}(x)$ the coefficient of the term containing z^{-p} and write

$$J_{-p}(x) = \sum_{j=0}^{\infty} \left[\frac{1}{j!}\left(\frac{x}{2}\right)^j\right]\left[\frac{(-1)^{p+j}}{(p+j)!}\left(\frac{x}{2}\right)^{p+j}\right]$$

$$= (-1)^p \sum_{j=0}^{\infty} \frac{(-1)^j}{j!\,(p+j)!}\left(\frac{x}{2}\right)^{p+2j}$$

$$= (-1)^p J_p(x) \tag{4.4.4a}$$

It follows directly from the definition of $J_{-p}(x)$ that

$$J_{-p}(-x) = J_p(x) \tag{4.4.4b}$$

Equation (4.4.2) can now be written as

$$u = \sum_{p=-\infty}^{\infty} J_p(x)\, z^p \tag{4.4.5}$$

The coefficients $J_p(x)$ that appear in (4.4.5) are called Bessel functions of the first kind of integral order p. For each p, the series for $J_p(x)$ converges for all values of x, as can readily be shown by the ratio test. Certain interesting properties of these functions will now be examined. If the partial derivative of u with respect to z is formed, we obtain, from the joint use of (4.4.1) and (4.4.5),

$$\frac{\partial u}{\partial z} = \frac{x}{2}\left(1 + \frac{1}{z^2}\right) u = \frac{x}{2}\left(1 + \frac{1}{z^2}\right) \sum_{p=-\infty}^{\infty} J_p(x)\, z^p$$

From (4.4.5) we obtain

$$\frac{\partial u}{\partial z} = \sum_{p=-\infty}^{\infty} pJ_p(x)\, z^{p-1} = \sum_{p=-\infty}^{\infty} (p+1)\, J_{p+1}(x)\, z^p$$

so that

$$\sum_{p=-\infty}^{\infty} \left[\frac{x}{2} J_p(x)\, z^p + \frac{x}{2} J_p(x)\, z^{p-2}\right] = \sum_{p=-\infty}^{\infty} (p+1)\, J_{p+1}(x)\, z^p$$

Equating coefficients of z^p in the preceding expression, we have

$$\frac{x}{2} J_p(x) + \frac{x}{2} J_{p+2}(x) = (p+1)\, J_{p+1}(x)$$

Writing $p - 1$ for p, we obtain the recurrence formula

$$\frac{x}{2}[J_{p-1}(x) + J_{p+1}(x)] = pJ_p(x) \tag{4.4.6}$$

If the partial derivative of u with respect to x is formed, we obtain, from the joint use of (4.4.1) and (4.4.5),

$$\frac{\partial u}{\partial x} = \frac{1}{2}(z - z^{-1})\,u = \frac{1}{2}(z - z^{-1}) \sum_{p=-\infty}^{\infty} J_p(x)\,z^p$$

From (4.4.5) we obtain

$$\frac{\partial u}{\partial x} - \sum_{p=-\infty}^{\infty} J_p'(x)\,z^p \qquad \left\{J_p'(x) \equiv \frac{d}{dx}[J_p(x)]\right\}$$

so that

$$\sum_{p=-\infty}^{\infty} \left[\frac{1}{2} J_p(x)\,z^{p+1} - \frac{1}{2} J_p(x)\,z^{p-1}\right] = \sum_{p=-\infty}^{\infty} J_p'(x)\,z^p$$

Equating coefficients of z^p in the preceding expression, we have

$$\tfrac{1}{2}[J_{p-1}(x) - J_{p+1}(x)] = J_p'(x) \tag{4.4.7}$$

For each p, the function J_p satisfies Bessel's second-order linear homogeneous differential equation

$$\frac{d^2y}{dx^2} + \frac{1}{x}\frac{dy}{dx} + \left(1 - \frac{p^2}{x^2}\right) y = 0 \tag{4.4.8}$$

Differentiating (4.4.7) with respect to x, we write

$$\begin{aligned} J_p'' &= \tfrac{1}{2}(J_{p-1}' - J_{p+1}') \\ &= \tfrac{1}{2}[\tfrac{1}{2}(J_{p-2} - J_p) - \tfrac{1}{2}(J_p - J_{p+2})] \\ &= \tfrac{1}{4}(J_{p-2} - 2J_p + J_{p+2}) \end{aligned}$$

If this result is added to $(1/x)\,J_p'$, we obtain

$$J_p'' + \frac{1}{x} J_p' = \frac{1}{4} J_{p-2} - \frac{1}{2} J_p + \frac{1}{4} J_{p+2} + \frac{1}{2x} J_{p-1} - \frac{1}{2x} J_{p+1}$$

The recurrence formula (4.4.6) can be used to write

$$\frac{1}{4} J_{p+2} = \frac{p}{2x} J_{p+1} + \frac{1}{2x} J_{p+1} - \frac{1}{4} J_p$$

$$\frac{1}{4} J_{p-2} = \frac{p}{2x} J_{p-1} - \frac{1}{2x} J_{p-1} - \frac{1}{4} J_p$$

whence

$$J_p'' + \frac{1}{x} J_p' = \frac{p}{2x}(J_{p+1} + J_{p-1}) - J_p$$

Again using (4.4.6), we write

$$J_p'' + \frac{1}{x} J_p' = \frac{p}{2x}\frac{2p}{x} J_p - J_p = J_p\left(\frac{p^2}{x^2} - 1\right)$$

showing that J_p is a solution of Bessel's equation (4.4.8).

The Bessel function of order p can be expressed in the form of an integral as follows. In (4.4.1) and (4.4.5), set $z = e^{i\theta}$, $i = \sqrt{-1}$, and write

$$u = \exp\left[\frac{x}{2}(e^{i\theta} - e^{-i\theta})\right] = e^{ix\sin\theta} = \sum_{p=-\infty}^{\infty} J_p(x)\, e^{ip\theta} \tag{4.4.9}$$

Clearly $e^{ix\sin\theta}$ is periodic, of smallest period 2π in θ, and satisfies all the conditions necessary to ensure a convergent Fourier expansion if the coefficients $J_p(x)$ are determined in the usual manner. Multiply both sides of (4.4.9) by $e^{-ik\theta}$, k an integer. Integrating between the limits of 0 and 2π, we have

$$\int_0^{2\pi} \exp[-i(k\theta - x\sin\theta)]\, d\theta = \int_0^{2\pi} \sum_{p=-\infty}^{\infty} J_p(x)\, e^{im\theta}\, d\theta$$

where $m = p - k$ is an integer. Since $\int_0^{2\pi} e^{im\theta}\, d\theta = 2\pi$, or 0, depending on whether $m = 0$ or $m \neq 0$, all terms on the right-hand side disappear except the term for which $m = 0$, that is, $k = p$. Thus we have

$$J_p(x) = \frac{1}{2\pi}\int_0^{2\pi} \exp[-i(p\theta - x\sin\theta)]\, d\theta \tag{4.4.10}$$

If the interval of integration is broken into two parts, then (4.4.10) can be written

$$J_p(x) = \frac{1}{2\pi}\int_0^{\pi} \exp[-i(p\theta - x\sin\theta)]\, d\theta + \frac{1}{2\pi}\int_{\pi}^{2\pi} \exp[-i(p\theta - x\sin\theta)]\, d\theta$$

When θ is replaced by $2\pi - \theta$ in the second integral on the right-hand side of the preceding equation, we can write

$$\begin{aligned} J_p(x) &= \frac{1}{2\pi}\int_0^{\pi} \exp[-i(p\theta - x\sin\theta)]\, d\theta + \frac{1}{2\pi} e^{-i2p\pi}\int_0^{\pi} \exp[i(p\theta - x\sin\theta)]\, d\theta \\ &= \frac{1}{2\pi}\int_0^{\pi} \{\exp[i(p\theta - x\sin\theta)] + \exp[-i(p\theta - x\sin\theta)]\}\, d\theta \\ &= \frac{1}{\pi}\int_0^{\pi} \cos(p\theta - x\sin\theta)\, d\theta \end{aligned} \tag{4.4.11}$$

4.5. FOURIER SERIES EXPANSIONS IN ELLIPTIC MOTION

Fourier series expansions for $\sin E$ and $\cos E$ in terms of the mean anomaly M will now be derived. These expansions will, in turn, permit us to express r as a series in M. The rectangular coordinates (ξ_0, η_0), referred to the center of the ellipse, and (ξ'', η''), referred to the focus of the ellipse, are then expressible in terms of M. It will be convenient to use complex variables in the development.

In Equation (3.2.2), we wrote Kepler's equation in the form

$$E = M + \psi$$

where $\psi = e \sin E = e \sin(M + \psi)$. As a function of M, ψ is periodic, of period 2π, since

$$\begin{aligned} e \sin(M + 2\pi + \psi) &= e[\sin(M + 2\pi) \cos \psi + \sin \psi \cos(M + 2\pi)] \\ &= e(\sin M \cos \psi + \sin \psi \cos M) \\ &= e \sin(M + \psi) \end{aligned}$$

The function (note the dual use of the symbol e),

$$f(M) = e^{ikE} = e^{ik(M+\psi)} \tag{4.5.1}$$

where k is an integer, is also periodic in M, with period 2π, for

$$\begin{aligned} f(M + 2\pi) &= e^{ik(M+2\pi+\psi)} \\ &= e^{2ik\pi} e^{ikM} e^{ik\psi} \\ &= e^{ik(M+\psi)} = f(M) \end{aligned}$$

If $f(M)$ is expanded in a Fourier series in M, expansions for $\sin E$ and $\cos E$ in Fourier series in M can be obtained by considering the real and imaginary parts of the expansion for $f(M)$. The Fourier expansion of $f(M)$ is given by

$$f(M) = e^{ikE} = \sum_{n=-\infty}^{\infty} C_n e^{inM} \tag{4.5.2}$$

where

$$C_n = \frac{1}{2\pi} \int_{-\pi}^{\pi} f(M)\, e^{-inM}\, dM = \frac{1}{2\pi} \int_{-\pi}^{\pi} e^{ikE} e^{-inM}\, dM$$

Considering first the coefficients for which n is not zero, we can write

$$C_n = \frac{i}{2\pi n} \int_{-\pi}^{\pi} e^{ikE} e^{-inM}\, d(-inM) = \frac{i}{2\pi n} \int_{-\pi}^{\pi} e^{ikE}\, d(e^{-inM}) \qquad (n \neq 0)$$

Then, integrating by parts, we obtain

$$C_n = \frac{i}{2\pi n}\left\{[e^{ikE}e^{-inM}]_{M=E=-\pi}^{M=E=\pi} - ik\int_{-\pi}^{\pi} e^{ikE}e^{-inM}\, dE\right\}$$

For any integer p, $e^{ip\pi} = e^{-ip\pi}$, so that

$$C_n = \frac{i}{2\pi n}\left\{[e^{ik\pi}e^{-in\pi} - e^{ik(-\pi)}e^{-in(-\pi)}] - ik\int_{-\pi}^{\pi} e^{ikE}e^{-inM}\, dE\right\}$$

$$= \frac{k}{2\pi n}\int_{-\pi}^{\pi} e^{i(kE-nM)}\, dE \tag{4.5.3}$$

since the quantity within the brackets vanishes. The integral in (4.5.3) can be represented as the sum of two integrals in the form

$$C_n = \frac{k}{2\pi n}\int_{0}^{\pi} e^{i(kE-nM)}\, dE - \frac{k}{2\pi n}\int_{0}^{-\pi} e^{i(kE-nM)}\, dE \tag{4.5.4}$$

Replace E by $-E$ in the second integral on the right-hand side of (4.5.4). In the same integral, M must be replaced by $-M$, for $M = E - e \sin E$. It then follows that

$$C_n = \frac{k}{2\pi n}\int_{0}^{\pi} [e^{i(kE-nM)} + e^{-i(kE-nM)}]\, dE = \frac{k}{\pi n}\int_{0}^{\pi} \cos(kE - nM)\, dE$$

$$= \frac{k}{\pi n}\int_{0}^{\pi} \cos[(k-n)E + ne \sin E]\, dE = \frac{k}{n} J_{n-k}(ne) \qquad (n \neq 0) \tag{4.5.5}$$

For $n = 0$, we return to (4.5.2) and write

$$C_0 = \frac{1}{2\pi}\int_{-\pi}^{\pi} e^{ikE}\, dM = \frac{1}{2\pi}\int_{-\pi}^{\pi} e^{ikE}\, d(E - e \sin E)$$

$$= \frac{1}{2\pi}\int_{-\pi}^{\pi} e^{ikE}(1 - e \cos E)\, dE$$

$$= \frac{1}{2\pi}\int_{-\pi}^{\pi} \left[e^{ikE} - \frac{e}{2} e^{ikE}(e^{iE} + e^{-iE})\right] dE$$

$$= \frac{1}{2\pi}\int_{-\pi}^{\pi} \left[e^{ikE} - \frac{e}{2} e^{i(k+1)E} - \frac{e}{2} e^{i(k-1)E}\right] dE \tag{4.5.6}$$

Clearly, C_0 vanishes unless $k = 0, \pm 1$, since $\int_{-\pi}^{\pi} e^{im\theta}\, d\theta = 0$ unless the integer $m = 0$. For the three cases where C_0 is not zero, we have

$$k = 0: \qquad C_0 = \frac{1}{2\pi} \int_{-\pi}^{\pi} dE = 1$$

$$k = +1: \qquad C_0 = \frac{1}{2\pi} \left(-\frac{e}{2}\right) \int_{-\pi}^{\pi} dE = -\frac{e}{2}$$

$$k = -1: \qquad C_0 = \frac{1}{2\pi} \left(-\frac{e}{2}\right) \int_{-\pi}^{\pi} dE = -\frac{e}{2}$$

The Fourier expansions of $f(M)$ for different values of k are summarized in the following equations:

$$k = 0: \qquad f(M) = 1 \tag{4.5.7}$$

$$k = 1: \qquad f(M) = C_0 + \sum_{\substack{n=-\infty \\ n\neq 0}}^{\infty} C_n e^{inM} = -\frac{e}{2} + \sum_{\substack{n=-\infty \\ n\neq 0}}^{\infty} \frac{1}{n} J_{n-1}(ne)\, e^{inM} \tag{4.5.8}$$

$$k = -1: \qquad f(M) = C_0 + \sum_{\substack{n=-\infty \\ n\neq 0}}^{\infty} C_n e^{inM} = -\frac{e}{2} + \sum_{\substack{n=-\infty \\ n\neq 0}}^{\infty} \frac{(-1)}{n} J_{n+1}(ne)\, e^{inM} \tag{4.5.9}$$

$$k \neq 0, \pm 1: \qquad f(M) = C_0 + \sum_{\substack{n=-\infty \\ n\neq 0}}^{\infty} C_n e^{inM} = \sum_{\substack{n=-\infty \\ n\neq 0}}^{\infty} \frac{k}{n} J_{n-k}(ne)\, e^{inM} \tag{4.5.10}$$

The Fourier expansions for $\cos E$ and $\sin E$ follow directly from (4.5.8). We have

$$\cos E + i \sin E = -\frac{e}{2} + \sum_{\substack{n=-\infty \\ n\neq 0}}^{\infty} \frac{1}{n} J_{n-1}(ne) \cos nM$$

$$+ i \sum_{\substack{n=-\infty \\ n\neq 0}}^{\infty} \frac{1}{n} J_{n-1}(ne) \sin nM$$

whence, equating real and imaginary parts,

$$\cos E = -\frac{e}{2} + \sum_{\substack{n=-\infty \\ n\neq 0}}^{\infty} \frac{1}{n} J_{n-1}(ne) \cos nM \tag{4.5.11}$$

$$\sin E = \sum_{\substack{n=-\infty \\ n\neq 0}}^{\infty} \frac{1}{n} J_{n-1}(ne) \sin nM \tag{4.5.12}$$

Equations (4.5.11) and (4.5.12) can be written in a somewhat more convenient form. If we first write

$$\cos E = -\frac{e}{2} + \sum_{n=-\infty}^{n=-1} \frac{1}{n} J_{n-1}(ne) \cos nM + \sum_{n=1}^{\infty} \frac{1}{n} J_{n-1}(ne) \cos nM$$

$$\sin E = \sum_{n=-\infty}^{n=-1} \frac{1}{n} J_{n-1}(ne) \sin nM + \sum_{n=1}^{\infty} \frac{1}{n} J_{n-1}(ne) \sin nM$$

and replace n by $-n$ in the first sum appearing in each of the two preceding equations, we obtain

$$\cos E = -\frac{e}{2} + \sum_{n=1}^{\infty} \frac{1}{n} [J_{n-1}(ne) - J_{-(n+1)}(-ne)] \cos nM \qquad (4.5.13)$$

$$\sin E = \sum_{n=1}^{\infty} \frac{1}{n} [J_{n-1}(ne) + J_{-(n+1)}(-ne)] \sin nM \qquad (4.5.14)$$

We can use (4.4.4b) to write

$$J_{-(n+1)}(-ne) = J_{n+1}(ne)$$

hence

$$\cos E = -\frac{e}{2} + \sum_{n=1}^{\infty} \frac{1}{n} [J_{n-1}(ne) - J_{n+1}(ne)] \cos nM \qquad (4.5.15)$$

$$\sin E = \sum_{n=1}^{\infty} \frac{1}{n} [J_{n-1}(ne) + J_{n+1}(ne)] \sin nM \qquad (4.5.16)$$

If the recurrence relationships (4.4.6) and (4.4.7) are now used, we can write

$$\cos E = -\frac{e}{2} + 2 \sum_{n=1}^{\infty} \frac{1}{n} \frac{d}{d(ne)} [J_n(ne)] \cos nM \qquad (4.5.17)$$

$$\sin E = 2 \sum_{n=1}^{\infty} \frac{1}{ne} J_n(ne) \sin nM \qquad (4.5.18)$$

For convenience, Appendix 4 gives a reference table of the Bessel coefficients and their derivatives, as they appear in the preceding equations, through the first six coefficients.

The expansions of E and r/a in terms of M follow directly from (4.5.17) and (4.5.18). They are given by

$$E = M + 2 \sum_{n=1}^{\infty} \frac{1}{n} J_n(ne) \sin nM \tag{4.5.19}$$

$$\frac{r}{a} = 1 + \frac{e^2}{2} - 2e \sum_{n=1}^{\infty} \frac{1}{n} \frac{d}{d(ne)} [J_n(ne)] \cos nM \tag{4.5.20}$$

The expansions of the coordinates ξ_0 and η_0, referred to the center of the ellipse, with ξ_0 taken to coincide with the major axis, positive from center to occupied focus, are

$$\xi_0 = a \cos E = -\frac{ae}{2} + 2a \sum_{n=1}^{\infty} \frac{1}{n} \frac{d}{d(ne)} [J_n(ne)] \cos nM \tag{4.5.21}$$

$$\eta_0 = b \sin E = 2b \sum_{n=1}^{\infty} \frac{1}{ne} J_n(ne) \sin nM \tag{4.5.22}$$

If we wish to use coordinates (ξ'', $\eta'' = \eta_0$) referred to the occupied focus of the ellipse, the expansion for ξ'' is given by

$$\begin{aligned} \xi'' &= \xi_0 - ae = a(\cos E - e) \\ &= -\frac{3ae}{2} + 2a \sum_{n=1}^{\infty} \frac{1}{n} \frac{d}{d(ne)} [J_n(ne)] \cos nM \end{aligned} \tag{4.5.23}$$

Expansions for the true anomaly w in terms of E and M will now be developed. We begin by rewriting (3.1.4). We have

$$\tan \frac{w}{2} = \frac{1}{\sqrt{\epsilon}} \tan \frac{E}{2} = \left(\frac{1+e}{1-e}\right)^{1/2} \tan \frac{E}{2}$$

Now introduce the substitutions

$$\log Z = iw, \qquad \log W = iE, \qquad \sqrt{\epsilon} = \frac{1-\beta}{1+\beta}$$

Since $Z = e^{iw}$, we can write

$$\tan \frac{w}{2} = \frac{\sin (w/2)}{\cos (w/2)} = \frac{e^{iw/2} - e^{-iw/2}}{i(e^{iw/2} + e^{-iw/2})} = \frac{Z-1}{i(Z+1)}$$

Similarly, we have

$$\tan \frac{E}{2} = \frac{W-1}{i(W+1)}$$

If these results are substituted into (3.1.4), we obtain

$$\frac{Z-1}{Z+1} = \frac{1+\beta}{1-\beta}\,\frac{W-1}{W+1}$$

When this last expression is solved for Z, we find

$$Z = \frac{W-\beta}{1-\beta W} = W(1-\beta W^{-1})(1-\beta W)^{-1}$$

If the logarithms of each side of the preceding equation are formed, we obtain

$$iw = iE + \log(1-\beta W^{-1}) - \log(1-\beta W) \tag{4.5.24a}$$

If γ is a complex number whose modulus is less than unity, the series expansion of $\log(1+\gamma)$ is given by

$$\log(1+\gamma) = \sum_{j=1}^{\infty} (-1)^{j+1}\frac{\gamma^j}{j}, \qquad |\gamma| < 1 \tag{4.5.24b}$$

Since

$$\beta = \frac{(1+e)^{1/2} - (1-e)^{1/2}}{(1+e)^{1/2} + (1-e)^{1/2}} < 1 \qquad \text{and} \qquad |W| = |e^{iE}| = 1$$

both $|\beta W|$ and $|\beta W^{-1}|$ are less than one. Thus (4.5.24b) can be applied to each of the last two terms on the right-hand side of (4.5.24a), in which γ is taken to be βW^{-1} and βW successively. It follows directly that

$$\log[1 + (-\beta W^{-1})] = -\sum_{j=1}^{\infty} \frac{\beta^j W^{-j}}{j}$$

$$\log[1 + (-\beta W)] = -\sum_{j=1}^{\infty} \frac{\beta^j W^{j}}{j}$$

Hence

$$\log[1 + (-\beta W^{-1})] - \log[1 + (-\beta W)] = \sum_{j=1}^{\infty} \frac{\beta^j}{j}(e^{iEj} - e^{-iEj})$$

$$= 2i \sum_{j=1}^{\infty} \frac{\beta^j \sin jE}{j}$$

After we eliminate i, Equation (4.5.24a) becomes

$$w = E + 2\sum_{j=1}^{\infty} \frac{\beta^j \sin jE}{j} \tag{4.5.25}$$

Equation (4.5.25) permits us to express w in terms of either E or M, for we can express $\sin jE$, for every j, in terms of M through (4.5.8)–(4.5.10). Equation (4.5.25) is not in a convenient form for calculation, however, and it is desirable to express β^j as a series in the eccentricity e by means of the Lagrange expansion theorem. To this end, we wish to cast the relationship between β and e in a form suitable for use in the expansion theorem. If the denominator of the expression

$$\beta = \frac{(1+e)^{1/2} - (1-e)^{1/2}}{(1+e)^{1/2} + (1-e)^{1/2}}$$

is rationalized, we can write

$$1 - e\beta = (1 - e^2)^{1/2}$$

Squaring both sides of this last expression and rearranging terms, we find

$$\beta = \frac{e}{2} + \frac{e}{2}\beta^2 \tag{4.5.26}$$

Comparing (4.5.26) with (4.2.1) and (4.2.2), we write

$$\beta^j = x^j + \sum_{n=1}^{\infty} \frac{\alpha^n}{n!} \frac{d^{n-1}}{dx^{n-1}} \left(x^{2n} \frac{d}{dx} x^j \right) \tag{4.5.27}$$

where $\beta = y$, $x = \alpha = e/2$, $\beta^2 = h(y)$, and $\beta^j = F(y)$. Since

$$\begin{aligned}
\frac{d^{n-1}}{dx^{n-1}} \left(x^{2n} \frac{d}{dx} x^j \right) &= j \frac{d^{n-1}}{dx^{n-1}} (x^{2n+j-1}) \\
&= j(2n+j-1)(2n+j-2) \cdots [2n+j-(n-1)]\, x^{2n+j-n} \\
&= j(2n+j-1)(2n+j-2) \cdots (n+j+1)\, x^{n+j} \\
&= \frac{j(2n+j-1)!}{(n+j)!} x^{n+j}
\end{aligned}$$

we can write (4.5.27) in the form

$$\beta^j = x^j \left[1 + j \sum_{n=1}^{\infty} \frac{\alpha^n}{n!} \frac{(2n+j-1)!}{(n+j)!} x^n \right]$$

or, in terms of e,

$$\beta^j = \left(\frac{e}{2}\right)^j \left[1 + j \sum_{n=1}^{\infty} \frac{(2n+j-1)!}{n!\,(n+j)!} \left(\frac{e}{2}\right)^{2n} \right] \tag{4.5.28}$$

The first few terms of the expansion (4.5.28) are

$$\beta^j = \left(\frac{e}{2}\right)^j \left[1 + \frac{j}{4}e^2 + \frac{j(j+3)}{4^2 \cdot 2!}e^4 + \frac{j(j+4)(j+5)}{4^3 \cdot 3!}e^6 + \cdots\right] \qquad (4.5.29)$$

If (4.5.29) is used in conjunction with (4.5.25), we obtain the so-called equation of the center, that is, the expansion of $w - M$. To terms of order e^4, the expansion can be written

$$w = M + \left(2e - \frac{1}{4}e^3\right)\sin M + \frac{5}{4}e^2 \sin 2M + \frac{13}{12}e^3 \sin 3M \qquad (4.5.30)$$

4.6. EXPANSION OF THE POSITION VECTOR IN TAYLOR SERIES

It is convenient, for certain purposes, to have a development of $\mathbf{r}$ in Taylor series. The second-order equation

$$\ddot{\mathbf{r}} = -\frac{\mu}{r^3}\mathbf{r} \qquad [\mu = k^2(m_0 + m)]$$

for the two-body orbit may be replaced by two first-order differential equations by introducing the substitution $\dot{\mathbf{r}} = \mathbf{v}$. Thus we obtain

$$\dot{\mathbf{r}} = \mathbf{v} \qquad (4.6.1)$$

$$\dot{\mathbf{v}} = -\frac{\mu}{r^3}\mathbf{r} \qquad (4.6.2)$$

If we differentiate (4.6.1) and (4.6.2) with respect to time, we find

$$\ddot{\mathbf{r}} = \dot{\mathbf{v}} = -\frac{\mu}{r^3}\mathbf{r}$$

$$\ddot{\mathbf{v}} = -\frac{\mu}{r^3}\dot{\mathbf{r}} + \frac{3\mu}{r^4}\dot{r}\mathbf{r} = -\frac{\mu}{r^3}\left(\mathbf{v} - \frac{3\dot{r}}{r}\mathbf{r}\right)$$

If the expression for $\ddot{\mathbf{v}}$ is differentiated with respect to time, we can write

$$\dddot{\mathbf{v}} = -\frac{\mu}{r^3}\left(-\frac{\mu}{r^3} + \frac{12\dot{r}^2}{r^2} - \frac{3\ddot{r}}{r}\right)\mathbf{r} + \frac{6\mu\dot{r}}{r^4}\mathbf{v} \qquad (4.6.3)$$

All the quantities in (4.6.3) are expressible directly in terms of the initial values $\mathbf{r}_0$ and $\mathbf{v}_0$, except $\ddot{r}$. To express $\ddot{r}$ in terms of position and speed, we note that $r^2 = \mathbf{r} \cdot \mathbf{r}$. Hence

$$\dot{r}r = \mathbf{r} \cdot \dot{\mathbf{r}} = \mathbf{v} \cdot \mathbf{r}$$

and

$$\dot{r}^2 + r\ddot{r} = \dot{\mathbf{v}} \cdot \mathbf{r} + \mathbf{v} \cdot \dot{\mathbf{r}} = -\frac{\mu}{r} + v^2 \qquad (v = |\mathbf{v}|)$$

Solving for $\ddot{r}$, we obtain the expression

$$\ddot{r} = -\frac{\mu}{r^2} + \frac{v^2}{r} - \frac{\dot{r}^2}{r}$$

A similar procedure may be used to express higher-order derivatives directly in terms of initial values.

The Taylor series expansion of $\mathbf{r}(t)$ in the neighborhood of $t = 0$ has the form

$$\mathbf{r}(t) = \mathbf{r}_0 + \dot{\mathbf{r}}_0 t + \frac{1}{2!}\ddot{\mathbf{r}}_0 t^2 + \frac{1}{3!}\dddot{\mathbf{r}}_0 t^3 + \frac{1}{4!}\ddddot{\mathbf{r}}_0 t^4 + \cdots \tag{4.6.4}$$

where the zero subscript denotes evaluation of the quantity at $t = 0$. We have

$$\dot{\mathbf{r}}_0 = \mathbf{v}_0$$

$$\ddot{\mathbf{r}}_0 = \dot{\mathbf{v}}_0 = -\frac{\mu}{r_0^3}\mathbf{r}_0$$

$$\dddot{\mathbf{r}}_0 = \ddot{\mathbf{v}}_0 = -\frac{\mu}{r_0^3}\left(\mathbf{v}_0 - \frac{3\dot{r}_0}{r_0}\mathbf{r}_0\right)$$

$$\ddddot{\mathbf{r}}_0 = \dddot{\mathbf{v}}_0 = -\frac{\mu}{r_0^5}\left(\frac{2\mu}{r_0} - 3v_0^2 + 15\dot{r}_0^2\right)\mathbf{r}_0 + \frac{6\mu\dot{r}_0}{r_0^4}\mathbf{v}_0$$

When the derivatives expressed in terms of the initial conditions are substituted into (4.6.4), we obtain the series

$$\mathbf{r}(t) = \mathbf{r}_0 + t\mathbf{v}_0 - \frac{1}{2!}\frac{\mu}{r_0^3}t^2\mathbf{r}_0 - \frac{1}{3!}\frac{\mu}{r_0^3}t^3\mathbf{v}_0 + \frac{1}{3!}\frac{3\mu\dot{r}_0}{r_0^4}t^3\mathbf{r}_0$$
$$+ \frac{1}{4!\,r_0^5}\left(-\frac{2\mu^2}{r_0} + 3\mu v_0^2 - 15\mu\dot{r}_0^2\right)t^4\mathbf{r}_0 + \frac{1}{4!}\frac{6\mu\dot{r}_0}{r_0^4}t^4\mathbf{v}_0 + \cdots \tag{4.6.5}$$

Collecting coefficients of $\mathbf{r}_0$ and $\mathbf{v}_0$, we obtain the series

$$f = 1 - \frac{1}{2!}\frac{\mu}{r_0^3}t^2 + \frac{1}{3!}\frac{3\mu\dot{r}_0}{r_0^4}t^3 + \frac{1}{4!\,r_0^5}\left(-\frac{2\mu^2}{r_0} + 3\mu v_0^2 - 15\mu\dot{r}_0^2\right)t^4 + \cdots \tag{4.6.6}$$

$$g = t - \frac{1}{3!}\frac{\mu}{r_0^3}t^3 + \frac{1}{4!}\frac{6\mu\dot{r}_0}{r_0^4}t^4 - \cdots \tag{4.6.7}$$

and (4.6.5) can be written in the form

$$\mathbf{r}(t) = f\mathbf{r}_0 + g\mathbf{v}_0 \tag{4.6.8}$$

If $\mathbf{r}_0$ and $\dot{\mathbf{r}}_0 = \mathbf{v}_0$ are prescribed at $t = 0$, approximate values of $\mathbf{r}(t)$ can be calculated from (4.6.8) in the neighborhood of $t = 0$. As we shall see, the notions of replacing a second-order differential equation with two first-order ones as in (4.6.1) and (4.6.2) and expanding in a Taylor series will find extensive application in our discussion of perturbations in later chapters.

Exercises

1. Consider our proof of the Lagrange expansion theorem. Assume that $h(z)$ is an integral function (i.e., it is analytic for every value of z). Further assume that $h(z)$ is neither a constant nor a polynomial of first degree. Let $M(r)$ be the maximum value of $h(z)$ along the circle C described about the point a as center with the radius r. Show that the quotient $r/M(r)$ passes through a maximum value, θ, for a value of r, say r_1, where $0 < r_1 < \infty$. Hence show that if $\phi(z)$ is analytic in the circle C of radius r_1 and $\alpha < \theta$, the Lagrange expansion converges. Are the cases excluded in the third sentence above trivial? Explain. (Refer to Appendix 3 for proof.)
2. Given the equation $z = a + \alpha h(z)$, we have shown that if a contour exists which encloses the complex point $z = a$ such that at every point of the boundary $|\alpha h(z)| < |z - a|$, then the equation has but one root within the area and the Lagrangian expansion for that root is convergent. Apply this theorem to Kepler's equation $M = E - e \sin E$ to show that the Lagrangian series development of E in powers of e converges for all values of e less than 0.663.
3. Evaluate the coefficients in the Laurent expansion of

$$u = \exp\left[\frac{x}{2}(z - z^{-1})\right], \qquad z = x + iy$$

about the singular point $z = 0$. Hence show that

$$J_p(x) = \frac{1}{\pi}\int_0^\pi \cos(p\theta - x\sin\theta)\, d\theta$$

where p is an integer.

4. Prove that

$$E = M + \sum_{n=1}^{\infty} B_n \sin nM$$

where

$$B_n = \frac{2}{\pi n^2} \int_0^{\pi} e \sin \theta (1 - e \cos \theta)^{-3} \sin n\Phi \, d\Phi$$

$$\Phi = \theta - e \sin \theta$$

It then follows that $B_n < 2k/n^2$, where k is the maximum value of $e \sin \theta/(1 - e \cos \theta)^{-3}$. Thus argue that the series is at least as convergent as $\sum (1/n^2)$.

5. Prove that

$$\frac{r}{a} = 1 + \frac{e^2}{2} + \sum_{n=1}^{\infty} A_n \cos nM$$

where

$$A_n = \frac{2e}{\pi n^2} \int_0^{\pi} (e - \cos \theta)(1 - e \cos \theta)^{-3} \cos n\Phi \, d\Phi$$

and Φ is defined in Exercise 4. Thus argue that the series is at least as convergent as $\sum (1/n^2)$.

6. Establish the expansions

$$\frac{1}{2}(w - E) = \sum_{n=1}^{\infty} \frac{\beta^n \sin nE}{n}, \qquad \frac{1}{2}(E - w) = \sum_{n=1}^{\infty} \frac{(-1)^n \beta^n \sin nw}{n}$$

where

$$\beta^{-1} = \frac{e}{1 - (1 - e^2)^{1/2}}$$

7. Show that

$$w = M + \sum_{n=1}^{\infty} C_n \sin nM$$

where

$$\pi n C_n = 2(1 - e^2)^{1/2} \int_0^{\pi} \cos nM (1 - e \cos E)^{-1} \, dE$$

Then show that

$$C_n = \frac{4(1 - e^2)^{1/2}}{\pi n^2} \int_0^{\pi} e \sin E (1 - e \cos E)^{-4} \sin nM \, dM$$

Thus argue that the series is at least as convergent as $\sum (1/n^2)$.

8. First show that $Z = W(1 - \beta W^{-1})(1 - \beta W)^{-1}$, where β is as in Exercise 6 and $Z = \exp(iw)$, $W = \exp(iE)$. Then show that

$$Z = -\beta + (1 - \beta^2) \sum_{n=1}^{\infty} \beta^{n-1} W^n$$

Hence deduce the expansions

$$\cos w = -\beta + (1 - \beta^2) \sum_{n=1}^{\infty} \beta^{n-1} \cos nE$$

$$\sin w = (1 - \beta^2) \sum_{n=1}^{\infty} \beta^{n-1} \sin nE$$

9. Proceeding as in Exercise 8, deduce the expansions

$$\cos E = \beta + (1 - \beta^2) \sum_{n=1}^{\infty} (-1)^{n-1} \beta^{n-1} \cos nw$$

$$\sin E = (1 - \beta^2) \sum_{n=1}^{\infty} (-1)^{n-1} \beta^{n-1} \sin nw$$

10. Show that the equation of center $(w - M)$ may be given in terms of w by the expression

$$\frac{w - M}{2} = \sum_{n=1}^{\infty} \frac{(-1)^{n-1} p^n[n + 1 - (n - 1) p^2] \sin nw}{n(1 + p^2)}, \qquad p = \frac{e}{2 - e\beta}$$

11. Show that

$$\frac{a}{r} = 1 + 2 \sum_{n=1}^{\infty} J_n(ne) \cos nM$$

$$e \cos w = -1 + (1 - e^2) \left[1 + 2 \sum_{n=1}^{\infty} J_n(ne) \cos nM\right]$$

12. Show that

$$J_n(x) = \frac{i^{-n}}{\pi} \int_0^{\pi} e^{ix\cos\theta} \cos(n\theta)\, d\theta$$

and that

$$I_n(x) \equiv i^{-n} J_n(ix) = \frac{1}{\pi} \int_0^{\pi} e^{x\cos\theta} \cos(n\theta)\, d\theta$$

The function $I_n(x)$ is called the modified Bessel function of the first kind of order n. Show that it satisfies the recurrence relations

$$I_n'(x) = I_{n-1}(x) - \frac{n}{x} I_n(x)$$

$$I_n'(x) = I_{n+1}(x) + \frac{n}{x} I_n(x)$$

Chapter 5

Systems of Astronomical Coordinates

5.1. DESCRIPTION OF THREE COMMONLY USED COORDINATE SYSTEMS

The choice of reference frame in which the motion of the body, or bodies, under study is to be described will be dictated by the problem under consideration. Certain coordinate systems having frequent application in celestial mechanics will now be introduced.

In order to specify a coordinate system, its origin and the directions (and senses) of its axes must be defined. Coordinate systems with origin at the center of the Sun are called *heliocentric systems*. If the origin is at the center of the Earth, the coordinate system is said to be *geocentric*; *planetocentric* if the origin is at the center of a particular planet; *selenocentric* if the origin is at the center of the Moon. If the coordinate system has its origin at the position of the observer on the Earth, it is called *topocentric*.

Since the main problem of celestial mechanics concerns the relative motion of one body about a second more massive body, a triad of coordinate systems with their common origin at the center of the massive body will be introduced. This triad will consist of an equatorial system, an ecliptic system, and an orbital system.

1. Equatorial System

For a fundamental axis, we choose the direction of the Earth's axis of rotation, reckoned positive from geocenter to the north. Coordinate systems with fundamental planes perpendicular to this fundamental axis are called *equatorial systems.* Rectangular coordinates in an equatorial system will be designated by (ξ_q, η_q, ζ_q); spherical coordinates by (r, l_q, b_q).

2. Ecliptic System

For a fundamental axis, we choose the direction of the normal to the plane of the Earth's orbit (ecliptic plane) about the Sun, reckoned positive in the sense of the Earth's angular momentum with respect to the Sun. Coordinate systems with fundamental planes perpendicular to this fundamental axis are called *ecliptic systems.* Rectangular coordinates in an ecliptic system will be designated by (ξ_c, η_c, ζ_c); spherical coordinates by (r, l_c, b_c).

3. Orbital System

For a fundamental axis, we choose the direction of the normal to the plane of the orbit of the body moving about the massive body reckoned positive in the sense of the orbiting body's angular momentum with respect to the massive body. Coordinate systems with fundamental planes perpendicular to this fundamental axis are called *orbital systems.* Rectangular coordinates in an orbital system will be designated by (ξ'', η'', ζ'') or (ξ', η', ζ'); spherical coordinates by (r, w, ψ) or (r, u, ψ).

A useful notion in astronomy is the concept of an imaginary celestial sphere assumed to surround the Earth at an infinite distance from it. Because this distance is infinite, all parallel lines converge at one point on the sphere. For example, a line parallel to the Earth's axis tangent to the surface of the Earth will intersect the celestial sphere at the same point as does an extension of the Earth's axis of rotation. Also, any family of parallel planes is represented by a single great circle on the celestial sphere.

The two points of intersection with the celestial sphere of a family of lines parallel to the Earth's axis of rotation are called the celestial poles. The point of intersection of the positive axis of rotation with the celestial sphere is called the celestial north pole (CNP). The antipoint is called the celestial south pole (CSP). The two points of intersection with the celestial sphere of a family of lines parallel to the fundamental axis of the ecliptic systems are called the ecliptic poles. The point of intersection of the positive fundamental axis of the ecliptic systems with the celestial sphere is called the ecliptic north pole (ENP). The antipoint is called the ecliptic south pole (ESP). The great circle representing the intersection with the celestial sphere of a family of planes parallel to the Earth's equator is called the *celestial equator*. The great circle representing the intersection with the celestial sphere of a family of planes parallel to the plane of the Earth's orbit about the Sun is called the *ecliptic*. It represents the apparent annual path of the Sun with respect to the Earth.

The celestial equator and the ecliptic intersect at two points on the celestial sphere. The point of intersection of the ecliptic and celestial equator where the Sun appears to cross the equator from south to north in its apparent annual motion along the ecliptic is called the *vernal equinox*, or the first point of Aries (symbol ♈; see Figure 5.1.1).

FIGURE 5.1.1

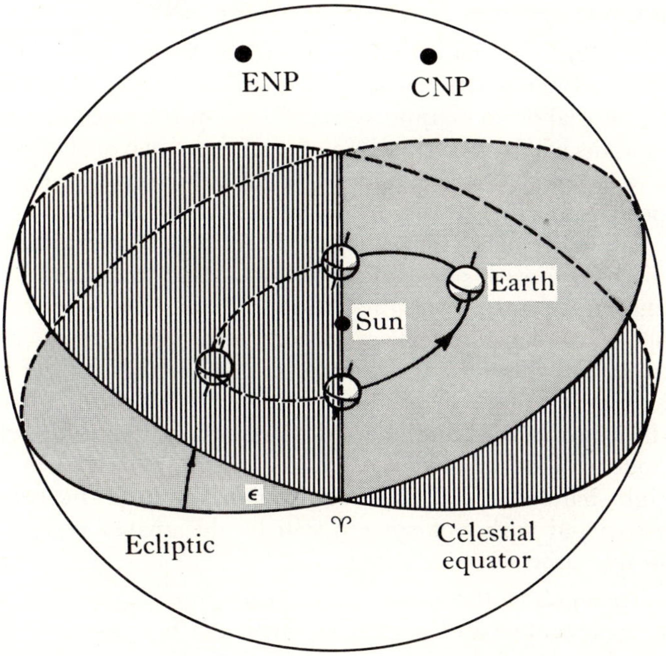

The angle ϵ between the ecliptic and the celestial equator is called the *obliquity of the ecliptic* ($\epsilon \approx 23^\circ.5$). Since the celestial sphere is hypothesized to be at infinity, its center can be thought of as coinciding with the origin of any coordinate system under discussion.

The specification of the principal directions (and senses) of the remaining two axes of each of the three types of coordinate system under discussion is given in what follows.

1. *Principal directions in the fundamental planes of equatorial and*

ecliptic systems. The principal direction in the fundamental planes of both equatorial and ecliptic systems is reckoned positive from the origin to ♈. Azimuthal angles in both systems are usually measured from this principal direction, clockwise as viewed from the CSP or ESP respectively. Azimuthal angles, so measured, are called *right ascensions*, l_q, *in equatorial systems* and *celestial longitudes*, l_c, *in ecliptic systems.* The complements of the polar angles are used to designate the other angular coordinates. These complements are called *declinations*, b_q, *in equatorial systems* and *celestial latitudes*, b_c, *in ecliptic systems* and they lie between -90 and $+90$ deg. Their signs must always be given. The third coordinate r is the distance from the origin. In equatorial systems, the positive ζ_q-axis is directed toward the CNP, the positive ξ_q-axis toward ♈, and the positive η_q-axis is chosen to form a right-handed system of coordinates. The same scheme is used for ecliptic systems except that the positive ζ_c-axis is directed toward the ENP.

2. *Principal directions in the fundamental plane of an orbital system.* The orbital system is usually employed in conjunction with either an equatorial or an ecliptic system. Two of the most commonly used orbital systems will be described. In each of these two orbital systems, the positive ζ'-axis and ζ''-axis coincide. This common axis has the sense of the positive normal to the orbital plane and passes through the origin at the occupied focus. The principal direction in the fundamental plane when directed toward pericenter will be labeled the ξ''-axis; the η''-axis is then chosen to form a right-handed rectangular coordinate system. The spherical coordinates associated with this system will be designated (r, w, ψ), where the third coordinate, ψ, is zero for the two-body problem and only enters in studies involving perturbing forces. The principal direction in the fundamental plane when directed toward the ascending node will be labeled the ξ'-axis; the η'-axis is then chosen to form a right-handed rectangular coordinate system. The spherical coordinates associated with this system will be designated (r, u, ψ), where ψ enters in the same sense as above.

In much of the following material, our discussions will be referenced to heliocentric and geocentric orbits. It is convenient in this connection to adopt the following notation:

Coordinate system	General origin	Geocenter	Heliocenter
Equatorial-rectangular	ξ_q, η_q, ζ_q	x_q, y_q, z_q	X_q, Y_q, Z_q
Ecliptic-rectangular	ξ_c, η_c, ζ_c	x_c, y_c, z_c	X_c, Y_c, Z_c
Equatorial-spherical	r, l_q, b_q	ρ, α, δ	r, A, Δ
Ecliptic-spherical	r, l_c, b_c	ρ, λ, β	r, Λ, B

5.2. GEOGRAPHIC AND TOPOCENTRIC COORDINATE SYSTEMS

In addition to the triad of coordinate systems discussed in the preceding section, two other coordinate systems commonly encountered will now be described briefly.

1. Geographic System

The geographic system of coordinates is an *equatorial system of coordinates* with origin at geocenter O. The half-plane bounded by the Earth's spin axis which contains the position vector $\boldsymbol{\rho}_P$ to a place P on the Earth's surface is called the *local meridian*; the half-plane bounded by the Earth's spin axis, which contains the position vector $\boldsymbol{\rho}_S$ to a celestial object S, is called the *hour-circle* of the object. In the geographic system of coordinates, the positive x_G-axis is taken to be the intersection of the meridian through G (the *Greenwich meridian*) with the Earth's equatorial plane; the y_G-axis is then chosen to form a right-handed system. The angle measured westward in the equatorial plane from the local meridian to the hour-circle of a celestial object is called the *local hour angle*, t, and is usually given in time measure. The conversion from seconds of arc (symbol: ″) to seconds of time (symbol: sec) is given by $15'' = 1$ sec. The *geographic longitude* of a place on the Earth's surface is the angle between the local meridian and the Greenwich meridian, designated longitude east l_E, or longitude west l_W, depending on whether the angle is measured from the Greenwich meridian, eastward or westward, along the equator. The *geocentric latitude*, ϕ', of a place on the Earth's surface is the angle at geocenter between the radius and its projection in the equatorial plane: ϕ' is positive if the place is north of the equatorial plane and negative if south of the equatorial plane. The geographic system is clearly not an inertial one as it rotates with the Earth through 360 deg in 24 mean solar hours (see Figure 5.2.1).

2. Topocentric System

Also shown on Figure 5.2.1 is a topocentric system of rectangular coordinates. This system, which has its origin at a place P on the Earth's surface, is often chosen to have its positive x_T-axis directed toward the south, along the curve of intersection of the local meridian through P with the surface of the Earth. The positive z_T-axis is directed toward the zenith; the positive y_T-axis is then directed toward the east to complete a right-handed system. The fundamental plane for a topocentric system is the horizon plane at P.

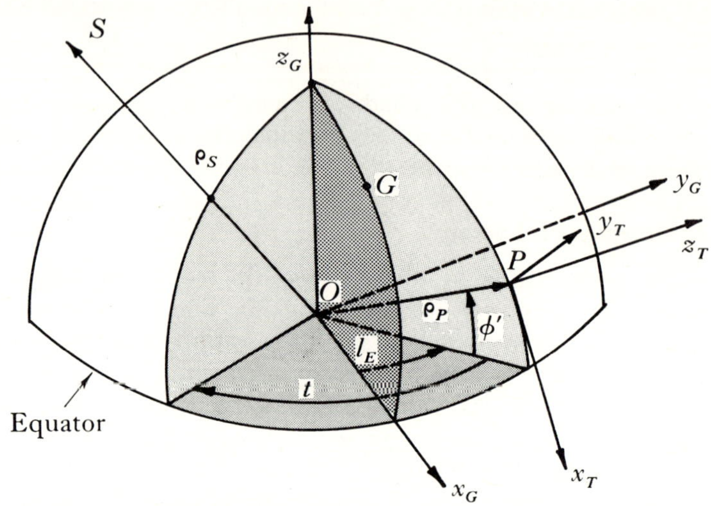

FIGURE 5.2.1

5.3. VARIATIONS IN THE PRINCIPAL DIRECTIONS AND FUNDAMENTAL PLANES

Thus far we have been discussing the principal directions and fundamental planes of the several coordinate systems as though they were fixed in space for all time. This, however, is not so. Certain motions of the principal directions and fundamental planes cause the reference axes of the coordinate systems to vary with time so that the reference system is not defined unless we specify the date to which it is referred. We will consider briefly how the separate motions arise and how they are taken into account.

At the outset, we remark that although the principal directions and fundamental planes of the coordinate systems used in the description of the motion of artificial Earth satellites are time dependent, thus far the precision of the observations has not been sufficient to warrant the highly refined specifications of the reference system we will now discuss. The applications of the concepts now considered are primarily to planetary motion in a heliocentric coordinate system.

As stated above, the fundamental planes for the ecliptic and equatorial systems do not remain fixed in space. They undergo small, but significant, variations with time. The ecliptic plane moves because of the attractions of the planets upon the Earth; the celestial equator moves because of torques exerted on a nonspherical, rotating Earth by the Sun and the Moon.

The attractions of the Moon and the planets, principally Venus and Jupiter, produce very small oscillations of geocenter perpendicular to the ecliptic plane. Thus, in essence, the concept of an ecliptic plane is a concept of a mean plane. Venus's effect is due to the large ($\approx 3°24'$) inclination of its orbital plane to the ecliptic plane. Jupiter's effect is due to its large mass. Another effect of planetary attraction is a motion, called *planetary precession*, in which the ecliptic plane slowly rotates in space at an almost constant rate of 47″ per annum.

In Figure 5.3.1 the symbols ENP and CNP refer to the ecliptic and

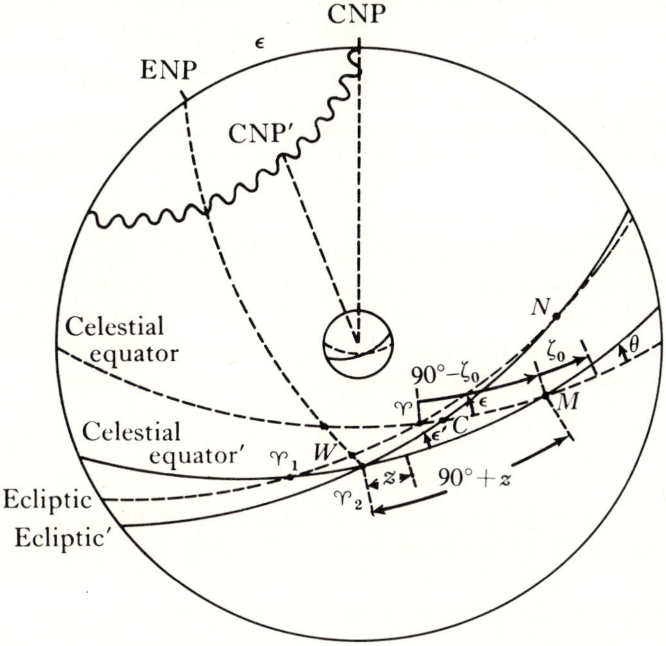

FIGURE 5.3.1

celestial north poles at some initial epoch and CNP′ refers to the celestial north pole at some later time. The motion of the ecliptic plane alone causes the equinox to move from ♈ to C. This motion, which is about 0″.13 per annum, is called *planetary precession in the equator*.

The rotation of the Earth on its axis has produced an equatorial bulge. Thus the figure of the Earth is more closely an oblate spheroid than a sphere. The forces exerted by the Sun and Moon on the two halves of the equatorial bulge result in a motion of the Earth's rotational axis in space that is composed of two parts. The first, and main part, called the *luni-solar precession*, is a conical motion of Earth's rotation axis that causes the CNP to describe a circle on the celestial sphere around the ENP, considered fixed. This precession, at a nearly uniform rate of $50''.37$ per annum, is a negative rotation, with apical angle equal to the obliquity ϵ and a period of about 26,000 years. The second part arises because the torques exerted by the Sun and Moon are not entirely uniform. The CNP does not move in a uniform circle around the ENP but rather in an oscillatory curve like the one depicted in Figure 5.3.1. This small disturbance in the smooth precessional motion is called *nutation.* It superimposes upon the uniform luni-solar precession an oscillatory motion of the equinox of amplitude of the order of $9''$ and period 18.6 years. If the effects of nutation are removed, the resulting fictitious equator and equinox are called the *mean equator* and *mean equinox*. The luni-solar precession is depicted on Figure 5.3.1 by the dashed arc $\Upsilon\Upsilon_1$.

Excluding nutation, the combined effect of planetary and luni-solar precession is to cause, in one year, say, the equinox to move about $50''$ from Υ to Υ_2. If we join ENP to Υ_2 with a great circle arc, it will cut the initial ecliptic in a point W. The quantity ΥW is called the *general precession in longitude*, χ. The combined planetary and luni-solar precessions (without nutation) also bring about a small ($0''.47$ per annum), nearly uniform decrease in the obliquity ϵ. Numerical expressions for calculating χ and ϵ, due to Newcomb [3, 4], are given below.

The general precession in longitude χ, per Julian year, is given by

$$\chi = 50''.2575 + 0''.0222T \tag{5.3.1}$$

where T is measured in Julian centuries of 36525 mean solar days from 1900 Jan. $0^d.50000$ U.T. [Universal time (U.T.) is the hour angle of the fictitious mean Sun plus 12 hours, measured from the Greenwich meridian.] The secular decrease in the obliquity ϵ is given by

$$\epsilon = 23°27'08''.26 - 46''.8645T - 0''.0059T^2 + 0''.00181T^3 \tag{5.3.2}$$

where T is measured in tropical centuries from the beginning of the Besselian fictitious year 1900.0 and

$$1 \text{ tropical year} = 365^d.242\,198\,79 - 0.000\,006\,14T$$

mean solar days, where T is the time in Julian centuries from 1900 Jan. $0^d.50000$ U.T. The tropical year is the time interval between successive apparent passages of the Sun through the vernal equinox. The

Besselian fictitious year is a tropical year that begins when the Sun's mean celestial longitude,† affected by aberration, is exactly 280 deg.

In Sections 5.1 and 5.2, we did not specify the reference frame other than to say it was defined by the position of a principal direction and a fundamental plane. Since the fundamental planes and principal directions are continuously changing, it must be understood that when we speak of an equatorial system, for example, we mean a system whose axes are defined by the orientation of the equator and equinox at some instant t. Thus all the formulas derived in the preceding chapters are correct in coordinate systems that are referenced to an instantaneous equator and equinox, (i.e., the true equator and equinox, including nutation). Ordinarily, however, the formulas are to be used with a mean equator and equinox. The instantaneous positions of the mean equator and equinox are called the *mean equator and mean equinox of date*. The adopted epoch is usually the beginning of some Besselian fictitious year. We then refer to the "mean equinox of 1959.0" or the "mean equinox of 1950.0," etc. For many purposes, a "standard equinox" is used. Those of 1950.0, 1925.0, 1900.0, and 1850.0 are especially useful. It should be pointed out that the beginning of the Besselian fictitious year must be carefully distinguished from the beginning of the calendar year. The time 1950.0 does not, for example, coincide with 1950 Jan. $0^d.0000$ U.T.

If observations of the right ascension and the declination of a celestial body are made, the observations are referenced to the equinox and equator at the instant of observation. The observed angles are called apparent right ascension and apparent declination because the effects of planetary aberration have not been taken into account. If the effects of aberration are removed, the angles are then called the true right ascension and the true declination. After the nutation effects that are present in the true right ascension and true declination are removed, the angles are then considered to be referenced to the mean equinox and the mean equator of date. It is not common practice, however, to refer the reduced coordinates to the mean equinox and equator of date but rather to the mean equator and equinox, at the beginning of the Besselian fictitious year, or to some standard equinox such as 1900.0 or 1950.0. Thus the problem of transforming from coordinates referenced to the mean equator and mean equinox of one date to coordinates referenced to a mean equator and mean equinox of another date arises. The transformations can be accomplished by the three separate rotations discussed in the following three sections.

† The mean celestial longitude is the angle $M + \bar{\omega}$, where M is the mean anomaly and $\bar{\omega}$ is the celestial longitude of pericenter (see Section 6.1).

5.4. TRANSFORMATIONS BETWEEN EQUATORIAL SYSTEMS REFERRED TO DIFFERENT EPOCHS

We represent the equatorial rectangular coordinates $(\xi_{q_0}, \eta_{q_0}, \zeta_{q_0})$ referred to a mean equinox at epoch t_0 by the column matrix $\mathbf{r}_{q_0}$ and the equatorial rectangular coordinates $(\xi_{q_f}, \eta_{q_f}, \zeta_{q_f})$ referred to a mean equinox at epoch t_f by the column matrix $\mathbf{r}_{q_f}$. The coordinates referred to the two epochs are related by the matrix equation

$$\mathbf{r}_{q_f} = F\mathbf{r}_{q_0} \tag{5.4.1}$$

where

$$F = \begin{pmatrix} F_{11} & F_{21} & F_{31} \\ F_{12} & F_{22} & F_{32} \\ F_{13} & F_{23} & F_{33} \end{pmatrix}$$

Formulas for calculating the elements F_{lm}, $l = 1, 2, 3$; $m = 1, 2, 3$, of F can be deduced as follows. An arbitrary origin O is assumed.

In Figure 5.3.1, let ♈ refer to the position of the mean equinox at epoch t_0 and ♈$_2$ refer to the position of the mean equinox at t_f. Since the precession of ♈ is retrograde, we consider a negative rotation of the equatorial rectangular axes $O\xi_{q_0}\eta_{q_0}\zeta_{q_0}$ through the angle $-\zeta_0$. This brings an intermediate η-axis into alignment with OM. The matrix C^* for this transformation is

$$C^* = \begin{pmatrix} \cos \zeta_0 & -\sin \zeta_0 & 0 \\ \sin \zeta_0 & \cos \zeta_0 & 0 \\ 0 & 0 & 1 \end{pmatrix} \tag{5.4.2}$$

Now rotate the intermediate equatorial rectangular axes positively through the angle θ about the intermediate η-axis. Although the rotation is positive, the matrix B^* for this transformation is

$$B^* = \begin{pmatrix} \cos \theta & 0 & -\sin \theta \\ 0 & 1 & 0 \\ \sin \theta & 0 & \cos \theta \end{pmatrix} \tag{5.4.3}$$

since the rotation is about an η-axis rather than a ξ-axis.

Finally, we rotate the intermediate η-axis in the plane of the celestial equator′ to the west through the angle $-z$ to align the ξ_{q_f}-axis with the line segment O♈$_2$. The matrix A^* for this transformation is

$$A^* = \begin{pmatrix} \cos z & -\sin z & 0 \\ \sin z & \cos z & 0 \\ 0 & 0 & 1 \end{pmatrix} \tag{5.4.4}$$

By forming the matrix product $A^*B^*C^*$, we can obtain the elements F_{lm}, $l = 1, 2, 3$; $m = 1, 2, 3$, of the F matrix. Explicit formulas for these elements are

$$
\begin{aligned}
F_{11} &= \cos \zeta_0 \cos z \cos \theta - \sin \zeta_0 \sin z \\
F_{12} &= \sin \zeta_0 \cos z + \cos \zeta_0 \sin z \cos \theta \\
F_{13} &= \cos \zeta_0 \sin \theta \\
F_{21} &= -\cos \zeta_0 \sin z - \sin \zeta_0 \cos z \cos \theta \\
F_{22} &= \cos \zeta_0 \cos z - \sin \zeta_0 \sin z \cos \theta \\
F_{23} &= -\sin \zeta_0 \sin \theta \\
F_{31} &= -\cos z \sin \theta \\
F_{32} &= -\sin z \sin \theta \\
F_{33} &= \cos \theta
\end{aligned}
\tag{5.4.5}
$$

The values of the F_{lm} are given in *Planetary Coordinates* [5] for reduction from the mean equinox at epoch $t_0 = 1950.0$ to the mean equinox at epoch t_f. They can also be used for reduction from the mean equinox at epoch t_f to the mean equinox of 1950.0. If $F_f(1950.0)$ indicates that the elements of the F matrix are to be evaluated for reduction from the mean equinox of 1950.0 to the mean equinox at epoch t_f, the coordinates at the two epochs are related by the matrix equation

$$\mathbf{r}_{q_f} = F_f(1950.0)\, \mathbf{r}_{q_{1950.0}}$$

Similarly, if $F_f(1950.0)^T$ represents the transpose of the $F_f(1950.0)$ matrix, we have

$$\mathbf{r}_{q_{1950.0}} = F_f(1950.0)^T\, \mathbf{r}_{q_f}$$

To transform coordinates from a mean equinox at epoch t_i to a mean equinox at epoch t_j, we have

$$\mathbf{r}_{q_{1950.0}} = F_i(1950.0)^T\, \mathbf{r}_{q_i}$$

and

$$\mathbf{r}_{q_j} = F_j(1950.0)\, \mathbf{r}_{q_{1950.0}} = F_j(1950.0)\, F_i(1950.0)^T\, \mathbf{r}_{q_i} = F\mathbf{r}_{q_i}$$

where

$$F = F_j(1950.0)\, F_i(1950.0)^T \tag{5.4.6}$$

For frequently used combinations, F can be computed and recorded. For years i and j not given in the tables, the angles ζ_0, z, and θ can be computed from the Newcomb formulas [3, 4]:

$$\begin{aligned}
\zeta_0 &= +2304''.25T + 0''.30T^2 + 0''.018T^3 \\
z &= +2304''.25T + 1''.09T^2 + 0''.019T^3 \\
\theta &= +2004''.298T - 0''.43T^2 - 0''.041T^3
\end{aligned} \tag{5.4.7}$$

where T is measured in tropical centuries from 1900.0.

If the coordinates $(\xi''_{q_i}, \eta''_{q_i}, \zeta''_{q_i})$ are given with respect to an orbital (equatorial) system† referenced to a mean equinox at epoch t_i and we want to transform to coordinates $\mathbf{r}_{q_j}$ in an equatorial system referenced to a mean equinox at epoch t_j, the transformation is given by

$$\mathbf{r}_{q_j} = F\mathbf{r}_{q_i} = H\mathbf{r}''_{q_i} \tag{5.4.8}$$

where

$$H = FD^T = \begin{pmatrix} H_{11} & H_{21} & H_{31} \\ H_{12} & H_{22} & H_{32} \\ H_{13} & H_{23} & H_{33} \end{pmatrix}$$

The matrix F is given by (5.4.6) and the matrix D^T by (2.5.6). The identity transformation for F (i.e., when no rotation of equinoxes is involved) is

$$F_I = \begin{pmatrix} 1 & 0 & 0 \\ 0 & 1 & 0 \\ 0 & 0 & 1 \end{pmatrix}$$

and, for this case, H reduces to the matrix D^T.

5.5. TRANSFORMATIONS BETWEEN ECLIPTIC AND EQUATORIAL SYSTEMS

If the coordinates are given with respect to an ecliptic reference system, it may be necessary to transform to an equatorial reference system and conversely. As indicated in Section 5.1, the subscripts q and c on the coordinates (ξ, η, ζ) will be used to refer to equatorial and ecliptic coordinates respectively (see Figure 5.5.1).

† By the notation orbital (equatorial) system, we mean that the orbital system is being used in conjunction with an equatorial system. A similar meaning is ascribed to the notation orbital (ecliptic) system.

The transformation matrix $\underset{\sim}{Q}$ from equatorial to ecliptic coordinates is given by

$$\underset{\sim}{Q} \equiv \begin{pmatrix} 1 & 0 & 0 \\ 0 & \cos\epsilon & \sin\epsilon \\ 0 & -\sin\epsilon & \cos\epsilon \end{pmatrix} \tag{5.5.1}$$

Hence

$$\mathbf{r}_c = \underset{\sim}{Q}\mathbf{r}_q \qquad \text{and} \qquad \mathbf{r}_q = \underset{\sim}{Q}^T\mathbf{r}_c \tag{5.5.2}$$

If the coordinates $\mathbf{r}''_{q_i}$ are given with respect to an orbital (equatorial) system referenced to a mean equinox at epoch t_i and we want to transform

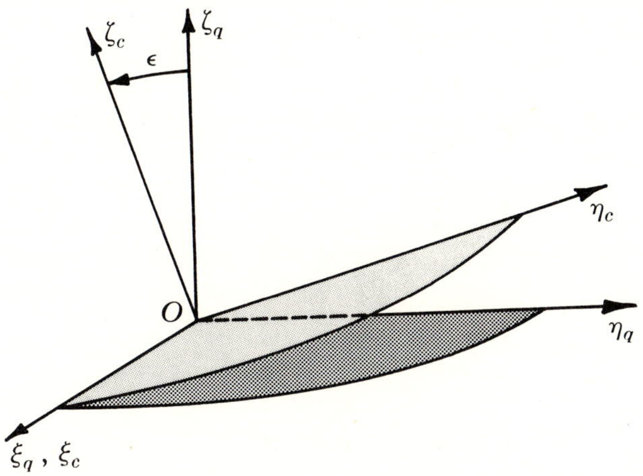

FIGURE 5.5.1

to coordinates $\mathbf{r}_{c_j}$ in an ecliptic system referenced to a mean equinox at epoch t_j , the transformation is given by

$$\mathbf{r}_{c_j} = P^*\mathbf{r}''_{q_i} \tag{5.5.3}$$

where

$$P^* \equiv \underset{\sim}{Q}H = \begin{pmatrix} 1 & 0 & 0 \\ 0 & \cos\epsilon & \sin\epsilon \\ 0 & -\sin\epsilon & \cos\epsilon \end{pmatrix} \begin{pmatrix} H_{11} & H_{21} & H_{31} \\ H_{12} & H_{22} & H_{32} \\ H_{13} & H_{23} & H_{33} \end{pmatrix}$$

and ϵ, for the epoch t_j , can be calculated from (5.3.2). The elements H_{lm} , $l = 1, 2, 3$; $m = 1, 2, 3$, are as defined earlier.

When the coordinates $\mathbf{r}''_{c_i}$ are given with respect to an orbital (ecliptic) system referenced to a mean equinox at epoch t_i and we want to transform to coordinates in an equatorial system referenced to a mean equinox at epoch t_j , the transformation is

$$\mathbf{r}_{q_j} = P\mathbf{r}''_{c_i} \tag{5.5.4}$$

where

$$P \equiv H\underset{\sim}{Q}^T = \begin{bmatrix} H_{11} & H_{21} & H_{31} \\ H_{12} & H_{22} & H_{32} \\ H_{13} & H_{23} & H_{33} \end{bmatrix} \begin{bmatrix} 1 & 0 & 0 \\ 0 & \cos\epsilon & -\sin\epsilon \\ 0 & \sin\epsilon & \cos\epsilon \end{bmatrix}$$

and ϵ is to be determined at epoch t_i.

5.6. TRANSFORMATIONS OF HELIOCENTRIC EQUATORIAL COORDINATES TO GEOCENTER OR TOPOCENTER

We have seen that coordinates given in either an ecliptic or equatorial system, referenced to a mean equinox at epoch t_i, can be transformed, by an appropriate rotation matrix, to coordinates in another reference system, either ecliptic or equatorial, with the same origin but referenced to a mean equinox at epoch t_f. We now wish to discuss a special type of transformation in which we transfer from a heliocentric coordinate system to a topocentric system. Since we have at our disposal techniques for transforming between ecliptic and equatorial systems, for different mean equinoxes, we may, for definiteness and purpose of exposition, restrict our discussion to a transfer from a heliocentric equatorial system to a topocentric equatorial system referred to the same equinox.

Let

$\mathbf{r}$ = position vector of celestial object referenced to heliocenter S

$\boldsymbol{\rho}$ = position vector of celestial object referenced to geocenter E

$\mathbf{r}_P$ = position vector of celestial object referenced to topocenter P

$\boldsymbol{\rho}'$ = position vector of topos P referenced to geocenter E

$\boldsymbol{\rho}_s$ = position vector of heliocenter referenced to geocenter E

ϕ' = geocentric latitude of topos P

θ = right ascension of topos P (local sidereal time)

(ξ_q, η_q, ζ_q) = equatorial rectangular coordinates with origin at P

Consider first the transformation from a heliocentric equatorial system to a geocentric equatorial system referenced to the same equinox. We have

$$\boldsymbol{\rho} = \mathbf{r} + \boldsymbol{\rho}_s$$

or, in component form (see Section 5.1 for notation),

$$\begin{aligned} x_q &= \rho \cos\delta \cos\alpha = X_q + x_{qs} \\ y_q &= \rho \cos\delta \sin\alpha = Y_q + y_{qs} \\ z_q &= \rho \sin\delta = Z_q + z_{qs} \end{aligned} \tag{5.6.1}$$

where the coordinates (x_{qs}, y_{qs}, z_{qs}) of the Sun referenced to geocenter are listed in the *Nautical Almanac.*

If we next consider the transfer from a heliocentric equatorial system to a topocentric equatorial system, we write

$$\mathbf{r}_P = \boldsymbol{\rho}_s + \mathbf{r} - \boldsymbol{\rho}'$$

or, in component form,

$$\begin{aligned}
\xi_q &= X_q + x_{qs} - x_{qP} = X_q + x_{qs} - \rho' \cos\phi' \cos\theta \\
\eta_q &= Y_q + y_{qs} - y_{qP} = Y_q + y_{qs} - \rho' \cos\phi' \sin\theta \\
\zeta_q &= Z_q + z_{qs} - z_{qP} = Z_q + z_{qs} - \rho' \sin\phi'
\end{aligned} \tag{5.6.2}$$

where ρ' is the radius of the Earth which, for most practical purposes, can be assumed constant in the treatment of planetary problems. The coordinates (x_{qP}, y_{qP}, z_{qP}) give the position of P referenced to geocenter.

Exercises

1. An artificial Earth satellite is at perigee ($q = 1.1$ gr) of a two-body orbit around the geocenter, for which each of the three orbital elements ω, Ω, and i, when referenced to the mean equator and equinox of 1950.0, is equal to $\pi/2$ radians. What are the coordinates of the satellite in the geocentric ecliptic system referenced to the ecliptic and mean equinox of 1955.0? You may use the following information: Equatorial coordinates, referred to the mean equinox 1950.0, may be reduced to the mean equinox 1955.0 by the transformation

$$\mathbf{r}_{q_{1955.0}} = F_{1955.0}(1950.0)\, \mathbf{r}_{q_{1950.0}}$$

where

$$F_{1955.0}(1950.0) = \begin{pmatrix} +1.0000 & -0.0010 & -0.0005 \\ +0.0010 & +1.0100 & -0.0000 \\ +0.0005 & -0.0000 & +1.0000 \end{pmatrix}$$

and the numerical values of the matrix elements have been rounded off for ease of computation. Assume that the obliquity of the ecliptic is constant and, for ease of computation, take $\cos\epsilon = 0.9000$ and $\sin\epsilon = 0.4000$.

2. Show that the angular distance between the mean equinox of 1900.0 and the mean equinox of 2150.0 is $12\,570''.98$.

Chapter 6

The Two-Body Orbit in Space

6.1. NOMENCLATURE

At various points in the preceding discussion, we introduced, as needed, some of the parameters that are used in describing the position in space of a body orbiting about another more massive body. Certain additional quantities will now be introduced. For convenience, we will give a recapitulation of the parameters already discussed as well as definitions of the new quantities that will be used to complete the description of the position of the body with respect to either an equatorial or ecliptic reference frame.

Consider the rectangular coordinate system $O\xi\eta\zeta$ (to be interpreted as $O\xi_q\eta_q\zeta_q$ or $O\xi_c\eta_c\zeta_c$, depending on whether the system is equatorial or ecliptic) shown in Figure 6.1.1. With reference to this figure, we have

True anomaly: w

Line of nodes: ON

Ascending node: N

Longitude of ascending node: $\Omega, \quad 0 \leqslant \Omega \leqslant 360$ (ecliptic system)[†]

[†] Throughout this section, replace word longitude by right ascension if system is equatorial.

Angle of inclination: i, $0 \leqslant i \leqslant 180°$ The motion is said to be direct if $0 \leqslant i \leqslant 90°$ and retrograde if $i > 90°$.

Argument of pericenter: ω The angle ω defines the orientation of the orbit in the orbital plane.

Time of pericenter passage: T The parameter T specifies some instant at which the orbiting body passed through pericenter.

Elements: a, e, q, P The size and shape of the orbit are specified by a (or q) and e. The period P introduces the masses of the two bodies.

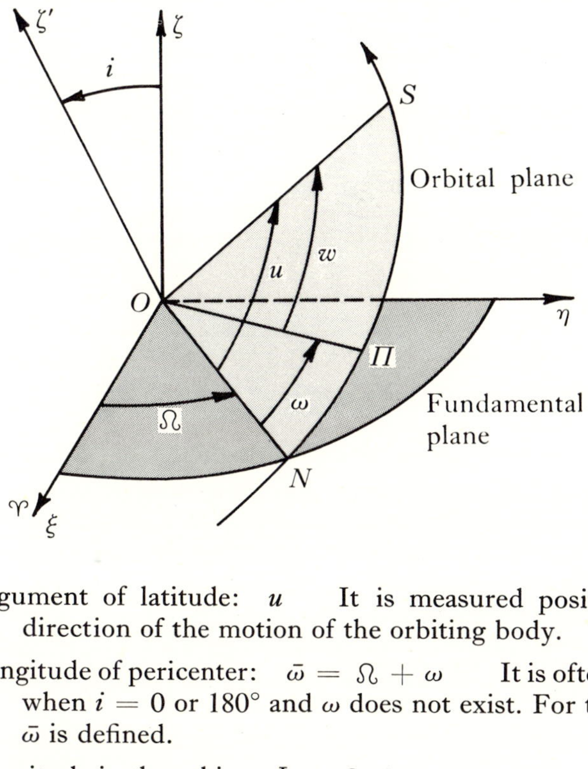

FIGURE 6.1.1

Argument of latitude: u It is measured positive (as is w) in the direction of the motion of the orbiting body.

Longitude of pericenter: $\varpi = \Omega + \omega$ It is often used as an element when $i = 0$ or $180°$ and ω does not exist. For these cases, the angle ϖ is defined.

Longitude in the orbit: $\Phi = \Omega + u = \varpi + w$ Even for $i = 0°$ or $180°$, the angle Φ is defined.

Mean longitude in the orbit: $L = \varpi + M = \varpi + n(t - T)$

Mean longitude in the orbit at epoch t_0: $L_0 = \varpi + n(t_0 - T)$ Clearly $L = L_0 + n(t - t_0)$. Sometimes the symbol ϵ is used for L_0 and ϵ is used as an orbital element instead of T.

6.2. SOME USEFUL RELATIONS FROM SPHERICAL TRIGONOMETRY

In Chapter 3 we examined the problem of determining the position coordinates (r, w) in the orbital plane when the orbital elements a or q, e, and T were given. The converse problem [that is, given the position and velocity at some instant, determine the orbital elements a (or q), e, and T] was also considered. In the next section, an extension of this problem is considered. We will examine the problem (and its converse) of determining the rectangular and spherical position coordinates in a given reference frame $O\xi\eta\zeta$ when a set of orbital elements $(a, e, i, \omega, \Omega, T)$, or some equivalent set of elements, has been specified. In this development, we will have occasion to draw on certain results from spherical trigonometry, which we now proceed to deduce, using some elementary notions of vector analysis.

Consider a sphere of unit radius centered at O. Let A, B, C represent three points on the surface of this sphere that define a spherical triangle ABC. The sides of the triangle are arcs of three great circles through the points (see Figure 6.2.1). The symbols a, b, and c are defined as

$$\angle BOC = a, \qquad \angle COA = b, \qquad \angle AOB = c \qquad (0\text{–}180^\circ)$$

The angle between the planes AOB and AOC is A, and B and C are similarly defined. The unit vectors **i**, **j**, and **k** used here are not necessarily orthogonal.

With reference to the figure, we write

$$(\mathbf{i} \times \mathbf{j}) \cdot (\mathbf{i} \times \mathbf{k}) = \sin c\mathbf{n} \cdot \sin b\mathbf{m}$$

where **n** and **m** are unit vectors perpendicular to the planes defined by

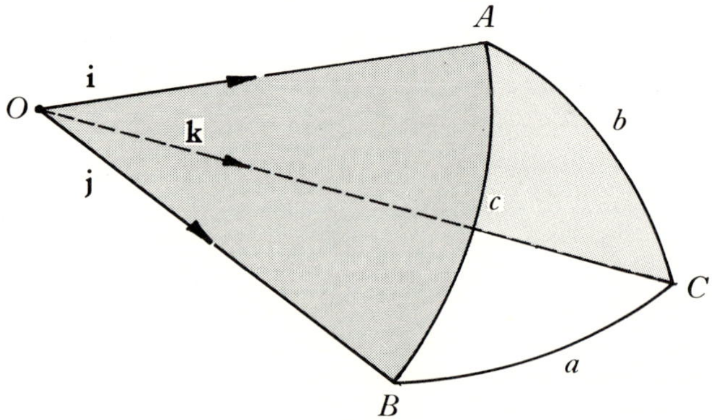

FIGURE 6.2.1

$\mathbf{i} \times \mathbf{j}$ and $\mathbf{i} \times \mathbf{k}$ in the usual sense of a cross product, and $\mathbf{n} \cdot \mathbf{m} = \cos A$. Thus

$$(\mathbf{i} \times \mathbf{j}) \cdot (\mathbf{i} \times \mathbf{k}) = \sin b \sin c \cos A \tag{6.2.1}$$

We can look at the dot product $(\mathbf{i} \times \mathbf{j}) \cdot (\mathbf{i} \times \mathbf{k})$ in a different light. For a triple scalar product of any three vectors $\boldsymbol{\alpha}$, $\boldsymbol{\beta}$, $\boldsymbol{\gamma}$, we can use the cyclic rule and write

$$\boldsymbol{\alpha} \cdot \boldsymbol{\beta} \times \boldsymbol{\gamma} = \boldsymbol{\gamma} \cdot \boldsymbol{\alpha} \times \boldsymbol{\beta} = \boldsymbol{\beta} \cdot \boldsymbol{\gamma} \times \boldsymbol{\alpha}$$

Setting $\boldsymbol{\alpha} = \mathbf{i} \times \mathbf{k}$, we have

$$\begin{aligned} \boldsymbol{\alpha} \cdot \mathbf{i} \times \mathbf{j} = \mathbf{j} \cdot \boldsymbol{\alpha} \times \mathbf{i} = \mathbf{i} \cdot \mathbf{j} \times \boldsymbol{\alpha} &= \mathbf{i} \cdot [\mathbf{j} \times (\mathbf{i} \times \mathbf{k})] \\ &= \mathbf{i} \cdot [(\mathbf{j} \cdot \mathbf{k})\, \mathbf{i} - (\mathbf{j} \cdot \mathbf{i})\, \mathbf{k}] \end{aligned}$$

where, in the last step, we used the vector identity

$$\boldsymbol{\alpha} \times (\boldsymbol{\beta} \times \boldsymbol{\gamma}) = (\boldsymbol{\alpha} \cdot \boldsymbol{\gamma})\, \boldsymbol{\beta} - (\boldsymbol{\alpha} \cdot \boldsymbol{\beta})\, \boldsymbol{\gamma} \tag{6.2.2}$$

It follows that

$$\begin{aligned} (\mathbf{i} \times \mathbf{j}) \cdot (\mathbf{i} \times \mathbf{k}) = (\mathbf{i} \times \mathbf{k}) \cdot (\mathbf{i} \times \mathbf{j}) &= \mathbf{i} \cdot [(\mathbf{j} \cdot \mathbf{k})\, \mathbf{i} - (\mathbf{j} \cdot \mathbf{i})\, \mathbf{k}] \\ &= \mathbf{j} \cdot \mathbf{k} - (\mathbf{i} \cdot \mathbf{k})(\mathbf{j} \cdot \mathbf{i}) \\ &= \cos a - \cos b \cos c \end{aligned} \tag{6.2.3}$$

Equating (6.2.1) and (6.2.3), we have

$$\cos a = \cos b \cos c + \sin b \sin c \cos A \tag{6.2.4}$$

Consider next the vector product $(\mathbf{i} \times \mathbf{j}) \times (\mathbf{i} \times \mathbf{k})$. Using the vector identity (6.2.2), we write

$$(\mathbf{i} \times \mathbf{j}) \times (\mathbf{i} \times \mathbf{k}) = (\mathbf{k} \cdot \mathbf{i} \times \mathbf{j})\, \mathbf{i} - (\mathbf{i} \cdot \mathbf{i} \times \mathbf{j})\, \mathbf{k} = (\mathbf{i} \cdot \mathbf{j} \times \mathbf{k})\, \mathbf{i} \tag{6.2.5}$$

since $\mathbf{i} \cdot \mathbf{i} \times \mathbf{j} = 0$. An alternative expression for $(\mathbf{i} \times \mathbf{j}) \times (\mathbf{i} \times \mathbf{k})$ is

$$\begin{aligned} (\mathbf{i} \times \mathbf{j}) \times (\mathbf{i} \times \mathbf{k}) = \sin c\mathbf{n} \times \sin b\mathbf{m} &= \sin c \sin b\mathbf{n} \times \mathbf{m} \\ &= \sin c \sin b \sin A\mathbf{i} \end{aligned} \tag{6.2.6}$$

where the last step in (6.2.6) follows from the fact that $\mathbf{n} \times \mathbf{m}$ is a vector perpendicular to the normals to each of the planes AOB and AOC and hence it is parallel to both planes. It is also parallel to the intersection of both planes and thus it is in the direction of $\mathbf{i}$. Equating (6.2.5) and (6.2.6), we write

$$\sin c \sin b \sin A = \mathbf{i} \cdot \mathbf{j} \times \mathbf{k} \tag{6.2.7}$$

Cyclic permutation of the vectors **i**, **j**, and **k** leads immediately to the two similar equations

$$\sin a \sin b \sin C = \mathbf{k} \cdot \mathbf{i} \times \mathbf{j} \tag{6.2.8}$$

$$\sin a \sin c \sin B = \mathbf{j} \cdot \mathbf{k} \times \mathbf{i} \tag{6.2.9}$$

Since

$$\mathbf{i} \cdot \mathbf{j} \times \mathbf{k} = \mathbf{k} \cdot \mathbf{i} \times \mathbf{j} = \mathbf{j} \cdot \mathbf{k} \times \mathbf{i}$$

we have

$$\sin c \sin b \sin A = \sin a \sin b \sin C = \sin a \sin c \sin B$$

from which it follows that

$$\frac{\sin A}{\sin a} = \frac{\sin B}{\sin b} = \frac{\sin C}{\sin c} \tag{6.2.10}$$

Equations (6.2.4) and (6.2.10) are the fundamental formulas of spherical trigonometry.

6.3. DETERMINATION OF THE COORDINATES WHEN THE ORBITAL ELEMENTS ARE GIVEN

We consider the problem of determining the rectangular coordinates (ξ, η, ζ) and the spherical coordinates (r, l, b) when the orbital elements a or q, e, i, ω, Ω, and T have been prescribed for the two-body problem. Since the discussion will apply to either equatorial or ecliptic systems, no subscripts will be needed on the symbols for the coordinates.

The principal direction in the fundamental plane is taken toward the vernal equinox (see Figure 6.3.1). Consider first an intermediate set of axes $O\xi'\eta^*\zeta$ obtained by rotating counterclockwise (as viewed from the positive ζ-axis), through an angle Ω about the ζ-axis. Using (2.5.2), we write

$$\xi = \xi' \cos \Omega - \eta^* \sin \Omega \tag{6.3.1a}$$

$$\eta = \xi' \sin \Omega + \eta^* \cos \Omega \tag{6.3.1b}$$

$$\zeta = \zeta \tag{6.3.1c}$$

The equations of transformation that relate the coordinates (ξ', η^*, ζ) and (r, l, b) are

$$\xi' = r \cos b \cos(l - \Omega) \tag{6.3.2a}$$

$$\eta^* = r \cos b \sin(l - \Omega) \tag{6.3.2b}$$

$$\zeta = r \sin b \tag{6.3.2c}$$

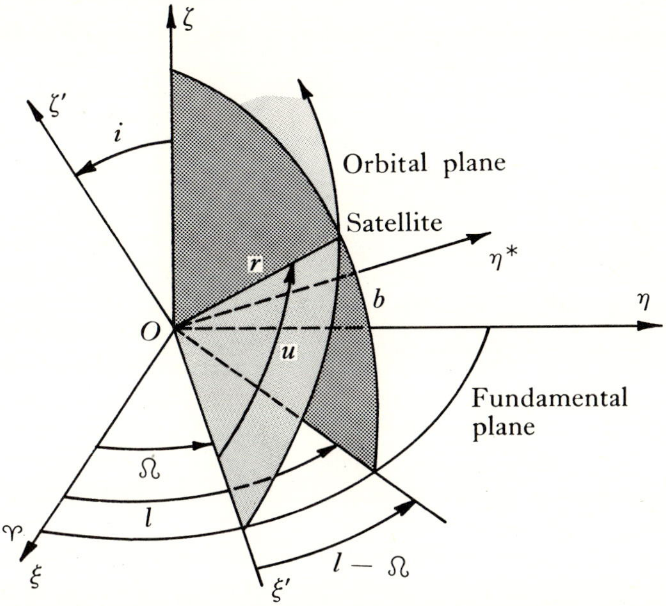

FIGURE 6.3.1

The spherical coordinates b and l may, in turn, be expressed in terms of i and Ω and the argument of latitude u if we use the fundamental formulas of spherical trigonometry derived in the preceding section.

Using (6.2.4), we can write

$$\cos u = \cos b \cos(l - \Omega)$$

and

$$\cos b = \cos(l - \Omega) \cos u + \sin u \sin(l - \Omega) \cos i$$

Eliminating $\cos u$ between the two preceding equations, we have

$$\cos b = \cos^2(l - \Omega) \cos b + \sin u \cos i \sin(l - \Omega)$$

from which it follows directly that

$$\cos b \sin(l - \Omega) = \sin u \cos i$$

Using (6.2.10), we can write

$$\sin b = \sin i \sin u$$

Recapitulating, we write

$$\cos b \cos(l - \Omega) = \cos u \tag{6.3.3a}$$

$$\cos b \sin(l - \Omega) = \sin u \cos i \tag{6.3.3b}$$

$$\sin b = \sin i \sin u \tag{6.3.3c}$$

from which it follows immediately that, for any conic section,

$$\xi = r(\cos u \cos \Omega - \sin u \cos i \sin \Omega) \tag{6.3.4a}$$

$$\eta = r(\cos u \sin \Omega + \sin u \cos i \cos \Omega) \tag{6.3.4b}$$

$$\zeta = r \sin i \sin u \tag{6.3.4c}$$

If the orbital elements, for any conic section, are given, the polar coordinates (r, w) at a given instant t may be determined from (3.3.4), (3.3.5), or (3.3.6), depending on whether the conic is a hyperbola, ellipse, or parabola. The argument of latitude is then determined since $u = w + \omega$. The spherical coordinates b and l can then be calculated using (6.3.3) and the rectangular coordinates can be calculated using (6.3.4).

The rectangular coordinates are uniquely determined by (6.3.4). We have already seen that w is uniquely determined by (3.3.4), (3.3.5), or (3.3.6). Thus, with ω given, $u = w + \omega$ is uniquely determined and hence the coordinates (ξ, η, ζ) are uniquely given by (6.3.4). Depending on whether $\sin u$ is positive, zero, or negative, $\sin b$, as given by (6.3.3c), is positive, zero, or negative. Thus, the spherical coordinate b $(0 \leqslant b \leqslant \pi)$ is uniquely determined. Equations (6.3.3a) and (6.3.3b) serve to fix the quadrant of $l - \Omega$ and hence l $(0 \leqslant l \leqslant 2\pi)$ is determined uniquely. Therefore, the spherical coordinates are determined uniquely when the orbital elements are given.

6.4. AN ALTERNATIVE DETERMINATION OF THE COORDINATES WHEN THE ORBITAL ELEMENTS ARE GIVEN

An alternative scheme for determining the coordinates when the orbital elements are given will be described briefly. Let (ξ'', η'', ζ'') represent the rectangular coordinates in the orbital system. The positive ξ''-axis is directed from the occupied focus toward pericenter. In this system, for any conic section, we have

$$\xi'' = r \cos w \tag{6.4.1a}$$

$$\eta'' = r \sin w \tag{6.4.1b}$$

$$\zeta'' = 0 \tag{6.4.1c}$$

Using the equations developed in Sections 3.1 and 3.2 of Chapter 3, we can write, for the various conic sections:

Hyperbolic orbit:

$$\xi'' = r \cos w = a(e - \cosh F) \tag{6.4.2a}$$

$$\eta'' = r \sin w = a(e^2 - 1)^{1/2} \sinh F \tag{6.4.2b}$$

$$\zeta'' = 0 \tag{6.4.2c}$$

Elliptic orbit:

$$\xi'' = r \cos w = a(\cos E - e) \tag{6.4.3a}$$

$$\eta'' = r \sin w = a(1 - e^2)^{1/2} \sin E \tag{6.4.3b}$$

$$\zeta'' = 0 \tag{6.4.3c}$$

Parabolic orbit:

$$\xi'' = r \cos w = q(1 - z^2) \tag{6.4.4a}$$

$$\eta'' = r \sin w = 2qz \tag{6.4.4b}$$

$$\zeta'' = 0 \tag{6.4.4c}$$

If the transformation (2.5.6) is used to introduce the unprimed coordinates (ξ, η, ζ), we can write

Hyperbolic orbit:

$$\xi = a[D_{11}(e - \cosh F) + D_{21}(e^2 - 1)^{1/2} \sinh F] \tag{6.4.5a}$$

$$\eta = a[D_{12}(e - \cosh F) + D_{22}(e^2 - 1)^{1/2} \sinh F] \tag{6.4.5b}$$

$$\zeta = a[D_{13}(e - \cosh F) + D_{23}(e^2 - 1)^{1/2} \sinh F] \tag{6.4.5c}$$

Elliptic orbit:

$$\xi = a[D_{11}(\cos E - e) + D_{21}(1 - e^2)^{1/2} \sin E] \tag{6.4.6a}$$

$$\eta = a[D_{12}(\cos E - e) + D_{22}(1 - e^2)^{1/2} \sin E] \tag{6.4.6b}$$

$$\zeta = a[D_{13}(\cos E - e) + D_{23}(1 - e^2)^{1/2} \sin E] \tag{6.4.6c}$$

Parabolic orbit:

$$\xi = q[D_{11}(1 - z^2) + 2D_{21}z] \tag{6.4.7a}$$

$$\eta = q[D_{12}(1 - z^2) + 2D_{22}z] \tag{6.4.7b}$$

$$\zeta = q[D_{13}(1 - z^2) + 2D_{23}z] \tag{6.4.7c}$$

When the orbital elements ω, Ω, and i are given, the D_{ij}, $i = 1, 2$; $j = 1, 2, 3$, can be calculated from (2.5.1) for all time t. Thus this particular scheme is quite useful for computing a table of positions.

When using Equations (6.4.5), (6.4.6), or (6.4.7), the procedure consists of the following four parts:

1. Given a or q, compute ν, n, or σ, using (3.3.4b), (3.3.5b), or (3.3.6b).
2. Given T, and the time t of interest, compute N, M, or Q from (3.3.4c), (3.3.5c), or (3.3.6c).
3. Determine F, E from (3.3.4d), (3.3.5d) or, in the parabolic case, determine w from Equations (3.3.6d) and (3.3.6e).
4. With F, E, or $z = \tan w/2$ known, use (6.4.5), (6.4.6), or (6.4.7) to obtain ξ, η, and ζ.

If coordinates (ξ, η, ζ) have been determined, the spherical coordinates can be obtained from the transformation equations (see Figure 6.3.1)

$$\xi = r \cos b \cos l \tag{6.4.8a}$$

$$\eta = r \cos b \sin l \tag{6.4.8b}$$

$$\zeta = r \sin b \tag{6.4.8c}$$

Suitable formulas for calculating the corresponding velocity components $\dot{\xi}$, $\dot{\eta}$, and $\dot{\zeta}$ can also be derived. To this end, we return to (2.5.6) and differentiate with respect to time to obtain the matrix equation

$$\begin{pmatrix} \dot{\xi} \\ \dot{\eta} \\ \dot{\zeta} \end{pmatrix} = D^T \begin{pmatrix} \xi'' \\ \eta'' \\ \zeta'' \end{pmatrix}^{\cdot} = D^T \begin{pmatrix} \dot{\xi}'' \\ \dot{\eta}'' \\ 0 \end{pmatrix} \tag{6.4.9}$$

where we have used the facts that $(D^T)^{\cdot} = 0$ since the orbital elements ω, Ω, and i are constant, and $\zeta'' = 0$.

To express the primed time derivatives in terms of the orbital elements and suitable anomalies, we return to (6.4.2), (6.4.3), and (6.4.4) and differentiate with respect to time. This procedure leads to the equations:

Hyperbolic orbit:

$$\dot{\xi}'' = -a(\sinh F)\dot{F} \tag{6.4.10a}$$

$$\dot{\eta}'' = a(e^2 - 1)^{1/2} (\cosh F)\dot{F} \tag{6.4.10b}$$

$$\dot{\zeta}'' = 0 \tag{6.4.10c}$$

Elliptic orbit:

$$\dot{\xi}'' = -a(\sin E)\dot{E} \tag{6.4.11a}$$

$$\dot{\eta}'' = a(1 - e^2)^{1/2} (\cos E)\dot{E} \tag{6.4.11b}$$

$$\dot{\zeta}'' = 0 \tag{6.4.11c}$$

Parabolic orbit:

$$\dot{\xi}'' = -2qz\dot{z} \tag{6.4.12a}$$

$$\dot{\eta}'' = 2q\dot{z} \tag{6.4.12b}$$

$$\dot{\zeta}'' = 0 \tag{6.4.12c}$$

The time derivatives of the anomalies are readily expressed in terms of the radial distance r and the orbital elements by differentiating Equations (3.3.4c, d), (3.3.5c, d), and (3.3.6e) with respect to time. We find that

$$\dot{F} = \left(\frac{\mu}{a}\right)^{1/2} \frac{1}{r} \tag{6.4.13a}$$

$$\dot{E} = \left(\frac{\mu}{a}\right)^{1/2} \frac{1}{r} \tag{6.4.13b}$$

$$\dot{z} = \left(\frac{\mu}{2q}\right)^{1/2} \frac{1}{r} \tag{6.4.13c}$$

Equations (6.4.9) may now be written in the form:

Hyperbolic orbit:

$$\begin{pmatrix} \dot{\xi} \\ \dot{\eta} \\ \dot{\zeta} \end{pmatrix} = D^T \begin{pmatrix} -r^{-1}(\mu a)^{1/2} \sinh F \\ r^{-1}[\mu a(e^2 - 1)]^{1/2} \cosh F \\ 0 \end{pmatrix} \tag{6.4.14}$$

Elliptic orbit:

$$\begin{pmatrix} \dot{\xi} \\ \dot{\eta} \\ \dot{\zeta} \end{pmatrix} = D^T \begin{pmatrix} -r^{-1}(\mu a)^{1/2} \sin E \\ r^{-1}[\mu a(1 - e^2)]^{1/2} \cos E \\ 0 \end{pmatrix} \tag{6.4.15}$$

Parabolic orbit:

$$\begin{pmatrix} \dot{\xi} \\ \dot{\eta} \\ \dot{\zeta} \end{pmatrix} = D^T \begin{pmatrix} -(2q\mu)^{1/2} zr^{-1} \\ (2q\mu)^{1/2} r^{-1} \\ 0 \end{pmatrix} \tag{6.4.16}$$

6.5. THE DETERMINATION OF THE ORBITAL ELEMENTS WHEN THE POSITION r AND THE VELOCITY v = ṙ ARE GIVEN

We have seen how the position and velocity of an orbiting body can be obtained when the elements of the two-body orbit have been prescribed. We now turn to the converse problem. If, at some instant t, the position $\mathbf{r}$ and the velocity $\mathbf{v}$ are given, how do we deduce the six elements of the two-body orbit?

We have already seen how three of the six elements (a or q, e, and T) can be obtained from Equations (3.3.1), (3.3.2), or (3.3.3) if $\mathbf{r}$ and $\mathbf{v}$ are prescribed at some instant t. We have also seen that the angular momentum per unit mass, $\mathbf{h} = c_1\mathbf{i} + c_2\mathbf{j} + c_3\mathbf{k} = \mathbf{r} \times \mathbf{v}$, which is prescribed when $\mathbf{r}$ and $\mathbf{v}$ are prescribed, permits us to determine Ω and i uniquely through Equations (2.5.7), namely,

$$h \sin \Omega \sin i = c_1$$

$$h \cos \Omega \sin i = -c_2$$

$$h \cos i = c_3$$

where

$$\begin{aligned} c_1 &= (\mathbf{r} \times \mathbf{v})_\xi = \eta\dot{\zeta} - \zeta\dot{\eta} \\ c_2 &= (\mathbf{r} \times \mathbf{v})_\eta = \zeta\dot{\xi} - \xi\dot{\zeta} \\ c_3 &= (\mathbf{r} \times \mathbf{v})_\zeta = \xi\dot{\eta} - \eta\dot{\xi} \end{aligned} \tag{6.5.1}$$

The preceding aspects of the calculations remain the same. We wish to examine the problem of determining the remaining element ω. From a knowledge of a or q, e,and T we may use (3.3.4), (3.3.5), or (3.3.6) to determine w for the particular conic section. To determine the argument of latitude, we can use (2.5.2) and write

$$\xi' = \xi \cos \Omega + \eta \sin \Omega = r \cos u \tag{6.5.2a}$$

$$\zeta' = \zeta = r \sin u \sin i \tag{6.5.2b}$$

Equations (6.5.2) permit us to determine an initial value of u, say u_0, uniquely. Then we may obtain ω from the relation

$$\omega = u_0 - w_0 \tag{6.5.3}$$

where w_0 is the corresponding initial value of w. This completes the determination of the orbital elements, for any conic section, when $\mathbf{r}$ and $\mathbf{v}$ are prescribed.

6.6. ALTERNATIVE DETERMINATION OF THE ORBITAL ELEMENTS WHEN THE POSITION AND THE VELOCITY ARE GIVEN

For some purposes, it is convenient to express the orbital elements Ω and i in terms of certain launch parameters. The determination of a or q, e, T, and ω proceed as before through Equations (3.3.1), (3.3.2), (3.3.3), and (6.5.3). We introduce a slightly different procedure, however, for determining the elements i and Ω.

Assume that an orbit is to be established about m_0 such that initially it passes through the origin $P(r, l, b)$ of a rectangular coordinate system $P\xi_T\eta_T\zeta_T$ with velocity $\mathbf{v}$ (see Figure 6.6.1). The positive ξ_T-axis is

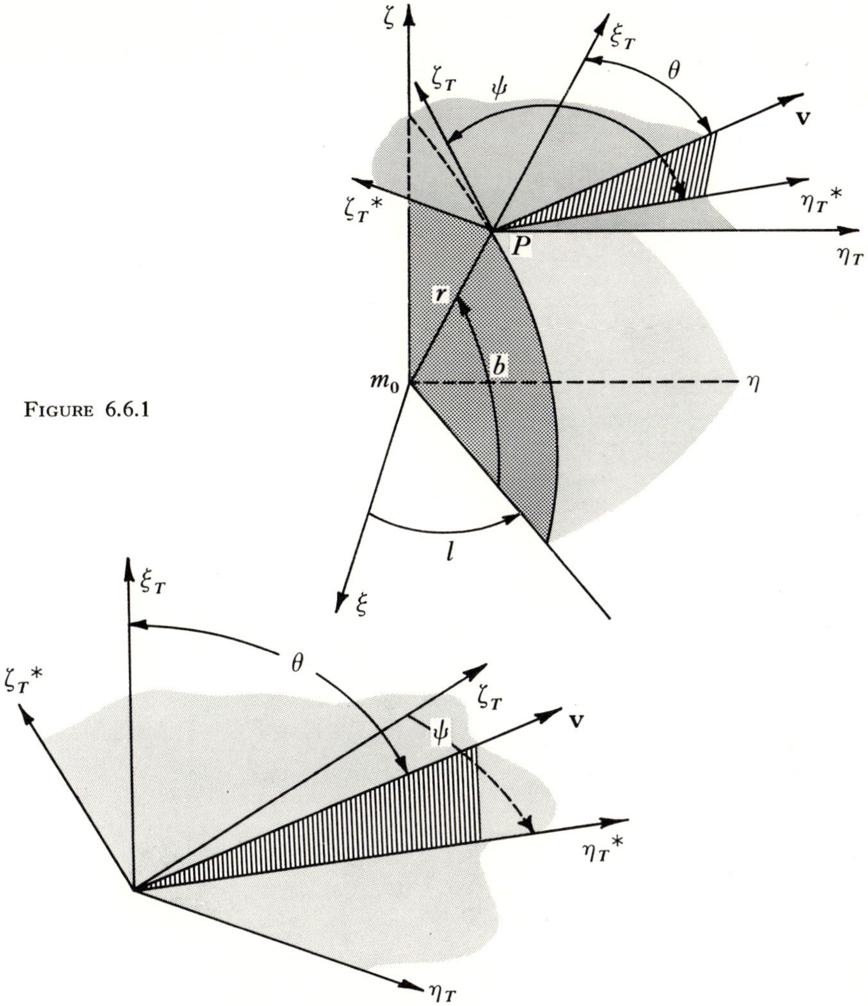

FIGURE 6.6.1

directed along **r**. The η_T- and ζ_T-axes lie in a plane perpendicular to ξ_T at P; the positive η_T-axis is directed toward the east, tangent to a circle of latitude through P. The ζ_T-axis is chosen to form a right-handed coordinate system. We define another rectangular coordinate system $P\xi_T\eta_T{}^*\zeta_T{}^*$ with origin at P. The $\eta_T{}^*$-axis is directed along the intersection of the $\eta_T\zeta_T$-plane and the plane determined by **r** and **v**; the positive sense of the $\eta_T{}^*$-axis is fixed by the condition that the angle between it and the velocity vector **v** be no greater than $\pi/2$ radians. The $\zeta_T{}^*$-axis is chosen to form a right-handed system. The azimuth angle ψ, $0 \leqslant \psi \leqslant 2\pi$, is measured in the $\eta_T\zeta_T$-plane, clockwise from ζ_T to $\eta_T{}^*$; ψ is known when **r** and **v** are given. The angle θ (depression angle) between **r** and **v** is also prescribed.

The coordinates $(\xi_T, \eta_T^*, \zeta_T^*)$ may be obtained from the coordinates (ξ, η, ζ) by performing, successively, the three rotations given by the matrix

$$R = \begin{pmatrix} R_{11} & R_{12} & R_{13} \\ R_{21} & R_{22} & R_{23} \\ R_{31} & R_{32} & R_{33} \end{pmatrix} = \begin{pmatrix} 1 & 0 & 0 \\ 0 & \sin\psi & \cos\psi \\ 0 & -\cos\psi & \sin\psi \end{pmatrix} \begin{pmatrix} \cos b & 0 & \sin b \\ 0 & 1 & 0 \\ -\sin b & 0 & \cos b \end{pmatrix}$$

$$\times \begin{pmatrix} \cos l & \sin l & 0 \\ -\sin l & \cos l & 0 \\ 0 & 0 & 1 \end{pmatrix} \tag{6.6.1}$$

where

$$\begin{aligned} R_{11} &= \cos b \cos l \\ R_{12} &= \cos b \sin l \\ R_{13} &= \sin b \\ R_{21} &= -\sin\psi \sin l - \cos\psi \sin b \cos l \\ R_{22} &= \sin\psi \cos l - \cos\psi \sin b \sin l \\ R_{23} &= \cos\psi \cos b \\ R_{31} &= \cos\psi \sin l - \sin\psi \sin b \cos l \\ R_{32} &= -\cos\psi \cos l - \sin\psi \sin b \sin l \\ R_{33} &= \sin\psi \cos b \end{aligned}$$

Associating the orthonormal bases $(\mathbf{i}, \mathbf{j}, \mathbf{k})$ and $(\mathbf{i}^*, \mathbf{j}^*, \mathbf{k}^*)$ with the coordinate systems $O\xi\eta\zeta$ and $P\xi_T\eta_T{}^*\zeta_T{}^*$ respectively, we write

$$\begin{pmatrix} \mathbf{i}^* \\ \mathbf{j}^* \\ \mathbf{k}^* \end{pmatrix} = \begin{pmatrix} R_{11} & R_{12} & R_{13} \\ R_{21} & R_{22} & R_{23} \\ R_{31} & R_{32} & R_{33} \end{pmatrix} \begin{pmatrix} \mathbf{i} \\ \mathbf{j} \\ \mathbf{k} \end{pmatrix}$$

We can express the angle of inclination uniquely in terms of the prescribed quantities ψ and b by noting that the vector

$$\mathbf{k}^* = R_{31}\mathbf{i} + R_{32}\mathbf{j} + R_{33}\mathbf{k}$$

is perpendicular to the orbital plane, so that

$$\mathbf{k} \cdot \mathbf{k}^* = \cos i = R_{33}\mathbf{k} \cdot \mathbf{k} = \sin\psi \cos b \tag{6.6.2}$$

The angle i is in the interval $[0, \pi/2]$ or $[\pi/2, \pi]$, depending on whether ψ is in the interval $[0, \pi]$ or $[\pi, 2\pi]$.

A unit vector $\mathbf{r}_\Omega$ directed along the line of nodes is given, in the $(\mathbf{i}, \mathbf{j}, \mathbf{k})$ system, by

$$\mathbf{r}_\Omega = \cos \Omega\mathbf{i} + \sin \Omega\mathbf{j} \tag{6.6.3}$$

If the symbols A, B, and C represent the components of $\mathbf{r}_\Omega$ when expressed in the $(\mathbf{i}^*, \mathbf{j}^*, \mathbf{k}^*)$ system, we write

$$\mathbf{r}_{\Omega} = A\mathbf{i}^* + B\mathbf{j}^* + C\mathbf{k}^* \tag{6.6.4}$$

The components of $\mathbf{r}_{\Omega}$ in the two systems are related by the matrix equation

$$\begin{pmatrix} A \\ B \\ C \end{pmatrix} = \begin{pmatrix} R_{11} & R_{12} & R_{13} \\ R_{21} & R_{22} & R_{23} \\ R_{31} & R_{32} & R_{33} \end{pmatrix} \begin{pmatrix} \cos \Omega \\ \sin \Omega \\ 0 \end{pmatrix} \tag{6.6.5}$$

where $C = 0$, since the line of nodes lies in the $\xi_T \eta_T{}^*$-plane. Thus we write

$$R_{31} \cos \Omega + R_{32} \sin \Omega = 0$$

The fact that $\mathbf{h} \cdot \mathbf{r}_{\Omega}$ equals zero yields the expression (see Section 2.5)

$$c_1 \cos \Omega + c_2 \sin \Omega = 0$$

Equating coefficients in the two preceding equations, we obtain

$$h \sin i \sin \Omega = R_{31} \tag{6.6.6a}$$

$$h \sin i \cos \Omega = -R_{32} \tag{6.6.6b}$$

Equations (6.6.6a) and (6.6.6b) serve to fix the quadrant of Ω. They can be expressed in simpler form. To do so, we first write their ratio in the form

$$\tan \Omega = \frac{\tan l - \tan \psi \sin b}{1 + \tan \psi \sin b \tan l} \tag{6.6.7}$$

and use (6.6.7) to substitute for $\tan \Omega$ in the identity

$$\tan(\Omega - l) = \frac{\tan \Omega - \tan l}{1 + \tan \Omega \tan l}$$

It follows directly that

$$\tan(\Omega - l) = -\sin b \tan \psi \tag{6.6.8}$$

Thus (6.6.2) and (6.6.8), when coupled with (3.3.1), (3.3.2), (3.3.3), (6.5.3), and (6.6.6), permit us to determine all six orbital elements, for any conic section, when **r** and **v** are prescribed.

6.7. DIFFERENTIAL CHANGES IN THE ORBITAL ELEMENTS

We are frequently interested in determining the effect on the orbital elements when small changes occur in the parameters r, v, ψ, and θ. Here we will consider only the case of elliptic orbits in detail. The scheme is readily extensible to both hyperbolic and parabolic orbits.

We will treat first a differential change in the semimajor axis. Since a depends only on the parameters r and v, as is clear from the equation

$$a = r\left(2 - \frac{v^2 r}{\mu}\right)^{-1} \qquad [3.3.2a]$$

the total differential of a takes the form

$$da = \frac{\partial a}{\partial r}\,dr + \frac{\partial a}{\partial v}\,dv \qquad (6.7.1)$$

The convenient expressions

$$\frac{\partial a}{\partial r} = \frac{2a}{r}\,\frac{1}{1 - e\cos E} \qquad (6.7.2a)$$

$$\frac{\partial a}{\partial v} = \frac{2a}{v}\,\frac{1 + e\cos E}{1 - e\cos E} \qquad (6.7.2b)$$

can readily be obtained from (3.3.2a) by partial differentiation and use of the relationship $r = a(1 - e\cos E)$. Thus (6.7.1) can be written

$$da = \frac{2a}{1 - e\cos E}\left[\frac{dr}{r} + (1 + e\cos E)\frac{dv}{v}\right] \qquad (6.7.3)$$

The eccentricity depends on the three parameters r, v, and θ through the relation

$$e^2 = 1 + \frac{v^2 r}{\mu}\left(\frac{v^2 r}{\mu} - 2\right)\sin^2\theta \qquad [3.3.2b]$$

The total differential of e is

$$de = \frac{\partial e}{\partial r}\,dr + \frac{\partial e}{\partial v}\,dv + \frac{\partial e}{\partial \theta}\,d\theta \qquad (6.7.4)$$

It is a straightforward matter to show that

$$\frac{\partial e}{\partial r} = \frac{(1 - e^2)\cos E}{1 - e\cos E}\,\frac{1}{r} \qquad (6.7.5a)$$

$$\frac{\partial e}{\partial v} = \frac{2(1 - e^2)\cos E}{1 - e\cos E}\,\frac{1}{v} \qquad (6.7.5b)$$

$$\frac{\partial e}{\partial \theta} = -\frac{1 - e^2}{e}\cot\theta \qquad (6.7.5c)$$

Equation (6.7.4) can be written as

$$de = \frac{(1 - e^2)\cos E}{1 - e\cos E}\left(\frac{dr}{r} + \frac{2\,dv}{v}\right) - \frac{1 - e^2}{e}\cot\theta\,d\theta \qquad (6.7.6)$$

The time of pericenter passage is given by the equation

$$T = t - \frac{1}{n} M \qquad \left[n = \left(\frac{a^3}{\mu} \right)^{1/2} \right] \tag{3.3.2e}$$

so that T depends on r, v, and θ. Its total differential is given by

$$dT = \frac{\partial T}{\partial r} dr + \frac{\partial T}{\partial v} dv + \frac{\partial T}{\partial \theta} d\theta \tag{6.7.7}$$

It is a straightforward but tedious matter to show that

$$\frac{\partial T}{\partial r} = \frac{1}{nr} \frac{(1 + 3e^2) \sin E - e^3 \sin E \cos E - 3eE}{e(1 - e \cos E)} \tag{6.7.8a}$$

$$\frac{\partial T}{\partial v} = \frac{1}{nv} \frac{(2 + 3e^2) \sin E + e^3 \sin E \cos E - 3e^2 E \cos E - 3eE}{e(1 - e \cos E)} \tag{6.7.8b}$$

$$\frac{\partial T}{\partial \theta} = \frac{1}{n} \frac{1 - e^2}{e^2} \frac{\cos E - e}{\sin E} \cot \theta \tag{6.7.8c}$$

In deriving Equations (6.7.8), certain intermediate but very useful expressions are obtained. The most important of these are

$$\frac{\partial E}{\partial r} = \frac{-\sin E}{re(1 - e \cos E)} \tag{6.7.9a}$$

$$\frac{\partial M}{\partial r} = \frac{-\sin E}{r} \left[\frac{1}{e} + \frac{(1 - e^2) \cos E}{1 - e \cos E} \right] \tag{6.7.9b}$$

$$\frac{\partial E}{\partial v} = \frac{-2 \sin E}{ve(1 - e \cos E)} \tag{6.7.9c}$$

$$\frac{\partial M}{\partial v} = \frac{-2 \sin E}{v} \left[\frac{1}{e} + \frac{(1 - e^2) \cos E}{1 - e \cos E} \right] \tag{6.7.9d}$$

$$\frac{\partial E}{\partial \theta} = \frac{-(1 - e^2)}{e^2} \cot E \cot \theta \tag{6.7.9e}$$

$$\frac{\partial M}{\partial \theta} = \frac{-(1 - e^2)}{e^2} \frac{\cos E - e}{\sin E} \cot \theta \tag{6.7.9f}$$

Equation (6.7.7) can be written

$$\begin{aligned} dT = {} & \frac{1}{nr} \frac{(1 + 3e^2) \sin E - e^3 \sin E \cos E - 3eE}{e(1 - e \cos E)} dr \\ & + \frac{1}{nv} \frac{(2 + 3e^2) \sin E - 3eE + e^3 \sin E \cos E - 3e^2 E \cos E}{e(1 - e \cos E)} dv \\ & + \frac{1}{n} \frac{(1 - e^2)(\cos E - e)}{e^2 \sin E} \cot \theta \, d\theta \end{aligned} \tag{6.7.10}$$

Next consider the inclination angle i. From (6.6.2) we see that i does not depend on r, v, or θ, so that

$$di = \frac{\partial i}{\partial \psi} d\psi \tag{6.7.11}$$

where

$$\frac{\partial i}{\partial \psi} = \frac{-\cos \psi \cos b}{(\cos^2 \psi \cos^2 b + \sin^2 b)^{1/2}} \tag{6.7.12}$$

From (6.6.8), it is seen that Ω does not depend on r, v, or θ, so that

$$d\Omega = \frac{\partial \Omega}{\partial \psi} d\psi \tag{6.7.13}$$

where

$$\frac{\partial \Omega}{\partial \psi} = \frac{-\sin b \cos^2(\Omega - l)}{\cos^2 \psi} \tag{6.7.14}$$

Finally, we consider the argument of pericenter ω. From the equations

$$\omega = u - w$$

$$\cos u = \cos b \cos(\Omega - l)$$

$$\cos w = \frac{1}{e}\left[\frac{a}{r}(1 - e^2) - 1\right]$$

derived earlier, it is clear that ω depends on all four parameters r, v, θ, and ψ. The total differential of ω is

$$d\omega = \frac{\partial \omega}{\partial r} dr + \frac{\partial \omega}{\partial v} dv + \frac{\partial \omega}{\partial \theta} d\theta + \frac{\partial \omega}{\partial \psi} d\psi \tag{6.7.15}$$

where it may be shown that

$$\frac{\partial \omega}{\partial r} = -\frac{\partial w}{\partial r} = \frac{\sin w}{re} \tag{6.7.16a}$$

$$\frac{\partial \omega}{\partial v} = -\frac{\partial w}{\partial v} = \frac{2 \sin w}{ve} \tag{6.7.16b}$$

$$\frac{\partial \omega}{\partial \theta} = -\frac{\partial w}{\partial \theta} = \frac{2e + (1 + e^2) \cos w}{e^2 \sin w} \cot \theta \tag{6.7.16c}$$

$$\frac{\partial \omega}{\partial \psi} = -\frac{\sin b \cos b \sin(\Omega - l) \cos^2(\Omega - l)}{\sin u \cos^2 \psi} \tag{6.7.16d}$$

6.8. THE GEOGRAPHIC TRACE OF A SATELLITE ORBIT

The usual procedure followed at a sensor control center of a satellite surveillance network is to issue sequential data reports on each satellite to the various observing stations that participate in the network. On these reports, the predicted times of passage of a satellite through a well-defined point in the orbit are usually given for perhaps three to seven days in advance of the date of issue. Suppose that the point chosen is the ascending node and that the subsatellite position $l_{\Omega}(T_N)$ is given in west longitude at time T_N, where T_N is the time the satellite passes through the ascending node at the beginning of the Nth revolution. The reports will then contain the orbital elements at the data of issue and, for a hundred revolutions or so in advance, the predicted values N, T_N, and $l_{\Omega}(T_N)$ calculated from these orbital elements.

The predicted times and longitudes of nodal passage appearing on a report are usually supplemented by a grid (see Exercise 11 at end of this chapter) that also appears on the report to permit an observer to compute a finding ephemeris for his station, for a given revolution. Alternatively, if a sheet of transparent plastic, containing a geographic trace of the satellite orbit during revolution N, is overlaid on a world map and oriented so that two successive points of equatorial crossing on the overlay coincide with the predicted values of $l_{\Omega}(T_N)$ and $l_{\Omega}(T_{N+1})$, a finding ephemeris for a station can be obtained graphically. Since this latter procedure is often used in practice, the problem of constructing a geographic trace of a satellite orbit for the Nth revolution, assuming that T_N and $l_{\Omega}(T_N)$ are given, is treated.

The trace of a satellite orbit on any world map is readily plotted when the latitude of the satellite is given as a function of its longitude. In Figure 6.8.1, the space-fixed reference frame is an equatorial system with origin O at geocenter. The x_q-axis points toward the vernal equinox and the z_q-axis toward the north pole. The sides of the spherical triangle NSA are arcs of three great circles on a sphere centered at geocenter. The symbol S refers to the position of the satellite at an instant Δt units after the time, T_N, at which it passed through the ascending node N to start its Nth revolution. The line segment OG lies on the intersection of the Greenwich meridian with the fundamental plane when the satellite is at S. The angle δ is the geocentric latitude of the satellite at $T_N + \Delta t$. The angle l_c, measured westward in the fundamental plane from the satellite meridian to the nodal line, is the amount that the geographic longitude l_S of the satellite differs from the geographic longitude of the node l_{Ω} when the satellite is at S.

The equations

$$\sin \delta = \sin i \sin u \tag{6.8.1}$$

$$\cos \delta \cos l_c = \cos u \tag{6.8.2}$$

hold for the spherical triangle NSA. The geographic longitude l_S, of the satellite at S, is given by

$$l_S = l_{\Omega}(T_N + \Delta t) - l_c = l_{\Omega}(T_N) + \dot{\psi}\, \Delta t - l_c \tag{6.8.3}$$

where $\dot{\psi}$ is the angular speed ($\dot{\psi} = 360°.985\,647\,24/$ mean solar day) of the Earth about its axis of rotation.

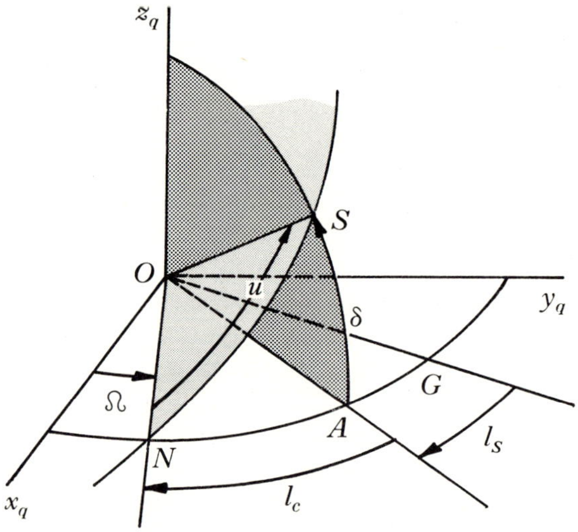

FIGURE 6.8.1

Only a rough estimate is needed in the type of calculation we are discussing. Thus, for small e, we can assume r is constant in the angular momentum condition, $\dot{w} = \dot{u} = h/r^2$, and write $\dot{u} = 2\pi/P$, where P is the anomalistic period, that is, perigee to perigee. With this assumption, u varies uniformly with time according to the equation

$$u = \frac{2\pi}{P} \Delta t \tag{6.8.4}$$

The Earth's angular speed $\dot{\psi}$ can be approximated by

$$\dot{\psi} \approx \frac{2\pi}{P_E} \tag{6.8.5}$$

where P_E is the period of the Earth's rotation ($\approx$1 mean solar day). We can substitute from (6.8.4) and (6.8.5) into (6.8.3) to obtain the equation

$$l_S = l_{\Omega}(T_N) - l_c + \frac{P}{P_E} u \tag{6.8.6}$$

Since i, P, and $l_{\Omega}(T_N)$ are prescribed, Equations (6.8.1), (6.8.2), and (6.8.6) give the required relationship between the geocentric latitude δ and the west longitude l_S of the satellite, for various choices of the parameter u.

For orbits of large eccentricity, the approximation for h/r^2 breaks down. We can then integrate the expression $\dot{u} = \dot{w} = h/r^2$ to obtain

$$I \equiv \int_{-\omega}^{u-\omega} \frac{dw}{(1 + e\cos w)^2} = \frac{h}{a^2(1 - e^2)^2} \Delta t \tag{6.8.7}$$

The integral I is elementary and it is readily found that

$$\begin{aligned} &\frac{h}{a^2(1-e^2)^{1/2}} \Delta t \\ &\quad = 2\left\{\tan^{-1}\left[\left(\frac{1-e}{1+e}\right)^{1/2} \tan\frac{u-\omega}{2}\right] + \tan^{-1}\left[\left(\frac{1-e}{1+e}\right)^{1/2} \tan\frac{\omega}{2}\right]\right\} \\ &\qquad - e(1-e^2)^{1/2}\left[\frac{\sin(u-\omega)}{1+e\cos(u-\omega)} + \frac{\sin\omega}{1+e\cos\omega}\right] = \frac{2\pi}{P}\Delta t \end{aligned} \tag{6.8.8}$$

where, in (6.8.8), we have used the fact that

$$\frac{h}{a^2(1-e^2)^{1/2}} = n = \frac{2\pi}{P}$$

Thus, substituting from (6.8.5) and (6.8.8) into (6.8.3), we find that

$$\begin{aligned} l_S &= l_{\Omega} - l_c + \frac{P}{P_E} \\ &\quad \times \left(2\left\{\tan^{-1}\left[\left(\frac{1-e}{1+e}\right)^{1/2} \tan\frac{u-\omega}{2}\right] + \tan^{-1}\left[\left(\frac{1-e}{1+e}\right)^{1/2} \tan\frac{\omega}{2}\right]\right\}\right. \\ &\qquad \left. - e(1-e^2)^{1/2}\left[\frac{\sin(u-\omega)}{1+e\cos(u-\omega)} + \frac{\sin\omega}{1+e\cos\omega}\right]\right) \end{aligned} \tag{6.8.9}$$

For $e = 0$, Equation (6.8.9) reduces to (6.8.6).

6.9. AN EXAMPLE OF HOW INITIAL INFORMATION ABOUT THE ORBITAL ELEMENTS MAY BE OBTAINED

In the discussions so far it has been assumed that a set of values of the orbital elements is available. We now inquire briefly into the means whereby such a set of orbital elements is obtained for an artificial Earth satellite.

Consider, for example, the planned launching of a satellite. In certain instances, orbital elements may be obtained from the prelaunch data supplied to the control center by the launching station. Orbital elements obtained in this way are called nominal elements. Another source of information is the launching agency, which may supply certain injection data. These data may appear in a variety of forms. Sometimes they will be given as orbital elements. Alternatively, they may be received in terms of the rectangular coordinates (ξ, η, ζ) and the velocities $(\dot{\xi}, \dot{\eta}, \dot{\zeta})$, which can be used to compute the orbital elements from the equations of transformation introduced in the preceding chapters. Still other possible forms exist. To illustrate the sort of thing that is involved, an example will be given of the form in which prelaunch injection data may be received at a control center.

Example 6.9.1. The following prelaunch information was received at a control center on July 7, 1960, for a satellite scheduled to be launched at 1960 July $19^{d}.0$ U.T. (The injection into orbit was to be parallel to the horizon plane at the subsatellite injection point.)

q = perigee height = 650 nautical miles
e = eccentricity = 0.0005
i = inclination = $28^{\circ}.318$
ϕ' = geocentric latitude of subsatellite injection point = $-17^{\circ}.015S$
l_S = longitude of subsatellite injection point = $33^{\circ}.905W$
T_i = time of injection into orbit = 241.6 sec after lift-off.

It would be quite instructive if the reader would take the trouble to calculate the four orbital elements a, ω, Ω, and T_i, not explicitly given in the prelaunch data. For this purpose, he will need the additional information that the sidereal time at Greenwich, $\alpha_G(\Upsilon_0)$, at 1960 January $0^{d}.0$, was $98^{\circ}.674\,008$.† It will be found that $a = 1.189\,63$ gr, $\omega = 218^{\circ}.088$, $\Omega = 49^{\circ}.291\,7$, and $T_i = 201.002\,796\,53$ msd from 1960 January $0^{d}.0$.

† The geocentric right ascension of a place on the Earth's surface is called the sidereal time at that place.

In carrying out the calculations suggested above, Figure 6.9.1 should prove helpful. In connection with this figure, the following notation is introduced:

O = geocenter

N, N' = ascending and descending nodes respectively

S = satellite injection point

z = angle at O between ON' and OS

Δl = angle at O between ON' and projection of OS into equatorial plane

NQS = arc on orbit between ascending node, just prior to injection, and injection point S (not traversed by satellite)

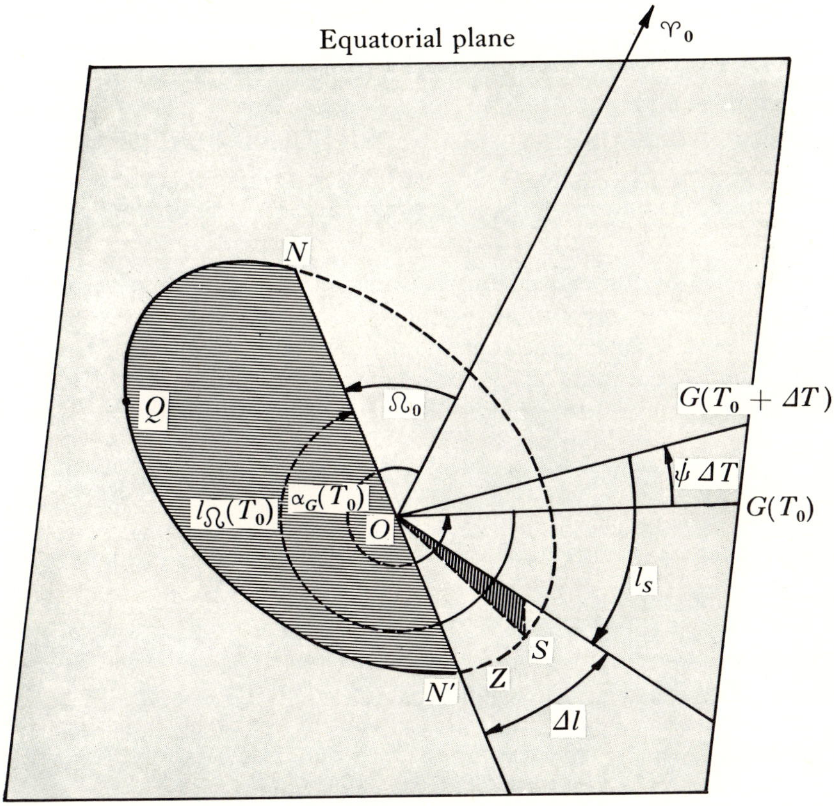

FIGURE 6.9.1

$\Delta T = T_i - T_0$, where T_i is the injection time (also time of first perigee passage) and T_0 is the time the satellite is imagined to have passed through the ascending node just prior to injection

$\alpha_G(T_0) =$ sidereal time at Greenwich at T_0 (measured in equatorial plane)

$l_{\Omega}(T_0) =$ longitude W of ascending node at T_0

$\dot{\psi}\Delta T =$ angle through which Earth rotates in time ΔT,

$OG(T_0) =$ intersection of Greenwich meridian with equator at T_0

$OG(T_0 + \Delta T) =$ intersection of Greenwich meridian with equator at $T_0 + \Delta T$

$\Upsilon_0 =$ vernal equinox 1960 January $0^d.0$. Note that 1960 January $0^d.0$ is the midnight that ushers in the day December 31, 1959.

6.10. METHODS USED TO MAINTAIN SURVEILLANCE OF ARTIFICIAL EARTH SATELLITES. PROCEDURES USED TO CALCULATE T_N AND $l_{\Omega}(T_N)$†

To introduce the central ideas, consider that at a time T^* of nodal passage, after N^* revolutions of the satellite, we enter the picture at a control center. We assume that the orbital elements a, e, i, ω, Ω at T^* are known. A computer program can then make use of the following empirical equations to predict values of T_N and node Ω_N, at the beginning of revolution number $N > N^*$:

$$T_N = T^* + P^*(N - N^*) + c(N - N^*)^2 + d(N - N^*)^3 \qquad (6.10.1)$$

$$\Omega_N = \Omega^* + \dot{\Omega}^*(T_N - T^*) + \frac{\ddot{\Omega}^*}{2}(T_N - T^*)^2 \qquad (6.10.2)$$

In (6.10.1), the nodal period P^* is the time between successive passages through the node [see Equation (6.10.6)]; c and d are coefficients, which are to be obtained from the reduced observations in a manner to be described. Formally, if we think of T_N as an analytic function of a real

† Only the underlying structure of the method is discussed. Refinements to the basic procedures described here are not treated.

variable N, then (6.10.1) is essentially the first four terms of the Taylor series expansion of T_N about the value N^*, which is given by

$$T_N = T^* + T_N'(T^*)(N - N^*) + T_N''(T^*)\frac{(N - N^*)^2}{2!} + T_N'''(T^*)\frac{(N - N^*)^3}{3!} + \cdots$$

where it is plausible to assume that $P^* = T_N'(T^*)$. Similarly, if we treat Ω_N as a function of a real variable T_N, then (6.10.2) represents the first three terms of a Taylor series expansion of Ω_N about the value $\Omega^* = \Omega_N(T^*)$.

The quantity $\dot{\Omega}^* = \dot{\Omega}(T^*)$ represents the regression of the line of nodes along the equator. It arises because the flattening of the Earth causes a departure of the satellite orbit from a true Keplerian ellipse. The most significant effect of the Earth's oblateness for our purposes here is that, through terms of first order in a small parameter corresponding to the Earth's ellipticity, it introduces linear time variations in the elements Ω and ω. These variations are given, in degrees per mean solar day, by the theoretical formulas†

$$\dot{\Omega} = \frac{-9.96^\circ \cos i}{a^{7/2}(1 - e^2)^2} \tag{6.10.3}$$

$$\dot{\omega} = \frac{-(5\cos^2 i - 1)\dot{\Omega}}{2\cos i} \tag{6.10.4}$$

when a is expressed in g-radii. The second time derivative $\ddot{\Omega}^* = \ddot{\Omega}(T^*)$, which appears in (6.10.2), is given by

$$\ddot{\Omega}^* = \frac{-\dot{a}(7 - e)\dot{\Omega}^*}{2a(1 + e)} \tag{6.10.5}$$

where, in deriving (6.10.5) from (6.10.3), it has been assumed that both $q = a(1 - e)$ and i are constants.‡

Using the value of a at T^*, we can obtain a good approximation to the anomalistic period P from (2.8.4). Then P^* is obtained from the proportion

$$P^* = \frac{360^\circ}{360^\circ + \dot{\omega}^* P} P \tag{6.10.6}$$

where $\dot{\omega}^* = \dot{\omega}(T^*)$ is calculated from (6.10.3) and (6.10.4).

† The derivation of these formulas is discussed in Chapter 13.

‡ Experience has shown that i varies but slightly over the entire lifetime of the satellite and that q is approximately constant until the satellite enters the decay phase.

Preliminary predictions of T_N and $l_{\Omega}(T_N)$ are then made by using the value of T_N, determined from the first two terms of (6.10.1), to compute Ω_N from the first two terms of (6.10.2). After Ω_N has been determined, $l_{\Omega}(T_N)$ is obtained from the formula

$$l_{\Omega}(T_N) = \alpha_G(T_N) - \Omega_N \tag{6.10.7}$$

where $\alpha_G(T_N)$ is the sidereal time at Greenwich at T_N.[†]

On the basis of these preliminary predictions, observations of the satellite can usually be obtained by the observing stations in the tracking network. These observations are then returned to the control center, where they are reduced to the node by means of special computer programs. After reduction to the node, the observations yield residuals ΔT_N and $\Delta \Omega_N$ for the various revolutions on which the observations were made, where

ΔT_N = difference between the observed and computed time of passage through the ascending node

$\Delta \Omega_N$ = difference between the observed and computed right ascension of the ascending node

Assume that a number of values of the residuals ΔT_N are available from the observations; in principle, four values will suffice. These residuals may then be used to obtain improved values of all the coefficients in (6.10.1). The differential of (6.10.1) is

$$\Delta T_N = \Delta T^* + (N - N^*)\,\Delta P^* + (N - N^*)^2\,\Delta c + (N - N^*)^3\,\Delta d \tag{6.10.8}$$

The left-hand side of (6.10.8) is known. Hence, for each ΔT_N, an equation is obtained in the unknowns ΔT^*, ΔP^*, Δc, and Δd. From four such equations we can, in principle, determine these unknowns, thus obtaining updated values of T^*, P^*, c, and d. With the new value of P^*, the semi-major axis can be updated by using the approximate formula[‡]

$$a \approx \left(\frac{P^*}{0.058\,672\,9}\right)^{2/3}$$

[†] The sidereal time at Greenwich at T_N can be calculated from the formula

$$\alpha_G(T_N) = \alpha_G(\Upsilon_0) + \dot{\psi}\,\Delta t$$

where $\alpha_G(\Upsilon_0)$, tabulated in the *American Ephemeris*, is the sidereal time at Greenwich at the beginning of the year; $\dot{\psi}$ is the Earth's angular speed; and Δt is the time elapsed since the beginning of the year.

[‡] This is formula (2.9.16), in which an accepted value of μ has been used. The semi-major axis a will be in g-radii when P^* is in mean solar mass.

If a has been computed and the perigee distance q is assumed to be constant, e can be updated since $q = a(1 - e)$. The inclination angle i is considered to remain constant.

Similarly, the node residuals may be used to update the coefficients Ω^*, $\dot{\Omega}^*$, and $\ddot{\Omega}^*/2$ appearing in (6.10.2) although, in practice, they are seldom used this way. Experience has shown that it is simpler to compute $\dot{\Omega}^*$ and $\ddot{\Omega}^*/2$ from (6.10.3) and (6.10.5), assuming a constant inclination i, and using a and e as determined above. Then a correction, deduced from an empirical analysis of the node residuals, is added algebraically to Ω^* in (6.10.2).

Exercises

1. An artificial Earth satellite is injected into an elliptic orbit at a height h above the Earth's surface and parallel to it with velocity $\mathbf{v}$. If $h = 186$ miles, $v = 26\,224$ ft/sec, decide whether the injection point is at the perigee or the apogee of the orbit.

2. Derive Equations (6.7.9) in the text and then verify Equation (6.7.10).

3. Show that

$$\frac{\partial \omega}{\partial v} = \frac{2 \sin w}{ev}$$

$$\frac{\partial \omega}{\partial \theta} = \frac{[2e + (1 + e^2) \cos w] \cot \theta}{e^2 \sin w}$$

$$\frac{\partial \omega}{\partial \psi} = \frac{-\sin b \cos b \sin(\Omega - l) \cos^2(\Omega - l)}{\sin u \cos^2 \psi}$$

for elliptic orbits.

4. If the velocity of a periodic comet is suddenly increased near its aphelion by a small amount $d\mathbf{v}$ without changing its direction, show that the changes produced in the eccentricity and semimajor axis are given by

$$de = -2 \left[\frac{a(1 - e^2)}{\mu}\right]^{1/2} dv$$

$$da = 2 \left[\frac{a^3(1 - e)}{\mu(1 + e)}\right]^{1/2} dv$$

Then show that the least distance from the heliocenter will be increased by the amount

$$\frac{4}{n}\left(\frac{1-e}{1+e}\right)^{1/2} dv$$

where n is the mean motion.†

5. A comet describes about the Sun an ellipse of eccentricity e nearly equal to unity. At a point where the radius vector makes an angle w with the major axis, the comet is instantaneously affected by a planet so that its velocity is increased in the ratio $n + 1 : n$, where n is great, without altering its direction. Show that if the new orbit is a parabola

$$e = 1 - \frac{4}{n}\cos^2\frac{w}{2}$$

6. Assume a spherical Earth of radius $R = 3963$ miles. An artificial Earth satellite is injected into an elliptic orbit at a height h above the Earth's surface and parallel to it with velocity $\mathbf{v}$. If $h = 186$ miles and $v = 26\,224$ ft/sec, determine the eccentricity of the orbit, the semimajor axis of the orbit in g-radii, the perigee and apogee distances in g-radii, and the period of revolution in minutes. If the same injection altitude is maintained, approximate the change in v required to decrease the apogee distance by 500 miles. If, at injection, h and v are as given, but the angle between the radius vector and the velocity vector is not precisely 90° but 88°.4 instead, approximate the change that this would make in the eccentricity e. How would the perigee and apogee be affected?

7. With reference to elliptic motion, show that

   (a) for a small tangential impulse in the orbital plane imparting a velocity $\delta\mathbf{v}$

$$da = \frac{2va^2\,\delta v}{\mu}, \qquad de = \frac{2(\cos w + e)\,\delta v}{v}, \qquad d\omega = \frac{2\sin w}{ev}\,\delta v$$

   (b) for a small impulse in the orbital plane normal to the orbit,

$$da = 0, \qquad de = \frac{-r\sin w}{av}\,\delta v, \qquad d\omega = \frac{(2ae + r\cos w)\,\delta v}{aev}$$

† Exercises 4, 5, 7, 9, and 10 are reprinted from A. E. H. Love, "Theoretical Mechanics," pp. 169, 161, 168, and 170. Copyright by Cambridge Univ. Press, London and New York. Used with permission of Cambridge Univ. Press.

(c) for a small radial impulse in the orbital plane,

$$da = \frac{2a^2 e \sin w}{h}\,\delta v, \qquad de = \frac{h \sin w}{\mu}\,\delta v, \qquad d\omega = \frac{-h \cos w}{e\mu}\,\delta v$$

(d) for a small transverse impulse in the orbital plane,

$$da = \frac{2a^2(1 + e \cos w)\,\delta v}{h}, \qquad de = \frac{[r(e + \cos w) + \mu^{-1} h^2 \cos w]\,\delta v}{h}$$

$$d\omega = \frac{\sin w(h^2 + r\mu)\,\delta v}{eh\mu}$$

8. (a) A particle is describing a given ellipse about a center of force in the focus. When at apogee its velocity is suddenly increased in the ratio $1 : n$. Show that, with the usual notation for elliptic motion,

$$a' = \frac{a(1 + e)}{2 - n^2(1 - e)}, \qquad e' = 1 - n^2(1 - e)$$

where the accented letters refer to the new orbit.

(b) Express in terms of equalities and inequalities the conditions on $n^2(1 - e)$ for which the new orbit is a

1. circle
2. ellipse
3. parabola
4. hyperbola

(c) Assume that $n = 1 + x$, where $x \ll 1$. Show that, neglecting terms of order x^2,

$$a' = a[1 - 2(1 - e)(1 - n)(1 + e)^{-1}], \qquad e' = e + 2(1 - e)(1 - n)$$

9. A particle describes an elliptic orbit about a focus and, when at the end of the minor axis, it receives a small impulse toward the center equal to $1/n$th of its momentum. Show that the eccentricity e is increased or diminished by $(1 - e^2)^{1/2}/n$ according to the direction of motion at the instant.

10. If a particle, describing an elliptic orbit about a focus, is at an end of the minor axis when the center of force is suddenly shifted a small distance da toward the particle, show that the eccentricity e of the orbit will be unaltered, but the major axis will be turned through an angle $a(e^{-2} - 1)^{1/2}$.

11. Make a geographic trace of the 1635th revolution of a near-Earth satellite based on the information

    1. $i = 81°.50$
    2. $P = 108.8$ min
    3. $l_{\Omega} = 4°$ at the beginning of 1635th revolution

    Compare your results with data below taken from the grid of a data report issued for this satellite.

REDUCTION TO OTHER LATITUDES FOR REVOLUTION 1635

	Lat N	(ΔT) Min plus	($-l_c$) Long corr		Lat S	(ΔT) Min plus	($-l_c$) Long corr
SN	00	.00	.00	NS	00	54.21	193.62
SN	10	3.13	359.29	NS	10	57.14	192.86
SN	20	6.27	358.48	NS	20	60.08	192.01
SN	30	9.41	357.46	NS	30	63.04	190.93
SN	40	12.58	356.01	NS	40	66.02	189.45
SN	50	15.71	353.78	NS	50	69.05	187.19
SN	60	18.90	349.88	NS	60	72.13	183.27
SN	70	22.17	341.57	NS	70	75.34	174.96
SN	80	26.07	309.51	NS	80	79.22	142.96
N	PT	27.71	277.19	S	PT	80.87	110.69
NS	80	29.34	244.73	SN	80	82.52	78.20
NS	70	33.18	212.41	SN	70	86.46	45.82
NS	60	36.36	204.03	SN	60	89.76	37.46
NS	50	39.42	200.09	SN	50	92.97	33.55
NS	40	42.43	197.81	SN	40	96.15	31.32
NS	30	45.40	196.32	SN	30	99.31	29.87
NS	20	48.34	195.24	SN	20	102.45	28.85
NS	10	51.28	194.38	SN	10	105.61	28.04
NS	00	54.21	193.62	SN	00	108.78	27.33

12. Calculate the orbital elements a, ω, Ω, and T_i referred to in connection with Example 6.9.1 in the text.

Chapter 7

The Lagrange Planetary Equations for General Perturbing Forces

7.1. INTRODUCTION TO THE THEORY OF PERTURBATIONS

Various methods are employed in celestial mechanics to approximate the true orbit of a body when, in addition to the inverse square attractive force, its motion is influenced by smaller perturbing forces that cause its trajectory to deviate from two-body motion. Following Moulton [6], we consider first a simple example from classical mechanics in which the vibrations of tuning fork are to be studied.†

We assume that the differential equation

$$\ddot{x} + k^2 x = -\mu \dot{x}^3 + \nu \cos lt \tag{7.1.1}$$

holds, where x is a linear coordinate and k^2, μ, ν, and l are positive constants not to be confused with any former use of these symbols for other quantities. The term $-\mu\dot{x}^3$ is considered to be a perturbing term that arises from the resistance of the medium; the term $\nu \cos lt$ is interpreted as a perturbation due to a driving force of angular frequency l.

† More modern and elegant introductions to the theory of perturbations are available. Moulton's example is used here because it serves to expose the underlying ideas in a simple and reasonably complete fashion without invoking undue mathematical sophistication (see p. 257).

Equation (7.1.1) is a second-order nonlinear differential equation that can be replaced by two first-order differential equations by means of the substitutions

$$x = x, \qquad \dot{x} = v \tag{7.1.2}$$

The first-order equations are

$$\dot{x} - v = 0 \tag{7.1.3a}$$

$$\dot{v} + k^2 x = -\mu v^3 + \nu \cos lt \tag{7.1.3b}$$

The method of variation of parameters will be used to obtain an approximate solution of (7.1.3). To use this method, we first solve the unperturbed system

$$\dot{x} - v = 0 \tag{7.1.4a}$$

$$\dot{v} + k^2 x = 0 \tag{7.1.4b}$$

exactly. To do this, differentiate (7.1.4a) with respect to t and substitute for $\dot{v}$ from (7.1.4b). This procedure yields the equation

$$\ddot{x} + k^2 x = 0$$

whose general solution is $x = \alpha \cos kt + \beta \sin kt$, where α and β are constants of integration. Thus the general solution of (7.1.4) is

$$x = \alpha \cos kt + \beta \sin kt \tag{7.1.5a}$$

$$\dot{x} = v = -k\alpha \sin kt + k\beta \cos kt \tag{7.1.5b}$$

The constants of integration α and β are analogous to the orbital elements that arise in the two-body problem. In a purely formal sense, we can interpret Equations (7.1.5a) and (7.1.5b) as equations of transformation that permit us to transform Equations (7.1.3) from the variables x and v to the variables α and β. With this point of view, Equation (7.1.3) can be rewritten in terms of the variables α and β. Since

$$\dot{x} = \dot{\alpha} \cos kt + \dot{\beta} \sin kt - k\alpha \sin kt + k\beta \cos kt$$

$$\dot{v} = -k\dot{\alpha} \sin kt + k\dot{\beta} \cos kt - k^2\alpha \cos kt - k^2\beta \sin kt$$

Equations (7.1.3) can be replaced by

$$(\cos kt)\, \dot{\alpha} + (\sin kt)\, \dot{\beta} = 0$$

$$-(\sin kt)\, \dot{\alpha} + (\cos kt)\, \dot{\beta} = \mu k^2 (\alpha \sin kt - \beta \cos kt)^3 + \frac{\nu}{k} \cos lt$$

After solving for $\dot{\alpha}$ and $\dot{\beta}$, we obtain

$$\dot{\alpha} = f(\alpha, \beta, t) \tag{7.1.6a}$$

$$\dot{\beta} = g(\alpha, \beta, t) \tag{7.1.6b}$$

where

$$f(\alpha, \beta, t) \equiv -\mu k^2(\alpha \sin kt - \beta \cos kt)^3 \sin kt - \frac{\nu}{k} \cos lt \sin kt$$

$$g(\alpha, \beta, t) \equiv \mu k^2(\alpha \sin kt - \beta \cos kt)^3 \cos kt + \frac{\nu}{k} \cos lt \cos kt$$

The solutions of the unperturbed problem have been used to transform Equations (7.1.3) in the variables x and v to the differential Equations (7.1.6) in the variables α and β. If the solutions of (7.1.6) for α and β have then been obtained, we can express the solution of (7.1.3) in terms of x and v through Equations (7.1.5).

At first glance, the transformed problem may seem more difficult to solve than the original problem. But if μ and ν are small parameters, the right-hand sides of (7.1.6a) and (7.1.6b) will, in general, be small. Hence α and β will vary but slowly with time, and we can approximate the solutions to (7.1.6) by means of a Taylor expansion in the sense described in what follows.

The Taylor series expansion of a function $f(x_1, x_2, \ldots, x_{2n}, t)$ of the $2n$ variables x_i, $i = 1, 2, \ldots, 2n$, and the time t, about the values $(x_1^0, x_2^0, \ldots, x_{2n}^0, t)$, can be written

$$f(x_1, x_2, \ldots, x_{2n}, t) = \sum_{j=0}^{m} \frac{1}{j!} \left[\sum_{s=1}^{2n} (x_s - x_s^0) \frac{\partial}{\partial x_s} \right]^j f^* + R_m \tag{7.1.7}$$

where

$$R_m = \frac{1}{(m+1)!} \left[\sum_{s=1}^{2n} (x_s - x_s^0) \frac{\partial}{\partial x_s} \right]^{m+1} f^{**}$$

The notation f^* indicates evaluation at $(x_1^0, x_2^0, \ldots, x_{2n}^0, t)$ after the requisite differential operations have been performed on f. The notation f^{**} has a similar interpretation except that the point of evaluation is $(x_1^0 + \theta_1 x_1, \ldots, x_{2n}^0 + \theta_{2n} x_{2n}, t)$, where $0 < \theta_i < 1$, $i = 1, 2, \ldots, 2n$.

If a subscript notation on f is used to indicate partial differentiation of f with respect to the variables used as the subscripts, the first few terms of (7.1.7) are $[f = f(x_1, x_2, \ldots, x_{2n}, t)]$:

$$\begin{aligned} f = f^* &+ f^*_{x_1}(x_1 - x_1^0) + f^*_{x_2}(x_2 - x_2^0) + \cdots + f^*_{x_{2n}}(x_{2n} - x_{2n}^0) \\ &+ \frac{1}{2!} [f^*_{x_1 x_1}(x_1 - x_1^0)^2 + f^*_{x_2 x_2}(x_2 - x_2^0)^2 + \cdots + f^*_{x_{2n} x_{2n}}(x_{2n} - x_{2n}^0)^2] \\ &+ f^*_{x_2 x_1}(x_2 - x_2^0)(x_1 - x_1^0) + f^*_{x_3 x_1}(x_3 - x_3^0)(x_1 - x_1^0) + \cdots \\ &+ f^*_{x_{2n} x_1}(x_{2n} - x_{2n}^0)(x_1 - x_1^0) + \cdots \end{aligned} \tag{7.1.8}$$

Let

$$\alpha - \alpha_0 = \sum_{n=1}^{\infty} \alpha_n \tag{7.1.9a}$$

$$\beta - \beta_0 = \sum_{n=1}^{\infty} \beta_n \tag{7.1.9b}$$

where α_0 and β_0 are fixed values of α and β, respectively, about which the expansion will be taken. Then

$$\dot{\alpha} = \sum_{n=1}^{\infty} \dot{\alpha}_n = \dot{\alpha}_1 + \dot{\alpha}_2 + \dot{\alpha}_3 + \cdots \tag{7.1.10a}$$

$$\dot{\beta} = \sum_{n=1}^{\infty} \dot{\beta}_n = \dot{\beta}_1 + \dot{\beta}_2 + \dot{\beta}_3 + \cdots \tag{7.1.10b}$$

If f and g in (7.1.6) are expanded in Taylor series about the values (α_0, β_0, t), we have, through the first three orders,

$$f(\alpha, \beta, t) = f^* + f_\alpha{}^*(\alpha - \alpha_0) + f_\beta{}^*(\beta - \beta_0) + \frac{1}{2!}[f_{\alpha\alpha}^*(\alpha - \alpha_0)^2 + 2f_{\beta\alpha}^*(\beta - \beta_0)(\alpha - \alpha_0) + f_{\beta\beta}^*(\beta - \beta_0)^2] \tag{7.1.11a}$$

$$g(\alpha, \beta, t) = g^* + g_\alpha{}^*(\alpha - \alpha_0) + g_\beta{}^*(\beta - \beta_0) + \frac{1}{2!}[g_{\alpha\alpha}^*(\alpha - \alpha_0)^2 + 2g_{\beta\alpha}^*(\beta - \beta_0)(\alpha - \alpha_0) + g_{\beta\beta}^*(\beta - \beta_0)^2] \tag{7.1.11b}$$

If, in (7.1.11), α, β, $\dot{\alpha}$, and $\dot{\beta}$ are replaced by their series representations, we can write

$$\sum_{n=1}^{\infty} \dot{\alpha}_n = f(\alpha, \beta, t) = f^* + f_\alpha{}^*\left(\sum_{n=1}^{\infty} \alpha_n\right) + f_\beta{}^*\left(\sum_{n=1}^{\infty} \beta_n\right) + \frac{1}{2!}\left[f_{\alpha\alpha}^*\left(\sum_{n=1}^{\infty} \alpha_n\right)^2 + 2f_{\beta\alpha}^*\left(\sum_{n=1}^{\infty} \alpha_n\right)\left(\sum_{n=1}^{\infty} \beta_n\right) + f_{\beta\beta}^*\left(\sum_{n=1}^{\infty} \beta_n\right)^2\right]$$

$$\sum_{n=1}^{\infty} \dot{\beta}_n = g(\alpha, \beta, t) = g^* + g_\alpha{}^*\left(\sum_{n=1}^{\infty} \alpha_n\right) + g_\beta{}^*\left(\sum_{n=1}^{\infty} \beta_n\right) + \frac{1}{2!}\left[g_{\alpha\alpha}^*\left(\sum_{n=1}^{\infty} \alpha_n\right)^2 + 2g_{\beta\alpha}^*\left(\sum_{n=1}^{\infty} \alpha_n\right)\left(\sum_{n=1}^{\infty} \beta_n\right) + g_{\beta\beta}^*\left(\sum_{n=1}^{\infty} \beta_n\right)^2\right]$$

The equation for $\dot\alpha$ can be written

$$\begin{aligned}\dot\alpha_1 + \dot\alpha_2 + \dot\alpha_3 + \cdots +\\ &= f^* + f_\alpha{}^*(\alpha_1 + \alpha_2 + \alpha_3 + \cdots)\\ &\quad + f_\beta{}^*(\beta_1 + \beta_2 + \beta_3 + \cdots) + \frac{1}{2!}[f_{\alpha\alpha}^*(\alpha_1 + \alpha_2 + \alpha_3 + \cdots)^2\\ &\quad + 2f_{\beta\alpha}^*(\alpha_1 + \alpha_2 + \alpha_3 + \cdots)(\beta_1 + \beta_2 + \beta_3 + \cdots)\\ &\quad + f_{\beta\beta}^*(\beta_1 + \beta_2 + \beta_3 + \cdots)^2] + \cdots\end{aligned}$$

The corresponding equation for $\dot\beta$ has a similar form.

We now set

$$\dot\alpha_1 = f^*, \qquad \dot\beta_1 = g^*$$

so that

$$\alpha_1 = \int f^*\,dt, \qquad \beta_1 = \int g^*\,dt$$

Both α_1 and β_1 are determined, for f^* and g^* are functions of t alone. With α_1 and β_1 so obtained, the next higher-order terms α_2 and β_2 can be obtained in similar fashion. We set

$$\begin{aligned}\dot\alpha_2 &= f_\alpha{}^*\alpha_1 + f_\beta{}^*\beta_1\\ \dot\beta_2 &= g_\alpha{}^*\alpha_1 + g_\beta{}^*\beta_1\end{aligned}$$

whence

$$\alpha_2 = \int (f_\alpha{}^*\alpha_1 + f_\beta{}^*\beta_1)\,dt$$

$$\beta_2 = \int (g_\alpha{}^*\alpha_1 + g_\beta{}^*\beta_1)\,dt$$

Since the integrands are known functions of t, we can obtain α_2 and β_2 from the two integrals just preceding. Assuming that α_2 and β_2 are determined as functions of t, we proceed to the next higher-order terms, namely, α_3 and β_3. We set

$$\dot\alpha_3 = f_\alpha{}^*\alpha_2 + f_\beta{}^*\beta_2 + \frac{1}{2!}(f_{\alpha\alpha}^*\alpha_1{}^2 + 2f_{\beta\alpha}^*\alpha_1\beta_1 + f_{\beta\beta}^*\beta_1{}^2)$$

$$\dot\beta_3 = g_\alpha{}^*\alpha_2 + g_\beta{}^*\beta_2 + \frac{1}{2!}(g_{\alpha\alpha}^*\alpha_1{}^2 + 2g_{\beta\alpha}^*\alpha_1\beta_1 + g_{\beta\beta}^*\beta_1{}^2)$$

whence

$$\alpha_3 = \int \left(f_\alpha{}^*\alpha_2 + f_\beta{}^*\beta_2 + \frac{1}{2!}f_{\alpha\alpha}^*\alpha_1{}^2 + f_{\beta\alpha}^*\alpha_1\beta_1 + \frac{1}{2!}f_{\beta\beta}^*\beta_1{}^2\right)dt$$

$$\beta_3 = \int \left(g_\alpha{}^*\alpha_2 + g_\beta{}^*\beta_2 + \frac{1}{2!}g_{\alpha\alpha}^*\alpha_1{}^2 + g_{\beta\alpha}^*\alpha_1\beta_1 + \frac{1}{2!}g_{\beta\beta}^*\beta_1{}^2\right)dt$$

Again α_3 and β_3 are determined because the integrands are functions of t alone. Proceeding in this way, we can develop the expression for α to any order desired. In practical problems arising in celestial mechanics, the mathematical complexities are so formidable that developments beyond the second order are rare and developments beyond the third order almost nonexistent.

An explicit development for α through terms of first order [see (7.1.6a)] will now be obtained; that is, we find α_1. It will be left as an exercise to develop a similar expression for β_1. We can write

$$\begin{aligned}\alpha_1 &= \int f(\alpha_0, \beta_0, t)\, dt \\ &= -\mu k^2 \int (\alpha_0 \sin kt - \beta_0 \cos kt)^3 \sin kt\, dt - \frac{\nu}{k} \int \cos lt \sin kt\, dt\end{aligned}$$

The integrals

$$I_1 = \int \cos lt \sin kt\, dt = -\frac{\cos(l+k)t}{2(l+k)} - \frac{\cos(k-l)t}{2(k-l)}$$

$$\begin{aligned}I_2 &= \int (\alpha_0 \sin kt - \beta_0 \cos kt)^3 \sin kt\, dt \\ &= \frac{\alpha_0^3}{k}\left(\frac{1}{32}\sin 4kt - \frac{1}{4}\sin 2kt + \frac{3}{8}kt\right) - \frac{3\alpha_0^2\beta_0}{k}\frac{\sin^4 kt}{4} \\ &\quad + \frac{3\alpha_0\beta_0^2}{k}\frac{1}{8}\left(kt - \frac{\sin 4kt}{4}\right) - \frac{\beta_0^3}{k}\left(-\frac{\cos^4 kt}{4}\right)\end{aligned}$$

may be obtained from any standard table of integrals. If the identities

$$\sin^4 kt = \tfrac{1}{8}(\cos 4kt - 4\cos 2kt + 3)$$

$$\cos^4 kt = \tfrac{1}{8}(\cos 4kt + 4\cos 2kt + 3)$$

are used, we can, after some tedious manipulation, write

$$\begin{aligned}I_2 = \frac{3}{8}\alpha_0(\alpha_0^2 + \beta_0^2)\, t - \frac{\alpha_0^3}{4k}\sin 2kt + \frac{\alpha_0}{32k}(\alpha_0^2 - 3\beta_0^2)\sin 4kt \\ - \frac{\beta_0}{32k}(3\alpha_0^2 - \beta_0^2)(\cos 4kt - 1) \\ + \frac{\beta_0}{8k}(3\alpha_0^2 + \beta_0^2)(\cos 2kt - 1) + \frac{\beta_0^3}{4k}\end{aligned}$$

A constant term C, which can be absorbed into the constant of integration, is to be added to I_1. We can write first

$$I_1 = -\frac{\cos(l+k)t}{2(l+k)} + \frac{\cos(l-k)t}{2(l-k)} + \frac{1}{2(l+k)} - \frac{1}{2(l-k)} - C$$

$$= -\frac{1}{2(l+k)}[\cos(l+k)\,t - 1] + \frac{1}{2(l-k)}[\cos(l-k)\,t - 1] - C$$

where

$$C = \frac{1}{2(l-k)} - \frac{1}{2(l+k)} = \frac{k}{l^2 - k^2}$$

Combining the preceding results, we can write the solution for α, through terms of first order, as

$$\alpha = -\mu k^2\left(I_2 - \frac{\beta_0^{\ 3}}{4k}\right) - \frac{\nu}{k}(I_1 + C) + \text{constant}$$

$$= K_1 - \mu k^2\left[\frac{3}{8}\alpha_0(\alpha_0^{\ 2} + \beta_0^{\ 2})\,t + \frac{\beta_0}{8k}(3\alpha_0^{\ 2} + \beta_0^{\ 2})(\cos 2kt - 1)\right.$$

$$- \frac{\beta_0}{32k}(3\alpha_0^{\ 2} - \beta_0^{\ 2})(\cos 4kt - 1) - \frac{\alpha_0^{\ 3}}{4k}\sin 2kt$$

$$\left. + \frac{\alpha_0}{32k}(\alpha_0^{\ 2} - 3\beta_0^{\ 2})\sin 4kt\right] + \frac{\nu}{2k(l+k)}[\cos(l+k)\,t - 1]$$

$$- \frac{\nu}{2k(l-k)}[\cos(l-k)\,t - 1] \qquad (7.1.12a)$$

where K_1 is an arbitrary constant of integration. In a similar fashion, the solution for β can be obtained through terms of first order. Where K_2 is an arbitrary constant of integration, the result is

$$\beta = K_2 + \mu k^2\left[-\frac{3\beta_0}{8}(\alpha_0^{\ 2} + \beta_0^{\ 2})\,t - \frac{\alpha_0}{8k}(\alpha_0^{\ 2} + 3\beta_0^{\ 2})(\cos 2kt - 1)\right.$$

$$+ \frac{\alpha_0}{32k}(\alpha_0^{\ 2} - 3\beta_0^{\ 2})(\cos 4kt - 1) - \frac{\beta_0^{\ 3}}{4k}\sin 2kt$$

$$\left. + \frac{\beta_0}{32k}(3\alpha_0^{\ 2} - \beta_0^{\ 2})\sin 4kt\right] + \frac{\nu}{2k(l+k)}\sin(l+k)t$$

$$+ \frac{\nu}{2k(l-k)}\sin(l-k)t \qquad (7.1.12b)$$

The expressions for α and β given in (7.1.12a) and (7.1.12b) contain terms in which t appears as a simple factor as well as terms in which t appears in the sine and cosine terms. The terms that change proportionally to t are called *secular terms*. In assessing the effect of secular terms, it is not correct to assume that α or β increase indefinitely with time. It is possible that if we had obtained rigorous expressions, higher powers of t would have appeared in such a way that the sums are bounded, just as

$$\sin t = t - \frac{t^3}{3!} + \frac{t^5}{3!} - \cdots$$

is a bounded function ($|\sin t| \leqslant 1$) for all values of t. The expressions for α and β also contain *periodic terms* of periods

$$\frac{\pi}{k}, \quad \frac{\pi}{2k}, \quad \frac{2\pi}{l+k}, \quad \frac{2\pi}{l-k}$$

If l and k are nearly equal, we speak of terms such as $\sin(l - k)t$ as *long-period terms*. If k is large and $l - k$ is small, we speak of terms like $\sin 2kt$ as *short-period terms*. Actually, the designation as long and short period is somewhat arbitrary and a relative matter. Terms involving the products of t and periodic terms are called *Poisson terms*.

7.2. CHANGE OF VARIABLES

The equations of motion of a particle of mass m_i, under the dominant attraction of a mass m_0, but acted upon by other gravitating masses m_j; $j = 1, 2, \ldots, n - 1$, $j \neq i$; and an external field intensity $\mathbf{E}$ appeared in Chapter 2. Specifically, it was found that for the rectangular reference frame $O\xi\eta\zeta$ with origin at m_0,

$$\ddot{\mathbf{r}}_i + k^2(m_0 + m_i) F(r_i)\, \mathbf{u}_{0i} = \nabla_i R_i + \mathbf{E}_i - \mathbf{E}_0 \qquad [2.3.4]$$

where

$$\nabla_i = \frac{\partial}{\partial \xi_i}\mathbf{i} + \frac{\partial}{\partial \eta_i}\mathbf{j} + \frac{\partial}{\partial \zeta_i}\mathbf{k}$$

$$R_i = -k^2 \sum_{\substack{j=1 \\ j\neq i}}^{n-1} m_j \left[P_{ij} + \frac{F(r_j)}{r_j}(\xi_i\xi_j + \eta_i\eta_j + \zeta_i\zeta_j)\right]$$

and P_{ij} satisfies the equation

$$\frac{dP_{ij}}{d\rho_{ij}} = F(\rho_{ij}) \qquad \text{for each } j$$

If the consideration is restricted to the case where, for each j, $F(\rho_{ij}) = \rho_{ij}^{-2}$, it follows that $F(r_j) = r_j^{-2}$ and we can take

$$P_{ij} = -\frac{1}{\rho_{ij}} \tag{7.2.1}$$

If we drop the subscript i, we can write

$$R = k^2 \sum_{j=1}^{n-2} m_j \left(\frac{1}{\rho_j} - \frac{\xi\xi_j + \eta\eta_j + \zeta\zeta_j}{r_j^3}\right) \tag{7.2.2}$$

where, in (7.2.2), it is to be understood that the summation extends over all particles other than m and m_0. The notation ρ_j is to be read "the distance from m to the jth particle," r_j is the distance from m_0 to the jth particle, and (ξ, η, ζ) are the coordinates of the mass m. In this notation, (2.3.4) becomes

$$\ddot{\mathbf{r}} + \frac{\mu\mathbf{r}}{r^3} = \nabla R + \mathbf{E} - \mathbf{E}_0 \tag{7.2.3}$$

where $\mu = k^2(m_0 + m)$. If the external field intensity is conservative, it can be represented as the gradient of a scalar function and combined with the perturbing function arising from the gravitating masses in the system of particles to form a total perturbing function, which can again be represented by the symbol R.

In some cases, it will be convenient to work with the total perturbing force intensity, $\mathbf{F}$, represented by the right-hand member of (7.2.3). We write

$$\mathbf{F} = \nabla R + \mathbf{E} - \mathbf{E}_0 = F_\xi\mathbf{i} + F_\eta\mathbf{j} + F_\zeta\mathbf{k} \tag{7.2.4}$$

where F_ξ, F_η, F_ζ denote the components of $\mathbf{F}$ along the axes indicated by the subscripts. Equation (7.2.3) can then by represented by the three scalar equations

$$\ddot{\xi} + \frac{\mu\xi}{r^3} = F_\xi \tag{7.2.5a}$$

$$\ddot{\eta} + \frac{\mu\eta}{r^3} = F_\eta \tag{7.2.5b}$$

$$\ddot{\zeta} + \frac{\mu\zeta}{r^3} = F_\zeta \tag{7.2.5c}$$

The substitutions $\dot{\xi} = v_\xi$, $\dot{\eta} = v_\eta$, and $\dot{\zeta} = v_\zeta$ will transform Equations (7.2.5) into the six first-order differential equations

$$\begin{aligned} \dot{\xi} &= v_\xi, & \dot{v}_\xi + \frac{\mu\xi}{r^3} &= F_\xi \\ \dot{\eta} &= v_\eta, & \dot{v}_\eta + \frac{\mu\eta}{r^3} &= F_\eta \\ \dot{\zeta} &= v_\zeta, & \dot{v}_\zeta + \frac{\mu\zeta}{r^3} &= F_\zeta \end{aligned} \tag{7.2.6}$$

In an earlier chapter it was seen that the coordinates (ξ, η, ζ) and the velocity $(\dot{\xi}, \dot{\eta}, \dot{\zeta})$ can, for unperturbed two-body motion, be expressed as

$$\xi = \xi(e_1, e_2, ..., e_6, t) \tag{7.2.7a}$$

$$\eta = \eta(e_1, e_2, ..., e_6, t) \tag{7.2.7b}$$

$$\zeta = \zeta(e_1, e_2, ..., e_6, t) \tag{7.2.7c}$$

$$\dot{\xi} = \dot{\xi}\ e_1, e_2, ..., e_6, t) \tag{7.2.7d}$$

$$\dot{\eta} = \dot{\eta}(e_1, e_2, ..., e_6, t) \tag{7.2.7e}$$

$$\dot{\zeta} = \dot{\zeta}(e_1, e_2, ..., e_6, t) \tag{7.2.7f}$$

where the e_j, $j = 1, 2, ..., 6$, represent the members of the set of six constant orbital elements that is being used in the discussion. Explicitly, Equations (7.2.7) are given by (6.4.5)–(6.4.7) and (6.4.14)–(6.4.16) after the anomalies have been replaced by functions of the constant orbital elements and the time.

In the spirit described in treating the vibrations of the tuning fork, Equations (7.2.7) will be regarded as equations of transformation that permit us to transform from the system of differential equations (7.2.6), which is to be solved for the variables $\xi, \eta, \zeta, \dot{\xi}, \dot{\eta}, \dot{\zeta}$, to an alternative system of six first-order differential equations which is to be solved for the variables e_j, $j = 1, 2, ..., 6$. It is this second system of six first-order differential equations that is to be determined. If the transformed set of differential equations in the $\dot{e}_j$ can be obtained and integrated so as to express the e_j as functions of the time, the results may be substituted back into (7.2.7) and thus the coordinates and velocity in the perturbed orbit will be determined as functions of time.

Assume that analytical expressions for the e_j as functions of time have been obtained for the true—that is, the perturbed—orbit, and suppose that these expressions are evaluated at some instant t. A set of values of the six orbital elements at t is thus obtained, and hence the coordinates and velocity of that instant t are determined for the true orbit. There exists an unperturbed two-body orbit corresponding to the same set of values of the six orbital elements, and Equations (7.2.7) also give the coordinates and velocity in it as functions of t. When t for this unperturbed orbit is chosen to be the same as the instant t at which we evaluated the orbital elements for the true orbit, then the unperturbed two-body orbit is called the *osculating orbit* at the point in question. The body moving in the perturbed orbit has, at this instant, the same coordinates and the same velocity as it would have if it were moving in the unperturbed (osculating) orbit at the same instant.

Clearly, corresponding to any point in the true orbit, there exists an osculating orbit at that point, with the property that the coordinates and velocity in the true orbit and the osculating orbit at the point are the same. Thus thinking of Equations (7.2.7) as applying to the true orbit, we write

$$\mathbf{r} = \xi\mathbf{i} + \eta\mathbf{j} + \zeta\mathbf{k} = \mathbf{r}(e_1, e_2, ..., e_6, t) \tag{7.2.8}$$

$$\dot{\mathbf{r}} = \frac{\partial \mathbf{r}}{\partial t} + \sum_{j=1}^{6} \frac{\partial \mathbf{r}}{\partial e_j} \dot{e}_j \qquad \text{(true orbit)} \tag{7.2.9}$$

where (7.2.9) is to be interpreted as applying to the true orbit, at any instant, say t_0. At the same instant, t_0, there is a particular osculating orbit and, thinking of (7.2.7) as applying to the osculating orbit, we write

$$\dot{\mathbf{r}}_0 = \frac{\partial \mathbf{r}}{\partial t} \qquad \text{(osculating orbit)} \tag{7.2.10}$$

The velocity $\dot{\mathbf{r}}_0$ in the osculating orbit at the point corresponding to $t = t_0$ is the same as the velocity in the true orbit at the same point. For each point on the true orbit, there is an osculating orbit at the point for which (7.2.10) must hold. Thus, since (7.2.9) and (7.2.10) are to hold simultaneously for each instant t, we must have for all time

$$\sum_{j=1}^{6} \frac{\partial \mathbf{r}}{\partial e_j} \dot{e}_j = 0 \tag{7.2.11}$$

Equation (7.2.11) represents three first-order differential equations for the time variations of the orbital elements. Three more equations are required.

The reasoning employed in the preceding two paragraphs may be applied to the velocity

$$\dot{\mathbf{r}} = \dot{\mathbf{r}}(e_1, e_2, ..., e_6, t) \tag{7.2.12}$$

Thinking of (7.2.12) as it applies to the true orbit and taking its time derivative, we have

$$\ddot{\mathbf{r}} = \frac{\partial \dot{\mathbf{r}}}{\partial t} + \sum_{j=1}^{6} \frac{\partial \dot{\mathbf{r}}}{\partial e_j} \dot{e}_j \qquad \text{(true orbit)}$$

or

$$\ddot{\mathbf{r}} = \frac{\partial^2 \mathbf{r}}{\partial t^2} + \sum_{j=1}^{6} \frac{\partial \dot{\mathbf{r}}}{\partial e_j} \dot{e}_j \qquad \text{(true orbit)} \tag{7.2.13}$$

since $\dot{\mathbf{r}} = \partial \mathbf{r}/\partial t$. Thinking of (7.2.12) as it applies to the osculating orbit and taking its time derivative, we have

$$\ddot{\mathbf{r}}_0 = \frac{\partial^2 \mathbf{r}}{\partial t^2} \qquad \text{(osculating orbit)} \tag{7.2.14}$$

where it is to be noted that the acceleration $\ddot{\mathbf{r}}_0$ in the osculating orbit is not the same as $\ddot{\mathbf{r}}$. If (7.2.14) is substituted into (7.2.13), it follows that

$$\ddot{\mathbf{r}} - \ddot{\mathbf{r}}_0 = \sum_{j=1}^{6} \frac{\partial \dot{\mathbf{r}}}{\partial e_j} \dot{e}_j \tag{7.2.15}$$

From the equations of motion

$$\ddot{\mathbf{r}} - \ddot{\mathbf{r}}_0 = -\frac{\mu \mathbf{r}}{r^3} + \mathbf{F} + \frac{\mu \mathbf{r}}{r^3} = \mathbf{F}$$

hence we have

$$\sum_{j=1}^{6} \frac{\partial \dot{\mathbf{r}}}{\partial e_j} \dot{e}_j = \mathbf{F} \tag{7.2.16}$$

Equations (7.2.16), together with (7.2.11), represent the six first-order differential equations in the $\dot{e}_j$ that we have been seeking. If these equations are to be useful for our purposes, however, they must be solved explicitly for the $\dot{e}_j$.

7.3. AN INTERMEDIATE SYSTEM OF DIFFERENTIAL EQUATIONS FOR $\dot{e}_j$, $j = 1, 2, \ldots, 6$

For the two-body problem, the coordinates (ξ, η, ζ) and velocity $(\dot{\xi}, \dot{\eta}, \dot{\zeta})$ are related to the coordinates $(\xi'', \eta'', 0)$ and velocity $(\dot{\xi}'', \dot{\eta}'', 0)$ in the orbital system by the equations

$$\begin{aligned}
\xi &= D_{11}\xi'' + D_{21}\eta'', & \dot{\xi} &= D_{11}\dot{\xi}'' + D_{21}\dot{\eta}'' \\
\eta &= D_{12}\xi'' + D_{22}\eta'', & \dot{\eta} &= D_{12}\dot{\xi}'' + D_{22}\dot{\eta}'' \\
\zeta &= D_{13}\xi'' + D_{23}\eta'', & \dot{\zeta} &= D_{13}\dot{\xi}'' + D_{23}\dot{\eta}''
\end{aligned} \tag{7.3.1}$$

which follow directly from (2.5.1) and (2.5.6), where the D_{ij}, $i = 1, 2$; $j = 1, 2, 3$, are functions only of the elements ω, Ω, and i.

Recall that, for an unperturbed conic orbit, we have

$$\xi'' = r \cos w, \qquad \eta'' = r \sin w$$

It is convenient to introduce the definition

$$p^* \equiv 2ep \tag{7.3.2}$$

and the convention of treating a as a negative number in hyperbolic orbits. Thus we write

$$p^* = 2q \qquad \text{(parabola)}$$

$$p^* = a(1 - e^2) \qquad \text{(ellipse and hyperbola)}$$

Since the true anomaly w depends only on a, e, T, and t and $r = p^*(1 + e \cos w)^{-1}$, the orbital coordinates $(\xi'', \eta'', 0)$ and velocity $(\dot{\xi}'', \dot{\eta}'', 0)$ depend only on p^*, e, T, and t. Since $r^2\dot{w} = (\mu p^*)^{1/2}$, it follows that

$$\dot{r} = \frac{p^* e \sin w}{(1 + e \cos w)^2}\, \dot{w} = \left(\frac{\mu}{p^*}\right)^{1/2} e \sin w$$

whence

$$\dot{\xi}'' = \dot{r} \cos w - r\,(\sin w)\dot{w} = -\left(\frac{\mu}{p^*}\right)^{1/2} \sin w \tag{7.3.3a}$$

$$\dot{\eta}'' = \dot{r} \sin w + r\,(\cos w)\dot{w} = \left(\frac{\mu}{p^*}\right)^{1/2} (\cos w + e) \tag{7.3.3b}$$

Although (7.3.1) and (7.3.3) have been developed for an unperturbed (two-body) orbit, they can be used to form the partial derivatives which appear in (7.2.11) and (7.2.16). The reason is that the partial derivatives of the coordinates and velocities have the same form in the perturbed orbit as they have in the osculating orbit at that instant.

Explicitly, we make the assignment

$$\begin{aligned} e_1 &\to \omega \\ e_2 &\to \Omega \\ e_3 &\to i \\ e_4 &\to p^* \quad \text{or} \quad a \\ e_5 &\to e \\ e_6 &\to T \end{aligned} \tag{7.3.4}$$

and, returning to (7.2.11), we replace the vector equation by the three scalar equations

$$\sum_{j=1}^{6} \frac{\partial A}{\partial e_j} \dot{e}_j = 0 \qquad (A = \xi, \eta, \zeta) \tag{7.3.5}$$

where the symbol A is to be interpreted as representing each of the variables ξ, η, and ζ successively.

The partial derivatives appearing in (7.3.5) are to be obtained from (7.3.1) by carrying out the indicated operations. In this connection it is convenient, here and elsewhere, to introduce the notation

$$b_1, b_2, \ldots, b_n: \quad \frac{\partial}{\partial}$$

to indicate the n expressions obtained by differentiating, successively, with respect to each of the variables b_k, $k = 1, 2, \ldots, n$. In this notation, the partial derivatives appearing in (7.3.5) can be written

$$e_1, e_2, e_3: \qquad \frac{\partial \xi}{\partial} = \xi'' \frac{\partial D_{11}}{\partial} + \eta'' \frac{\partial D_{21}}{\partial}$$

$$e_4, e_5, e_6: \qquad \frac{\partial \xi}{\partial} = D_{11} \frac{\partial \xi''}{\partial} + D_{21} \frac{\partial \eta''}{\partial}$$

with similar expressions for the coordinates η and ζ. An entirely analogous result can be derived from (7.2.16). The six scalar differential equations corresponding to (7.2.11) and (7.2.16) then become

$$\sum_{j=1}^{3} \left(\xi'' \frac{\partial D_{11}}{\partial e_j} + \eta'' \frac{\partial D_{21}}{\partial e_j}\right) \dot{e}_j + \sum_{j=4}^{6} \left(D_{11} \frac{\partial \xi''}{\partial e_j} + D_{21} \frac{\partial \eta''}{\partial e_j}\right) \dot{e}_j = 0 \tag{7.3.6a}$$

$$\sum_{j=1}^{3} \left(\xi'' \frac{\partial D_{12}}{\partial e_j} + \eta'' \frac{\partial D_{22}}{\partial e_j}\right) \dot{e}_j + \sum_{j=4}^{6} \left(D_{12} \frac{\partial \xi''}{\partial e_j} + D_{22} \frac{\partial \eta''}{\partial e_j}\right) \dot{e}_j = 0 \tag{7.3.6b}$$

$$\sum_{j=1}^{3} \left(\xi'' \frac{\partial D_{13}}{\partial e_j} + \eta'' \frac{\partial D_{23}}{\partial e_j}\right) \dot{e}_j + \sum_{j=4}^{6} \left(D_{13} \frac{\partial \xi''}{\partial e_j} + D_{23} \frac{\partial \eta''}{\partial e_j}\right) \dot{e}_j = 0 \tag{7.3.6c}$$

$$\sum_{j=1}^{3} \left(\dot{\xi}'' \frac{\partial D_{11}}{\partial e_j} + \dot{\eta}'' \frac{\partial D_{21}}{\partial e_j}\right) \dot{e}_j + \sum_{j=4}^{6} \left(D_{11} \frac{\partial \dot{\xi}''}{\partial e_j} + D_{21} \frac{\partial \dot{\eta}''}{\partial e_j}\right) \dot{e}_j = F_\xi \tag{7.3.6d}$$

$$\sum_{j=1}^{3} \left(\dot{\xi}'' \frac{\partial D_{12}}{\partial e_j} + \dot{\eta}'' \frac{\partial D_{22}}{\partial e_j}\right) \dot{e}_j + \sum_{j=4}^{6} \left(D_{12} \frac{\partial \dot{\xi}''}{\partial e_j} + D_{22} \frac{\partial \dot{\eta}''}{\partial e_j}\right) \dot{e}_j = F_\eta \tag{7.3.6e}$$

$$\sum_{j=1}^{3} \left(\dot{\xi}'' \frac{\partial D_{13}}{\partial e_j} + \dot{\eta}'' \frac{\partial D_{23}}{\partial e_j}\right) \dot{e}_j + \sum_{j=4}^{6} \left(D_{13} \frac{\partial \dot{\xi}''}{\partial e_j} + D_{23} \frac{\partial \dot{\eta}''}{\partial e_j}\right) \dot{e}_j = F_\zeta \tag{7.3.6f}$$

Certain relationships based on the fact that the D_{ij} are the elements of a matrix associated with an orthogonal transformation can be used to simplify (7.3.6). Specifically, the D_{ij} satisfy the orthogonality conditions

$$\sum_{j=1}^{3} D_{ij}D_{kj} = \delta_{ik} \qquad (i = 1, 2, 3;\ k = 1, 2, 3) \tag{7.3.7}$$

where the Kronecker delta δ_{ik} is equal to one when $i = k$ and zero otherwise. Since, from (2.5.6), we have

$$\begin{aligned} \mathbf{i}'' &= D_{11}\mathbf{i} + D_{12}\mathbf{j} + D_{13}\mathbf{k} \\ \mathbf{j}'' &= D_{21}\mathbf{i} + D_{22}\mathbf{j} + D_{23}\mathbf{k} \\ \mathbf{k}'' &= D_{31}\mathbf{i} + D_{32}\mathbf{j} + D_{33}\mathbf{k} \end{aligned}$$

Equations (7.3.7) and certain useful relations derivable from them can be written in terms of inner products of the unit vectors $\mathbf{i}'', \mathbf{j}'', \mathbf{k}''$ with themselves and with their partial derivatives with respect to the orbital elements e_1, e_2, and e_3. Specifically, we have

$$\mathbf{i}'' \cdot \mathbf{i}'' = \mathbf{j}'' \cdot \mathbf{j}'' = \mathbf{k}'' \cdot \mathbf{k}'' = 1 \tag{7.3.8a}$$

$$\mathbf{i}'' \cdot \mathbf{j}'' = \mathbf{j}'' \cdot \mathbf{k}'' = \mathbf{i}'' \cdot \mathbf{k}'' = 0 \tag{7.3.8b}$$

$$e_1, e_2, e_3: \qquad \mathbf{i}'' \cdot \frac{\partial \mathbf{i}''}{\partial} = \mathbf{j}'' \cdot \frac{\partial \mathbf{j}''}{\partial} = \mathbf{k}'' \cdot \frac{\partial \mathbf{k}''}{\partial} = 0 \tag{7.3.8c}$$

In addition, it follows directly from (7.3.8b) that

$$e_1, e_2, e_3: \qquad \mathbf{i}'' \cdot \frac{\partial \mathbf{j}''}{\partial} = -\mathbf{j}'' \cdot \frac{\partial \mathbf{i}''}{\partial} \tag{7.3.9a}$$

$$e_1, e_2, e_3: \qquad \mathbf{j}'' \cdot \frac{\partial \mathbf{k}''}{\partial} = -\mathbf{k}'' \cdot \frac{\partial \mathbf{j}''}{\partial} \tag{7.3.9b}$$

$$e_1, e_2, e_3: \qquad \mathbf{i}'' \cdot \frac{\partial \mathbf{k}''}{\partial} = -\mathbf{k}'' \cdot \frac{\partial \mathbf{i}''}{\partial} \tag{7.3.9c}$$

Equations (7.3.9) can be evaluated in a straightforward manner. To illustrate, we have

$$\begin{aligned} \mathbf{i}'' \cdot \frac{\partial \mathbf{j}''}{\partial e_1} &= \mathbf{i}'' \cdot \frac{\partial \mathbf{j}''}{\partial \omega} = D_{11}\frac{\partial D_{21}}{\partial \omega} + D_{12}\frac{\partial D_{22}}{\partial \omega} + D_{13}\frac{\partial D_{23}}{\partial \omega} \\ &= D_{11}(-\cos\omega\cos\Omega + \sin\omega\sin\Omega\cos i) \\ &\quad + D_{12}(-\cos\omega\sin\Omega - \sin\omega\cos\Omega\cos i) - D_{13}\sin\omega\sin i \\ &= -D_{11}^2 - D_{12}^2 - D_{13}^2 = -1 \end{aligned} \tag{7.3.10}$$

In a similar fashion, the remaining expressions appearing in (7.3.9) can be determined. It is found that

$$\begin{aligned}
&\mathbf{i}'' \cdot \frac{\partial \mathbf{j}''}{\partial e_1} = -1, && \mathbf{j}'' \cdot \frac{\partial \mathbf{i}''}{\partial e_1} = 1 \\
&\mathbf{i}'' \cdot \frac{\partial \mathbf{j}''}{\partial e_2} = -D_{33}, && \mathbf{j}'' \cdot \frac{\partial \mathbf{i}''}{\partial e_2} = D_{33} \\
&\mathbf{i}'' \cdot \frac{\partial \mathbf{j}''}{\partial e_3} = 0, && \mathbf{j}'' \cdot \frac{\partial \mathbf{i}''}{\partial e_3} = 0 \\
&\mathbf{j}'' \cdot \frac{\partial \mathbf{k}''}{\partial e_1} = 0, && \mathbf{k}'' \cdot \frac{\partial \mathbf{j}''}{\partial e_1} = 0 \\
&\mathbf{j}'' \cdot \frac{\partial \mathbf{k}''}{\partial e_2} = -D_{13}, && \mathbf{k}'' \cdot \frac{\partial \mathbf{j}''}{\partial e_2} = D_{13} \\
&\mathbf{j}'' \cdot \frac{\partial \mathbf{k}''}{\partial e_3} = -\cos \omega, && \mathbf{k}'' \cdot \frac{\partial \mathbf{j}''}{\partial e_3} = \cos \omega \\
&\mathbf{i}'' \cdot \frac{\partial \mathbf{k}''}{\partial e_1} = 0, && \mathbf{k}'' \cdot \frac{\partial \mathbf{i}''}{\partial e_1} = 0 \\
&\mathbf{i}'' \cdot \frac{\partial \mathbf{k}''}{\partial e_2} = D_{23}, && \mathbf{k}'' \cdot \frac{\partial \mathbf{i}''}{\partial e_2} = -D_{23} \\
&\mathbf{i}'' \cdot \frac{\partial \mathbf{k}''}{\partial e_3} = -\sin \omega, && \mathbf{k}'' \cdot \frac{\partial \mathbf{i}''}{\partial e_3} = \sin \omega
\end{aligned} \tag{7.3.11}$$

Equations (7.3.6) can be replaced by the reduced system

$$-\eta''\dot{\omega} - \eta''\dot{\Omega} + \sum_{j=4}^{6} \frac{\partial \xi''}{\partial e_j} \dot{e}_j = 0 \tag{7.3.12a}$$

$$\xi''\dot{\omega} + \xi''\dot{\Omega} + \sum_{j=4}^{6} \frac{\partial \eta''}{\partial e_j} \dot{e}_j = 0 \tag{7.3.12b}$$

$$-\dot{\eta}''\dot{\omega} - \dot{\eta}''\dot{\Omega} + \sum_{j=4}^{6} \frac{\partial \dot{\xi}''}{\partial e_j} \dot{e}_j = F''_\xi \tag{7.3.12c}$$

$$\dot{\xi}''\dot{\omega} + \dot{\xi}''\dot{\Omega} + \sum_{j=4}^{6} \frac{\partial \dot{\eta}''}{\partial e_j} \dot{e}_j = F''_\eta \tag{7.3.12d}$$

$$(-\xi''D_{23} + \eta''D_{13})\,\dot{\Omega} + (\xi'' \sin \omega + \eta'' \cos \omega)(i)^{\cdot} = 0 \tag{7.3.12e}$$

$$(-\dot{\xi}''D_{23} + \dot{\eta}''D_{13})\,\dot{\Omega} + (\dot{\xi}'' \sin \omega + \dot{\eta}'' \cos \omega)(i)^{\cdot} = F''_\zeta \tag{7.3.12f}$$

where Equations (7.3.12) are obtained by performing the following operations on (a)–(f) of (7.3.6), in conjunction with the use of (7.3.8)–(7.3.11):

$$D_{11}(a) + D_{12}(b) + D_{13}(c) = (7.3.12\text{a})$$

$$D_{21}(a) + D_{22}(b) + D_{23}(c) = (7.3.12\text{b})$$

$$D_{31}(a) + D_{32}(b) + D_{33}(c) = (7.3.12\text{e})$$

$$D_{11}(d) + D_{12}(e) + D_{13}(f) = (7.3.12\text{c})$$

$$D_{21}(d) + D_{22}(e) + D_{23}(f) = (7.3.12\text{d})$$

$$D_{31}(d) + D_{32}(e) + D_{33}(f) = (7.3.12\text{f})$$

The quantities F''_ξ, F''_η, and F''_ζ appearing in (7.3.12) are defined by the relations

$$F''_\xi \equiv \mathbf{F}\cdot\mathbf{i}'' = D_{11}F_\xi + D_{12}F_\eta + D_{13}F_\zeta \tag{7.3.13a}$$

$$F''_\eta \equiv \mathbf{F}\cdot\mathbf{j}'' = D_{21}F_\xi + D_{22}F_\eta + D_{23}F_\zeta \tag{7.3.13b}$$

$$F''_\zeta \equiv \mathbf{F}\cdot\mathbf{k}'' = D_{31}F_\xi + D_{32}F_\eta + D_{33}F_\zeta \tag{7.3.13c}$$

7.4. DERIVATION OF PERTURBATION EQUATIONS FOR $(i)^{\cdot}$ AND $\dot{\Omega}$

A rectangular coordinate system $O\xi^u\eta^u$ (in the instantaneous orbital plane) is introduced in Figure 7.4.1. The new coordinates are defined by the relations

$$\xi^u = r\cos u, \qquad \eta^u = r\sin u$$

where u is the argument of latitude. At any instant t, the transformations from the coordinates and velocity in the $O\xi''\eta''$-system to the coordinates and velocity in the $O\xi^u\eta^u$-system are given by the matrix relations

$$\begin{aligned}\begin{pmatrix}\xi^u\\ \eta^u\end{pmatrix} &= \begin{pmatrix}\cos\omega & -\sin\omega\\ \sin\omega & \cos\omega\end{pmatrix}\begin{pmatrix}\xi''\\ \eta''\end{pmatrix}\\ \begin{pmatrix}\dot{\xi}^u\\ \dot{\eta}^u\end{pmatrix} &= \begin{pmatrix}\cos\omega & -\sin\omega\\ \sin\omega & \cos\omega\end{pmatrix}\begin{pmatrix}\dot{\xi}''\\ \dot{\eta}''\end{pmatrix}\end{aligned} \tag{7.4.1}$$

where ω is to be evaluated at t.

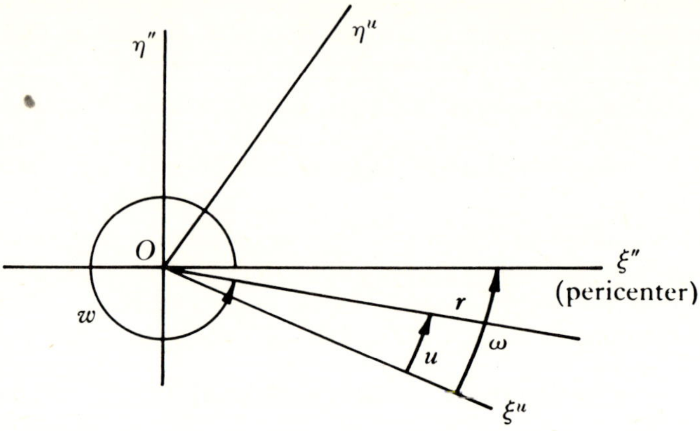

FIGURE 7.4.1

Since $D_{23} = \cos\omega \sin i$ and $D_{13} = \sin\omega\sin i$, Equations (7.3.12e) and (7.3.12f) can be written

$$(-\xi'' \cos\omega + \eta'' \sin\omega)(\sin i)\dot{\Omega} + (\xi'' \sin\omega + \eta'' \cos\omega)(i)^{\cdot} = 0 \tag{7.4.2a}$$

$$(-\dot{\xi}'' \cos\omega + \dot{\eta}'' \sin\omega)(\sin i)\dot{\Omega} + (\dot{\xi}'' \sin\omega + \dot{\eta}'' \cos\omega)(i)^{\cdot} = F''_\zeta \tag{7.4.2b}$$

If (7.4.1) is used to introduce the new coordinates and velocity into (7.4.2), the equations become

$$\begin{aligned} -\xi^u (\sin i)\dot{\Omega} + \eta^u(i)^{\cdot} &= 0 \\ -\dot{\xi}^u (\sin i)\dot{\Omega} + \dot{\eta}^u(i)^{\cdot} &= F''_\zeta \end{aligned} \tag{7.4.3}$$

whence

$$(i)^{\cdot} = \frac{-\xi^u F''_\zeta}{\eta^u\dot{\xi}^u - \xi^u\dot{\eta}^u} \tag{7.4.4a}$$

$$\dot{\Omega} = \frac{-\eta^u F''_\zeta}{\sin i(\eta^u\dot{\xi}^u - \xi^u\dot{\eta}^u)} \tag{7.4.4b}$$

Equations (7.4.4) can be further simplified if use is made of the fact that, in the instantaneous orbital plane, $\zeta^u = \dot{\zeta}^u = 0$. Thus the instantaneous angular momentum per unit mass is given by

$$\mathbf{h} = \mathbf{r} \times \mathbf{v} = (\xi^u\dot{\eta}^u - \eta^u\dot{\xi}^u)\,\mathbf{k}''$$

so that

$$h = \xi^u\dot{\eta}^u - \eta^u\dot{\xi}^u = (\mu p^*)^{1/2}$$

If this result is introduced into (7.4.4), we obtain

$$(i)^{\cdot} = \frac{r(\cos u)\, F''_\zeta}{(\mu p^*)^{1/2}} \tag{7.4.5}$$

$$\dot{\Omega} = \frac{r(\sin u)\, F''_\zeta}{\sin i(\mu p^*)^{1/2}} \tag{7.4.6}$$

Equations (7.4.5) and (7.4.6) apply to any conic orbit. The difficulty arising for small i due to the appearance of $\sin i$ in the denominator of (7.4.6) can be removed by introducing the new variables α_1 and β_1, defined by the relations

$$\alpha_1 = \tan i \sin \Omega, \qquad \beta_1 = \tan i \cos \Omega \tag{7.4.7}$$

It can be shown that, under this transformation, the expressions for the time derivatives of α_1 and β_1 will not contain a singularity at $i = 0$.

7.5. DERIVATION OF PERTURBATION EQUATIONS FOR $\dot{a}$ AND $\dot{p}^*$

We obtain the three equations

$$\sum_{j=4}^{6} \left(\xi'' \frac{\partial \xi''}{\partial e_j} + \eta'' \frac{\partial \eta''}{\partial e_j}\right) \dot{e}_j = 0 \tag{7.5.1a}$$

$$\sum_{j=4}^{6} \left(\dot{\xi}'' \frac{\partial \dot{\xi}''}{\partial e_j} + \dot{\eta}'' \frac{\partial \dot{\eta}''}{\partial e_j}\right) \dot{e}_j = \dot{\xi}'' F''_\xi + \dot{\eta}'' F''_\eta \tag{7.5.1b}$$

$$\sum_{j=4}^{6} \left(\dot{\xi}'' \frac{\partial \xi''}{\partial e_j} + \dot{\eta}'' \frac{\partial \eta''}{\partial e_j} + \xi'' \frac{\partial \dot{\xi}''}{\partial e_j} + \eta'' \frac{\partial \dot{\eta}''}{\partial e_j}\right) \dot{e}_j = \xi'' F''_\xi + \eta'' F''_\eta \tag{7.5.1c}$$

by operating on (a), (b), (c), and (d) of (7.3.12) according to the scheme

$$\xi''(a) + \eta''(b) = (7.5.1\text{a})$$

$$\dot{\xi}''(c) + \dot{\eta}''(d) = (7.5.1\text{b})$$

$$\dot{\xi}''(a) + \dot{\eta}''(b) + \xi''(c) + \eta''(d) = (7.5.1\text{c})$$

The terms within the parentheses in (7.5.1) can be further simplified. Since the radial distance r and the speed v are related to the double-primed coordinates and velocity components by the relations

$$r^2 = (\xi'')^2 + (\eta'')^2, \qquad v^2 = (\dot{\xi}'')^2 + (\dot{\eta}'')^2$$

it follows directly that

$$e_4, e_5, e_6: \qquad r\frac{\partial r}{\partial} = \xi''\frac{\partial \xi''}{\partial} + \eta''\frac{\partial \eta''}{\partial} \tag{7.5.2a}$$

$$e_4, e_5, e_6: \qquad v\frac{\partial v}{\partial} = \dot{\xi}''\frac{\partial \dot{\xi}''}{\partial} + \dot{\eta}''\frac{\partial \dot{\eta}''}{\partial} \tag{7.5.2b}$$

$$e_4, e_5, e_6: \qquad \frac{\partial(r\dot{r})}{\partial} = \xi''\frac{\partial \dot{\xi}''}{\partial} + \eta''\frac{\partial \dot{\eta}''}{\partial} + \dot{\xi}''\frac{\partial \xi''}{\partial} + \dot{\eta}''\frac{\partial \eta}{\partial} \tag{7.5.2c}$$

If Equations (7.5.2) are introduced into (7.5.1), we can write

$$\sum_{j=4}^{6} r\frac{\partial r}{\partial e_j}\dot{e}_j = 0 \tag{7.5.3a}$$

$$\sum_{j=4}^{6} v\frac{\partial v}{\partial e_j}\dot{e}_j = \dot{\xi}''F''_\xi + \dot{\eta}''F''_\eta \tag{7.5.3b}$$

$$\sum_{j=4}^{6} \frac{\partial(r\dot{r})}{\partial e_j}\dot{e}_j = \xi''F''_\xi + \eta''F''_\eta \tag{7.5.3c}$$

The relation

$$v^2 = \mu\left(\frac{2}{r} - \frac{1}{a}\right)$$

which has already been deduced for two-body orbits, can be used to obtain the expressions

$$v\frac{\partial v}{\partial e_4} = -\frac{\mu}{r^2}\frac{\partial r}{\partial e_4} + \frac{\mu}{2a^2}\frac{\partial a}{\partial e_4} \tag{7.5.4a}$$

$$v\frac{\partial v}{\partial e_5} = -\frac{\mu}{r^2}\frac{\partial r}{\partial e_5} + \frac{\mu}{2a^2}\frac{\partial a}{\partial e_5} \tag{7.5.4b}$$

$$v\frac{\partial v}{\partial e_6} = -\frac{\mu}{r^2}\frac{\partial r}{\partial e_6} \tag{7.5.4c}$$

for the coefficients of the $\dot{e}_j$ appearing in (7.5.3b). If (7.5.3a) is multiplied by μr^{-3} and Equations (7.5.4) are introduced into (7.5.3b), we obtain the two equations

$$\frac{\mu}{r^2}\sum_{j=4}^{6}\frac{\partial r}{\partial e_j}\dot{e}_j = 0 \tag{7.5.5a}$$

$$-\frac{\mu}{r^2}\sum_{j=4}^{6}\frac{\partial r}{\partial e_j}\dot{e}_j + \frac{\mu}{2a^2}\left(\frac{\partial a}{\partial e_4}\dot{e}_4 + \frac{\partial a}{\partial e_5}\dot{e}_5\right) = \dot{\xi}''F''_\xi + \dot{\eta}''F''_\eta \tag{7.5.5b}$$

which, when added, yield the equation

$$\dot{a} = \frac{2a^2}{\mu}(\dot{\xi}''F''_\xi + \dot{\eta}''F''_\eta) \tag{7.5.6}$$

if we take $e_4 = a$. When the expressions for $\dot{\xi}''$ and $\dot{\eta}''$ given by (7.3.3) are introduced into (7.5.6), we have

$$\dot{a} = \frac{2a^2}{(\mu p^*)^{1/2}}[-(\sin w)F''_\xi + (\cos w + e)F''_\eta] \tag{7.5.7}$$

Equation (7.5.7) expresses $\dot{a}$ in terms of the components of acceleration F''_ξ and F''_η in the instantaneous orbital plane. This is not, however, the most convenient decomposition of acceleration. Particularly useful decompositions are given either in terms of the components R along the radius and S perpendicular to R in the direction of advance of w, or in terms of the components T along the tangent, positive in the direction of advance of w and N normal to the tangent and directed away from the center of curvature.† With reference to Figure 7.5.1, the unit vectors

FIGURE 7.5.1

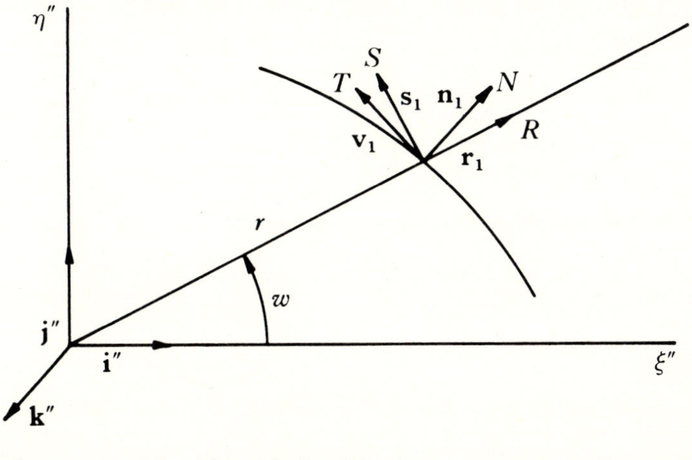

$\mathbf{r}_1$, $\mathbf{s}_1$, $\mathbf{v}_1$, and $\mathbf{n}_1$ are to be identified with the components R, S, T, and N respectively, in the senses already described. The unit vectors $\mathbf{i}''$, $\mathbf{j}''$, $\mathbf{k}''$ appear in the usual sense.

Since the ($\mathbf{r}_1$, $\mathbf{s}_1$, $\mathbf{k}''$) basis is obtained from the ($\mathbf{i}''$, $\mathbf{j}''$, $\mathbf{k}''$) basis by a rotation through the angle w, we can write the matrix relation

$$\begin{pmatrix}\mathbf{i}''\\ \mathbf{j}''\end{pmatrix} = \begin{pmatrix}\cos w & -\sin w\\ \sin w & \cos w\end{pmatrix}\begin{pmatrix}\mathbf{r}_1\\ \mathbf{s}_1\end{pmatrix} \tag{7.5.8}$$

† It will be clear from the context when the alternative use of R for the perturbing function and T for a time of pericenter passage is intended.

In terms of the osculating elements, the velocity $\mathbf{v}$, in the perturbed orbit, can be written as

$$\mathbf{v} = \dot{\xi}''\mathbf{i}'' + \dot{\eta}''\mathbf{j}'' = -\left(\frac{\mu}{p^*}\right)^{1/2} [\sin w\mathbf{i}'' - (\cos w + e)\,\mathbf{j}''] \tag{7.5.9}$$

so that

$$v = \left(\frac{\mu}{p^*}\right)^{1/2} f \tag{7.5.10}$$

where

$$f \equiv (1 + 2e \cos w + e^2)^{1/2} \tag{7.5.11}$$

It follows directly that

$$\begin{pmatrix} \mathbf{i}'' \\ \mathbf{j}'' \end{pmatrix} = \begin{pmatrix} \dfrac{\cos w + e}{f} & -\dfrac{\sin w}{f} \\ \dfrac{\sin w}{f} & \dfrac{\cos w + e}{f} \end{pmatrix} \begin{pmatrix} \mathbf{n}_1 \\ \mathbf{v}_1 \end{pmatrix} \tag{7.5.12}$$

The inverse transformations corresponding to (7.5.8) and Equation (7.5.12) can be used to relate the $(\mathbf{r}_1, \mathbf{s}_1)$ and $(\mathbf{n}_1, \mathbf{v}_1)$ bases. Explicitly, we find

$$\begin{pmatrix} \mathbf{r}_1 \\ \mathbf{s}_1 \end{pmatrix} = \begin{pmatrix} \dfrac{1 + e \cos w}{f} & \dfrac{e \sin w}{f} \\ -\dfrac{e \sin w}{f} & \dfrac{1 + e \cos w}{f} \end{pmatrix} \begin{pmatrix} \mathbf{n}_1 \\ \mathbf{v}_1 \end{pmatrix} \tag{7.5.13}$$

In terms of the components, we write

$$fR = (1 + e \cos w)\, N + e(\sin w)T \tag{7.5.14a}$$

$$fS = -e(\sin w)\, N + (1 + e \cos w)T \tag{7.5.14b}$$

and

$$fN = (1 + e \cos w)\, R - e(\sin w)S \tag{7.5.15a}$$

$$fT = e(\sin w)\, R + (1 + e \cos w)S \tag{7.5.15b}$$

If expressions for F''_ξ and F''_η obtained from (7.5.8) and (7.5.14) are substituted into (7.5.7), we have

$$\dot{a} = \frac{2a^2}{(\mu p^*)^{1/2}} \left[e(\sin w)\, R + \frac{p^*}{r} S\right] \tag{7.5.16}$$

or

$$\dot{a} = \frac{2a^2}{(\mu p^*)^{1/2}} fT \tag{7.5.17}$$

The equation

$$\sum_{j=4}^{6} \frac{\partial h}{\partial e_j} \dot{e}_j = \xi'' F''_\eta - \eta'' F''_\xi \tag{7.5.18}$$

can be obtained from (a), (b), (c), and (d) of (7.3.12) by performing the following sequence of operations:

$$\dot{\eta}''(a) - \dot{\xi}''(b) - \eta''(c) + \xi''(d) = (7.5.18)$$

In obtaining (7.5.18), we have made use of the fact that the instantaneous angular momentum h is given by

$$h = \xi''\dot{\eta}'' - \eta''\dot{\xi}'' = (\mu p^*)^{1/2}$$

If we take $e_4 = p^*$, it follows that

$$\dot{p}^* = 2\left(\frac{p^*}{\mu}\right)^{1/2} rS \tag{7.5.19}$$

or

$$\dot{p}^* = 2\left(\frac{p^*}{\mu}\right)^{1/2} \frac{r}{f} [(1 + e\cos w)\, T - e(\sin w) N] \tag{7.5.20}$$

7.6. DERIVATION OF PERTURBATION EQUATIONS FOR $\dot{e}$ AND $\dot{\omega}$

Returning to the first four equations of (7.3.12), we note that the partial derivatives of ξ'', η'', $\dot{\xi}''$, and $\dot{\eta}''$ with respect to the orbital elements e_j, $j = 4, 5, 6$, are to be expressed in terms of the orbital elements and the time. Since

$$\xi'' = r\cos w, \qquad \dot{\xi}'' = -\left(\frac{\mu}{p^*}\right)^{1/2} \sin w$$

$$\eta'' = r\sin w, \qquad \dot{\eta}'' = \left(\frac{\mu}{p^*}\right)^{1/2} (\cos w + e) \tag{7.3.3}$$

we have

$$e_4, e_5, e_6: \qquad \frac{\partial \xi''}{\partial} = \cos w \frac{\partial r}{\partial} - \frac{\sin w}{r} r^2 \frac{\partial w}{\partial}$$

with similar equations for η'', $\dot{\xi}''$, and $\dot{\eta}''$. Thus we can, after some manipulation, replace (a), (b), (c), and (d) of (7.3.12) by the four equations

$$\sum_{j=4}^{6} \frac{\partial r}{\partial e_j} \dot{e}_j = 0 \tag{7.6.1a}$$

$$\sum_{j=4}^{6} \frac{\partial w}{\partial e_j} \dot{e}_j + \dot{\chi} = 0 \tag{7.6.1b}$$

$$\sum_{j=4}^{6} \left(\sin w \frac{\partial e}{\partial e_j} - \frac{e \sin w}{2p^*} \frac{\partial p^*}{\partial e_j} - \frac{\partial w}{\partial e_j}\right) \dot{e}_j - (1 + e \cos w)\dot{\chi} = \left(\frac{p^*}{\mu}\right)^{1/2} (F_\xi'' \cos w + F_\eta'' \sin w) \tag{7.6.1c}$$

$$\sum_{j=4}^{6} \left(\cos w \frac{\partial e}{\partial e_j} - \frac{1 + e \cos w}{2p^*} \frac{\partial p^*}{\partial e_j}\right) \dot{e}_j + e(\sin w)\dot{\chi} = \left(\frac{p^*}{\mu}\right)^{1/2} (-F_\xi'' \sin w + F_\eta'' \cos w) \tag{7.6.1d}$$

where

$$\dot{\chi} = \dot{\omega} + D_{33}\dot{\Omega} \qquad (D_{33} = \cos i) \tag{7.6.2}$$

If (7.6.1b) and (7.6.1c) are added, we obtain a fifth equation

$$\sum_{j=4}^{6} \left(\sin w \frac{\partial e}{\partial e_j} - \frac{e \sin w}{2p^*} \frac{\partial p^*}{\partial e_j}\right) \dot{e}_j - e(\cos w)\dot{\chi} = \left(\frac{p^*}{\mu}\right)^{1/2} (F_\xi'' \cos w + F_\eta'' \sin w) \tag{7.6.3}$$

We can now obtain two new equations from (7.6.1d) and (7.6.3), one in $\dot{p}^*$ and $\dot{e}$ alone, the other in $\dot{p}^*$ and $\dot{\chi}$ alone, by taking $e_4 = p^*$ and performing the following operations:

$$\cos w(d) + \sin w(7.6.3) = (7.6.4a)$$

$$-\sin w(d) + \cos w(7.6.3) = (7.6.4b)$$

Explicitly, we find

$$-\frac{\cos w + e}{2p^*} \dot{p}^* + \dot{e} = \left(\frac{p^*}{\mu}\right)^{1/2} F_\eta'' \tag{7.6.4a}$$

$$\frac{\sin w}{2p^*} \dot{p}^* - e\dot{\chi} = \left(\frac{p^*}{\mu}\right)^{1/2} F_\xi'' \tag{7.6.4b}$$

Equations (7.6.4) can be solved for $\dot{e}$ and $\dot{\chi}$. When this is done and use is made of (7.4.4b), (7.5.19), (7.5.20), (7.6.2), (7.5.8), (7.5.14), and (7.5.15), we obtain, in a straightforward manner, the following equations:

$$\dot{e} = \left(\frac{p^*}{\mu}\right)^{1/2} \left[(\sin w)\, R + \frac{(r + p^*) \cos w + er}{p^*}\, S\right] \tag{7.6.5}$$

$$e\dot{\chi} = \left(\frac{p^*}{\mu}\right)^{1/2} \left[-(\cos w)\, R + \frac{(r + p^*) \sin w}{p^*}\, S\right] \tag{7.6.6}$$

$$e\dot{\omega} = \left(\frac{p^*}{\mu}\right)^{1/2} \left[-(\cos w)\, R + \frac{(r + p)^* \sin w}{p^*}\, S - \frac{re \sin u\, (\cot i)}{p^*}\, F_\zeta''\right] \tag{7.6.7}$$

or

$$\dot{e} = \frac{1}{f}\left(\frac{p^*}{\mu}\right)^{1/2} \left[2(\cos w + e)T + \frac{r(1 - e^2) \sin w}{p^*}\, N\right] \tag{7.6.8}$$

$$e\dot{\chi} = \frac{1}{f}\left(\frac{p^*}{\mu}\right)^{1/2} \left[2(\sin w)T - \frac{r}{p^*}(2e + \cos w + e^2 \cos w)N\right] \tag{7.6.9}$$

$$e\dot{\omega} = \frac{1}{f}\left(\frac{p^*}{\mu}\right)^{1/2} \left[2(\sin w)T - \frac{r}{p^*}(2e + \cos w + e^2 \cos w)N - \frac{efr \sin u\, (\cot i)}{p^*}\, F_\zeta''\right] \tag{7.6.10}$$

The six equations just preceding apply to any conic orbit. The difficulty arising for small e due to its presence in the denominators of the expressions for $\dot{\chi}$ and $\dot{\omega}$ can be removed by introducing the new variables

$$\alpha_2 = e \sin \omega, \qquad \beta_2 = e \cos \omega \tag{7.6.11}$$

It can be shown that, under this transformation, the expressions for the time derivatives of $\dot{\alpha}_2$ and $\dot{\beta}_2$ will not contain a singularity at $e = 0$.

7.7. DERIVATION OF A PERTURBATION EQUATION FOR $\dot{T}$

Since expressions for $\dot{e}_4$, $\dot{e}$, and $\dot{\chi}$ have been determined, the perturbation equation for $\dot{T}$ could now be obtained from either (a) or (b) of (7.6.1) if we had explicit expressions for the partial derivatives of w and r with respect to the elements e_4, e, T appearing in the two equations. Actually, the problem reduces to that of determining suitable expressions for $r^2(\partial w/\partial)$ with respect to these elements. If such expressions were available, we could readily obtain relations for $\partial r/\partial$ with respect to the three elements by differentiating the equation

$$r = p^*(1 + e \cos w)^{-1}$$

for a conic orbit. To illustrate, we take $e_4 = p^*$ and write

$$p^*, e, T: \qquad \frac{\partial r}{\partial} = \frac{r}{p^*}\frac{\partial p^*}{\partial} - \frac{r^2 \cos w}{p^*}\frac{\partial e}{\partial} + \frac{e \sin w}{p^*}\left(r^2 \frac{\partial w}{\partial}\right) \tag{7.7.1}$$

We have expressed the coordinates and velocity in terms of the true anomaly because it is a unifying variable; that is, it is a common variable for each of the three conic orbits. To examine the way in which w changes at some fixed instant t as the orbital elements p^*, e, and T change, we return to the equation

$$\int_0^w r^2\, dw = h(t - T) = (\mu p^*)^{1/2}\,(t - T) \qquad [3.2.3]$$

To obtain expressions for $r^2(\partial w/\partial)$ with respect to the elements p^*, e, T, we will apply Leibnitz' rule for differentiating under the integral sign in (3.2.3). We can do this because the change in w as the orbital elements p^*, e, T change has the same form in the perturbed orbit as it does in the osculating orbit at that instant. For our purposes, the rule can be stated as follows:

LEIBNITZ' RULE: Let

$$f(\phi, p^*, e), \qquad \frac{\partial f(\phi, p^*, e)}{\partial p^*}, \qquad \frac{\partial f(\phi, p^*, e)}{\partial e}$$

be continuous functions of ϕ, p^*, and e in some domain, $w_1 \leqslant \phi \leqslant w_2$, $p_1^* \leqslant p^* \leqslant p_2^*$, $e_1 \leqslant e \leqslant e_2$. In addition, let w_1 and w_2 and their first partial derivatives with respect to p^*, e, T be continuous functions of p^*, e, and T for $p_1^* \leqslant p^* \leqslant p_2^*$, $e_1 \leqslant e \leqslant e_2$, $T_1 \leqslant T \leqslant T_2$. Then if we write

$$\psi(p^*, e, T) = \int_{w_1}^{w_2} f(\phi, p^*, e)\, d\phi$$

it is valid to differentiate under the integral sign to obtain

$$p^*, e, T: \qquad \frac{\partial \psi(p^*, e, T)}{\partial} = f(w_2, p^*, e)\frac{\partial w_2}{\partial} + \int_{w_1}^{w_2} \frac{\partial f(\phi, p^*, e)}{\partial}\, d\phi - f(w_1, p^*, e)\frac{\partial w_1}{\partial}$$

If, in the integrand of (3.2.3), w is replaced by the dummy variable ϕ, we have

$$\int_0^w r^2\, d\phi = \int_0^w p^{*2}(1 + e \cos\phi)^{-2}\, d\phi = (\mu p^*)^{1/2}\,(t - T) \qquad (7.7.2)$$

Certain restrictions on the domain of ϕ arise, depending on the type of conic orbit under study, but, for any of the three conic orbits, the integral in (7.7.2) satisfies all the criteria necessary for the application of Leibnitz' rule. Thus we obtain

$$p^*, e, T: \qquad r^2 \frac{\partial w}{\partial} + 2\int_0^w r \frac{\partial r}{\partial}\, d\phi = \frac{1}{2}\left(\frac{\mu}{p^*}\right)^{1/2} (t - T) \frac{\partial p^*}{\partial} - (\mu p^*)^{1/2} \frac{\partial T}{\partial} \qquad (7.7.3)$$

Explicitly, the three equations corresponding to (7.7.3) are

$$r^2 \frac{\partial w}{\partial p^*} + 2p^* I_1 = \frac{1}{2}\left(\frac{\mu}{p^*}\right)^{1/2} (t - T) \qquad (7.7.4a)$$

$$r^2 \frac{\partial w}{\partial e} - 2p^{*2} I_2 = 0 \qquad (7.7.4b)$$

$$r^2 \frac{\partial w}{\partial T} = -(\mu p^*)^{1/2} \qquad (7.7.4c)$$

where

$$I_1 \equiv \int_0^w (1 + e \cos\phi)^{-2}\, d\phi$$

$$I_2 \equiv \int_0^w (\cos\phi)(1 + e\cos\phi)^{-3}\, d\phi$$

The evaluation of I_1 follows directly from inspection of (7.7.2). Specifically, we have

$$I_1 \equiv \frac{\mu^{1/2}}{p^{*3/2}} (t - T) \qquad (7.7.5)$$

It can also be shown that

$$I_2 = \frac{1}{2(1 - e^2)\, p^{*3/2}} \left[\frac{r \sin w (r + p^*)}{p^{*1/2}} - 3e\mu^{1/2}(t - T)\right] \qquad (7.7.6)$$

When the expressions for I_1 and I_2 are substituted into (7.7.4), we obtain

$$r^2 \left[\frac{\partial w}{\partial p^*}\right]_{(p^*,e,T)} = -\frac{3}{2}\left(\frac{\mu}{p^*}\right)^{1/2}(t - T) \tag{7.7.7a}$$

$$r^2 \left[\frac{\partial w}{\partial e}\right]_{(p^*,e,T)} = \frac{1}{1-e^2}\,[(r + p^*)\, r \sin w - 3(\mu p^*)^{1/2}\, e(t - T)] \qquad (e \neq 1) \tag{7.7.7b}$$

$$r^2 \left[\frac{\partial w}{\partial T}\right]_{(p^*,e,T)} = -(\mu p^*)^{1/2} \tag{7.7.7c}$$

where the square bracket subscript notation has been introduced to indicate the variables that were taken to be independent in applying Leibnitz' rule. The corresponding equations for the partial derivatives of r follow directly from (7.7.7) and (7.7.1). Explicitly, they are

$$\left[\frac{\partial r}{\partial p^*}\right]_{(p^*,e,T)} = \frac{1}{p^*}\left[r - \frac{3}{2}\left(\frac{\mu}{p^*}\right)^{1/2}(t - T)\, e \sin w\right] \tag{7.7.8a}$$

$$\left[\frac{\partial r}{\partial e}\right]_{(p^*,e,T)} = \frac{1}{p^*}\left\{-r^2 \cos w + \frac{e \sin w}{1-e^2}\,[(r + p^*)\, r \sin w - 3(\mu p^*)^{1/2}\, e(t - T)]\right\} \qquad (e \neq 1) \tag{7.7.8b}$$

$$\left[\frac{\partial r}{\partial T}\right]_{(p^*,e,T)} = -\left(\frac{\mu}{p^*}\right)^{1/2} e \sin w \tag{7.7.8c}$$

The perturbation equation for $\dot{T}$ can be obtained by solving (7.6.1b) for $\dot{T}$. We find that

$$(\mu p^*)^{1/2}\, \dot{T} = r^2 \dot{\chi} + r^2 \frac{\partial w}{\partial p^*}\, \dot{p}^* + r^2 \frac{\partial w}{\partial e}\, \dot{e}$$

which leads to the equations

$$\mu \dot{T} = a\left[\left(2r - \frac{p^*}{e}\cos w\right) - 3\left(\frac{\mu}{p^*}\right)^{1/2} e(t - T)\sin w\right] R + a\left[\frac{(r + p^*)\sin w}{e} - \frac{3}{r}(\mu p^*)^{1/2}(t - T)\right] S \qquad (e \neq 1) \tag{7.7.9}$$

$$\mu \dot{T} = a\left[2\left(\frac{p^*}{e} + re\right)\frac{\sin w}{f} - 3\left(\frac{\mu}{p^*}\right)^{1/2} f(t - T)\right] T - \frac{rp^* \cos w}{ef} N \qquad (e \neq 1) \tag{7.7.10}$$

Equations (7.7.9) and (7.7.10) are not in a convenient form for use with nearly parabolic cases for which e is close to unity. For such nearly parabolic orbits, however, it can be shown that, for the triad of elements p^*, e, T, the equations

$$r^2 \frac{\partial w}{\partial e} = \frac{2p^{*2}B}{(1+e)^3} \tag{7.7.11a}$$

$$\frac{\partial r}{\partial e} = -\frac{r^2 \cos w}{p^*} + \frac{4ezp^*B}{(1+z^2)(1+e)^3} \tag{7.7.11b}$$

hold, where

$$z \equiv \tan \frac{w}{2}$$

$$B \equiv \sum_{n=0}^{\infty} (-1)^n (n+1)(n+2) \left(\frac{1-e}{1+e}\right)^n \left(\frac{z^{2n+1}}{2n+1} - \frac{z^{2n+5}}{2n+5}\right) \tag{7.7.12}$$

It then follows directly that

$$\begin{aligned} \mu\dot{T} = {} & \left[-\frac{r^2 \cos w}{e} + \frac{2p^{*2}B \sin w}{(1+e)^3}\right] R \\ & + \left\{\frac{r^2(r+p^*)}{ep^*} \sin w - 3r \left(\frac{\mu}{p^*}\right)^{1/2} (t-T)\right. \\ & \left. + \frac{[(r+p^*) \cos w + er]\, 2p^*B}{(1+e)^3}\right\} S \end{aligned} \tag{7.7.13}$$

For the parabolic case, $e = 1$, and we have

$$\mu\dot{T} = \frac{q^2}{5(1+z^2)} [(z^6 + 5z^4 + 15z^2 - 5)\, R + 4(5 - z^4)\, zS] \qquad (e = 1) \tag{7.7.14}$$

A perturbation equation for the mean anomaly M is now readily derived by differentiating the equation

$$M = n(t - T) \qquad (n = \mu^{1/2}a^{-3/2})$$

with respect to time. Specifically, we find

$$\dot{M} = n - \frac{1}{(\mu a)^{1/2}} \left[\left(2r - \frac{p^*}{e} \cos w\right) R + \frac{(r+p^*) \sin w}{e} S\right] \tag{7.7.15}$$

or

$$\dot{M} = n - \frac{1}{f(\mu a)^{1/2}} \left[2 \sin w \left(\frac{p^*}{e} + re\right) T - \frac{r}{e}(1 - e^2)(\cos w) N\right] \tag{7.7.16}$$

7.8. SUMMARY

For convenience, most of the perturbation equations derived in the preceding sections are repeated below and renumbered. They are referred to as Lagrange planetary equations.

1. The Planetary Equations in Terms of the R, S, F''_ζ Components of Acceleration

$$\dot{p}^* = 2\left(\frac{p^*}{\mu}\right)^{1/2} rS \tag{7.8.1a}$$

$$(i)^{\cdot} - \frac{r\cos u}{(\mu p^*)^{1/2}} F''_\zeta \tag{7.8.1b}$$

$$\dot{\Omega} = \frac{r\sin u}{\sin i(\mu p^*)^{1/2}} F''_\zeta \tag{7.8.1c}$$

$$\dot{a} = \frac{2a^2}{(\mu p^*)^{1/2}}\left[e(\sin w)\,R + \frac{p^*}{r}\,S\right] \tag{7.8.1d}$$

$$\dot{e} = \left(\frac{p^*}{\mu}\right)^{1/2}\left[(\sin w)\,R + \frac{(r+p^*)\cos w + er}{p^*}\,S\right] \tag{7.8.1e}$$

$$e\dot{\chi} = \left(\frac{p^*}{\mu}\right)^{1/2}\left[-(\cos w)\,R + \frac{(r+p^*)\sin w}{p^*}\,S\right] \tag{7.8.1f}$$

$$e\dot{\omega} = \left(\frac{p^*}{\mu}\right)^{1/2}\left[-(\cos w)\,R + \frac{(r+p^*)\sin w}{p^*}\,S - \frac{re\sin u\,(\cot i)}{p^*}F''_\zeta\right] \tag{7.8.1g}$$

$$\begin{aligned}\mu\dot{T} &= a\left[\left(2r - \frac{p^*}{e}\cos w\right) - 3\left(\frac{\mu}{p^*}\right)^{1/2} e(t-T)\sin w\right] R \\ &\quad + a\left[\frac{(r+p^*)\sin w}{e} - \frac{3(\mu p^*)^{1/2}(t-T)}{r}\right] S \qquad (e \neq 1)\end{aligned} \tag{7.8.1h}$$

$$\mu\dot{T} = \frac{q^2}{5(1+z^2)}\left[(z^6 + 5z^4 + 15z^2 - 5)\,R + 4(5z - z^5)S\right] \qquad \left(e = 1,\ z = \tan\frac{w}{2}\right) \tag{7.8.1i}$$

$$\dot{M} = n - \frac{1}{(\mu a)^{1/2}}\left[\left(2r - \frac{p^*}{e}\cos w\right) R + \frac{(r+p^*)\sin w}{e}\,S\right] \tag{7.8.1j}$$

$$\dot{M} = n - \frac{2r}{(\mu a)^{1/2}}\,R - (1-e^2)^{1/2}\,\dot{\chi} \tag{7.8.1k}$$

2. The Planetary Equations in Terms of the T, N, F''_ζ Components of Acceleration

$$\dot{p}^* = 2\left(\frac{p^*}{\mu}\right)^{1/2}\frac{r}{f}[(1 + e\cos w)\,T - e(\sin w)N] \tag{7.8.2a}$$

$$(i)^{\cdot} = \frac{r\cos u}{(\mu p^*)^{1/2}}F''_\zeta \tag{7.8.2b}$$

$$\dot{\Omega} = \frac{r\sin u}{(\mu p^*)^{1/2}\sin i}F''_\zeta \tag{7.8.2c}$$

$$\dot{a} = \frac{2a^2}{(\mu p^*)^{1/2}}fT \tag{7.8.2d}$$

$$\dot{e} = \left(\frac{p^*}{\mu}\right)^{1/2}\frac{1}{f}\left[2(\cos w + e)\,T + \frac{r(1-e^2)\sin w}{p^*}N\right] \tag{7.8.2e}$$

$$e\dot{\chi} = \left(\frac{p^*}{\mu}\right)^{1/2}\frac{1}{f}\left[2(\sin w)\,T - \frac{r}{p^*}(2e + \cos w + e^2\cos w)N\right] \tag{7.8.2f}$$

$$e\dot{\omega} = \left(\frac{p^*}{\mu}\right)^{1/2}\left[\frac{2(\sin w)\,T - (r/p^*)(2e + \cos w + e^2\cos w)N}{f} - \frac{re\sin u\,(\cot i)}{p^*}F''_\zeta\right] \tag{7.8.2g}$$

$$\mu\dot{T} = a\left[2\left(\frac{p^*}{e} + re\right)\frac{\sin w}{f} - 3\left(\frac{\mu}{p^*}\right)^{1/2}f(t-T)\right]T - \frac{rp^*\cos w}{ef}N \qquad (e \neq 1) \tag{7.8.2h}$$

$$\dot{M} = n - \frac{1}{f(\mu a)^{1/2}}\left[2\sin w\left(\frac{p^*}{e} + re\right)T - \frac{r}{e}(1-e^2)(\cos w)N\right] \tag{7.8.2i}$$

If it is desired, Equations (7.8.2) can be expressed in terms of the speed v through the relation

$$v = \left(\frac{\mu}{p^*}\right)^{1/2}f$$

Exercises

1. Show that, through first order, β is given by (7.1.12b) in the problem of the vibrating tuning fork described in the text.
2. Obtain the first-order approximation to the solution of the differential equation

$$\ddot{x} + x = cx^2, \qquad |c| \ll 1$$

3. Prove or disprove analytically the following hypotheses for perturbed orbits:

$$\text{(a)}\quad \dot{r} = \frac{\partial r}{\partial t}, \qquad \text{(b)}\quad \dot{w} = \frac{\partial w}{\partial t}, \qquad \text{(c)}\quad \dot{\xi}'' = \frac{\partial \xi''}{\partial t}$$

4. Verify Equations (7.3.11) in the text.
5. Verify the statement immediately following (7.4.7) in the text.
6. Verify Equations (7.6.1) and (7.6.2) in the text.
7. Verify the statement immediately following (7.6.11) in the text.
8. In the text, expressions for $r^2(\partial w/\partial)$ with respect to the elements p^*, e, T in the perturbed orbit were obtained by using Equation (3.2.3). Can you reconcile this procedure with the result you obtained in answering Exercise 3(b)?
9. Show that $\int_0^w r^2\, d\phi$ satisfies the conditions necessary for the application of Leibnitz' rule.
10. Show that

$$\begin{aligned} I_2 &= \int_0^w (\cos\phi)(1 + e\cos\phi)^{-3}\, d\phi \\ &= \frac{1}{2(1-e^2)}\left[\frac{r\sin w}{p^{*2}}(r+p^*) - \frac{3e\mu^{1/2}(t-T)}{p^{*3/2}}\right] \qquad (e \neq 1) \end{aligned}$$

11. Verify that, for the triad of orbital elements a, e, T,

$$\begin{aligned} r^2\left[\frac{\partial w}{\partial e}\right]_{(a,e,T)} &= \frac{1}{1-e^2}\, r(r+p^*)\sin w \\ \left[\frac{\partial r}{\partial e}\right]_{(a,e,T)} &= -a\cos w \\ r^2\left[\frac{\partial w}{\partial a}\right]_{(a,e,T)} &= -\frac{3}{2}(1-e^2)\left(\frac{\mu}{p^*}\right)^{1/2}(t-T) \\ \left[\frac{\partial r}{\partial a}\right]_{(a,e,T)} &= \frac{r}{a} - \frac{3}{2}\left(\frac{\mu}{p^*}\right)^{1/2}(t-T)\frac{e\sin w}{a} \\ r^2\left[\frac{\partial w}{\partial T}\right]_{(a,e,T)} &= -(\mu p^*)^{1/2} \\ \left[\frac{\partial r}{\partial T}\right]_{(a,e,T)} &= -\left(\frac{\mu}{p^*}\right)^{1/2} e\sin w \end{aligned}$$

12. Verify that, for the triad of orbital elements a, e, M,

$$r^2\left[\frac{\partial w}{\partial e}\right]_{(a,e,M)} = \frac{r(r+p^*)\sin w}{1-e^2}, \qquad \left[\frac{\partial r}{\partial e}\right]_{(a,e,M)} = -a\cos w$$

$$r^2\left[\frac{\partial w}{\partial a}\right]_{(a,e,M)} = 0, \qquad \left[\frac{\partial r}{\partial a}\right]_{(a,e,M)} = \frac{r}{a}$$

$$r^2\left[\frac{\partial w}{\partial M}\right]_{(a,e,M)} = a^2(1-e^2)^{1/2}, \qquad \left[\frac{\partial r}{\partial M}\right]_{(a,e,M)} = \frac{ae\sin w}{(1-e^2)^{1/2}}$$

13. Verify that, for a nearly parabolic orbit, for the triad of elements p^*, e, T,

$$r^2\frac{\partial w}{\partial e} = \frac{2p^{*2}B}{(1+e)^3}$$

and

$$\frac{\partial r}{\partial e} = -\frac{r^2\cos w}{p^*} + \frac{4ezp^*}{(1+z^2)(1+e)^3}B$$

where

$$z = \tan\frac{w}{2}$$

$$B = \sum_{n=0}^{\infty}(-1)^n(n+1)(n+2)\,\epsilon^n\left(\frac{z^{2n+1}}{2n+1} - \frac{z^{2n+5}}{2n+5}\right) \qquad \left(\epsilon = \frac{1-e}{1+e}\right)$$

Hence deduce formula (7.7.13) in the text.

14. Show that for $e = 1$, $z = \tan(w/2)$, the perturbation equation for T is

$$5(1+z^2)\,\mu\dot{T} = q^2[(z^6 + 5z^4 + 15z^2 - 5)\,R + 4(5z - z^5)S]$$

15. (a) A small spherical satellite experiences a resisting force given by $\mathbf{F} = -cv^2\mathbf{v}_1/r^2$, where c is a positive constant. Show that during the time t the semimajor axis contracts by an amount

$$\Delta a = \frac{-2cnKt}{(1-e^2)^2} + \text{periodic terms}$$

where

$$K = \frac{1}{\pi}\int_0^{\pi} f^3\,dw, \qquad f = [1 + e^2 + 2e\cos w]^{1/2}$$

(b) Examine the sign of

$$I = \frac{e}{\pi}\int_0^\pi f\,dw + \frac{1}{\pi}\int_0^\pi f\cos w\,dw$$

Now show that during time t the eccentricity decreases by an amount

$$\Delta e = -\frac{2cnIt}{p^*} + \text{periodic terms}$$

Chapter 8

The Lagrange Planetary Equations for Conservative Perturbing Forces

8.1. LAGRANGE BRACKETS

Return to Equation (7.2.4) and consider the case where the total perturbing acceleration **F** is expressible as the gradient of a scalar potential, say R'. For the planetary problem, $R' = R$ as given by (7.2.2), but, more generally, the only restriction on R' is that the total perturbing acceleration be derivable from it by taking its gradient. With this understanding, we will drop the primed notation and write

$$\ddot{\mathbf{r}} + \frac{\mu \mathbf{r}}{r^3} = \nabla R \qquad (\mathbf{F} = \nabla R) \tag{8.1.1}$$

If we rewrite Equations (7.2.11) and (7.2.16) as

$$\sum_{k=1}^{6} \frac{\partial \mathbf{r}}{\partial e_k} \dot{e}_k = \mathbf{0} \tag{8.1.2}$$

$$\sum_{k=1}^{6} \frac{\partial \dot{\mathbf{r}}}{\partial e_k} \dot{e}_k = \nabla R \tag{8.1.3}$$

form the dot product of (8.1.3) and (8.1.2) by $\partial\mathbf{r}/\partial e_1$ and $\partial\dot{\mathbf{r}}/\partial e_1$, respectively, and then subtract the resultant equations, we obtain a new equation. Repeating the procedure, successively, for $\partial\mathbf{r}/\partial e_j$, $\partial\dot{\mathbf{r}}/\partial e_j$, $j = 2, 3, 4, 5, 6$, we obtain the system of six equations,

$$\sum_{k=1}^{6} \left\{ \frac{\partial\mathbf{r}}{\partial e_j} \cdot \frac{\partial\dot{\mathbf{r}}}{\partial e_k} - \frac{\partial\mathbf{r}}{\partial e_k} \cdot \frac{\partial\dot{\mathbf{r}}}{\partial e_j} \right\} \dot{e}_k = \nabla R \cdot \frac{\partial\mathbf{r}}{\partial e_j} \qquad (j = 1, 2, \ldots, 6) \tag{8.1.4}$$

The quantity within the brace in (8.1.4) will be designated by the symbols $[e_j, e_k]$. It is called the Lagrange bracket with reference to e_j and e_k. Explicitly, we have

$$[e_j, e_k] = \frac{\partial\mathbf{r}}{\partial e_j} \cdot \frac{\partial\dot{\mathbf{r}}}{\partial e_k} - \frac{\partial\mathbf{r}}{\partial e_k} \cdot \frac{\partial\dot{\mathbf{r}}}{\partial e_j} \tag{8.1.5}$$

or, in Jacobian notation, where

$$\frac{\partial(A, \dot{A})}{\partial(e_j, e_k)} = \begin{vmatrix} \dfrac{\partial A}{\partial e_j} & \dfrac{\partial A}{\partial e_k} \\ \dfrac{\partial \dot{A}}{\partial e_j} & \dfrac{\partial \dot{A}}{\partial e_k} \end{vmatrix} \qquad (A = \xi, \eta, \zeta)$$

we write

$$[e_j, e_k] = \frac{\partial(\xi, \dot{\xi})}{\partial(e_j, e_k)} + \frac{\partial(\eta, \dot{\eta})}{\partial(e_j, e_k)} + \frac{\partial(\zeta, \dot{\zeta})}{\partial(e_j, e_k)} \tag{8.1.6}$$

It follows directly from the definition that

$$[e_j, e_j] = 0, \qquad [e_j, e_k] = -[e_k, e_j] \tag{8.1.7}$$

Since

$$\nabla R \cdot \frac{\partial\mathbf{r}}{\partial e_j} = \frac{\partial R}{\partial e_j} \tag{8.1.8}$$

Equations (8.1.4) can be written

$$\sum_{k=1}^{6} \{[e_j, e_k]\}\, \dot{e}_k = \frac{\partial R}{\partial e_j} \qquad (j = 1, 2, \ldots, 6) \tag{8.1.9}$$

A 6×6 array of Lagrange brackets corresponds to the coefficients of the $\dot{e}_k$ in (8.1.9) but, in view of the properties expressed in (8.1.7), only the 15 brackets above the diagonal need to be computed. In the next section, a convenient formula for evaluating a Lagrange bracket will be derived.

8.2. CAMPBELL'S FORMULA

Consider a two-body orbit for which $\phi = -\mu r^{-1}$ is the potential of the acceleration field $\ddot{\mathbf{r}}$ at the point $\mathbf{r}$, so that

$$\ddot{\mathbf{r}} = -\nabla\phi = -\frac{\partial\phi}{\partial\xi}\mathbf{i} - \frac{\partial\phi}{\partial\eta}\mathbf{j} - \frac{\partial\phi}{\partial\zeta}\mathbf{k}$$

In component form, we have the three equations

$$\ddot{A} = -\frac{\partial\phi}{\partial A} \qquad (A = \xi, \eta, \zeta) \tag{8.2.1}$$

If $\bar{T}$ represents the kinetic energy per unit mass, we can write

$$\bar{T} = \frac{1}{2}v^2 = \frac{\mu}{2p^*}f^2 = \frac{\mu}{p^*}\left(\frac{1}{2} + \frac{e^2}{2} + e\cos w\right) \tag{8.2.2}$$

The constant α of (2.7.2) is given by

$$\begin{aligned}\alpha = \bar{T} + \phi &= \frac{\mu}{p^*}\left(\frac{1}{2} + \frac{e^2}{2} + e\cos w\right) - \frac{\mu}{p^*}(1 + e\cos w)\\ &= -\frac{\mu(1 - e^2)}{2p^*}\end{aligned} \tag{8.2.3}$$

Equation (8.2.3) is a restatement of a previous result given by (2.7.11).

Let P, Q be any two of the orbital elements, or any two independent functions of the orbital elements. It follows from (8.2.3) that

$$\frac{\partial\bar{T}}{\partial P} = \frac{\partial\alpha}{\partial P} - \frac{\partial\phi}{\partial P}$$

Since ϕ is a function of ξ, η, ζ, we can use (8.2.1) and write

$$\frac{\partial\bar{T}}{\partial P} = \frac{\partial\alpha}{\partial P} + \ddot{\xi}\frac{\partial\xi}{\partial P} + \ddot{\eta}\frac{\partial\eta}{\partial P} + \ddot{\zeta}\frac{\partial\zeta}{\partial P}$$

The alternative expression

$$\frac{\partial\bar{T}}{\partial P} = \dot{\xi}\frac{\partial\dot{\xi}}{\partial P} + \dot{\eta}\frac{\partial\dot{\eta}}{\partial P} + \dot{\zeta}\frac{\partial\dot{\zeta}}{\partial P}$$

follows from (8.2.2). If the two expressions for $\partial\bar{T}/\partial P$ are added, we obtain the equation

$$2\frac{\partial\bar{T}}{\partial P} = \frac{\partial\alpha}{\partial P} + \frac{d}{dt}\left(\dot{\xi}\frac{\partial\xi}{\partial P} + \dot{\eta}\frac{\partial\eta}{\partial P} + \dot{\zeta}\frac{\partial\zeta}{\partial P}\right) \tag{8.2.4}$$

If (8.2.4) is integrated between limits extending from a time of pericentric passage T to some instant of time t, we have

$$2\int_T^t \frac{\partial \bar{T}}{\partial P}\,dt = \frac{\partial \alpha}{\partial P}(t - T) + \left[\dot{\xi}\frac{\partial \xi}{\partial P} + \dot{\eta}\frac{\partial \eta}{\partial P} + \dot{\zeta}\frac{\partial \zeta}{\partial P}\right]_t - \left[\dot{\xi}\frac{\partial \xi}{\partial P} + \dot{\eta}\frac{\partial \eta}{\partial P} + \dot{\zeta}\frac{\partial \zeta}{\partial P}\right]_T \tag{8.2.5}$$

where the subscript on a bracket indicates that the quantity within the bracket is to be evaluated at the subscripted value. Noting the fact that

$$\int_T^t \frac{\partial \bar{T}}{\partial P}\,dt = \frac{\partial}{\partial P}\int_T^t \bar{T}\,dt + \bar{T}(T)\frac{\partial T}{\partial P}$$

we can rearrange the terms in (8.2.5) and write

$$2\frac{\partial}{\partial P}\int_T^t \bar{T}\,dt - t\frac{\partial \alpha}{\partial P} - \left[\dot{\xi}\frac{\partial \xi}{\partial P} + \dot{\eta}\frac{\partial \eta}{\partial P} + \dot{\zeta}\frac{\partial \zeta}{\partial P}\right]_t = C_P \tag{8.2.6}$$

where

$$C_P \equiv -2\bar{T}(T)\frac{\partial T}{\partial P} - T\frac{\partial \alpha}{\partial P} - \left[\dot{\xi}\frac{\partial \xi}{\partial P} + \dot{\eta}\frac{\partial \eta}{\partial P} + \dot{\zeta}\frac{\partial \zeta}{\partial P}\right]_T \tag{8.2.7}$$

It is clear that if the partial derivative of $\bar{T}$ is taken with respect to Q, we obtain the expression

$$2\frac{\partial}{\partial Q}\int_T^t \bar{T}\,dt - t\frac{\partial \alpha}{\partial Q} - \left[\dot{\xi}\frac{\partial \xi}{\partial Q} + \dot{\eta}\frac{\partial \eta}{\partial Q} + \dot{\zeta}\frac{\partial \zeta}{\partial Q}\right]_t = C_Q \tag{8.2.8}$$

where

$$C_Q \equiv -2\bar{T}(T)\frac{\partial T}{\partial Q} - T\frac{\partial \alpha}{\partial Q} - \left[\dot{\xi}\frac{\partial \xi}{\partial Q} + \dot{\eta}\frac{\partial \eta}{\partial Q} + \dot{\zeta}\frac{\partial \zeta}{\partial Q}\right]_T \tag{8.2.9}$$

It follows directly from the definitions of C_P and C_Q that both quantities depend only on the elements and do not contain the time explicitly.

If Equations (8.2.6) and (8.2.8) are differentiated partially with respect to Q and P respectively, we obtain two expressions:

$$2\frac{\partial^2}{\partial Q\,\partial P}\int_T^t \bar{T}\,dt - t\frac{\partial^2 \alpha}{\partial Q\,\partial P} - \frac{\partial}{\partial Q}\left[\dot{\xi}\frac{\partial \xi}{\partial P} + \dot{\eta}\frac{\partial \eta}{\partial P} + \dot{\zeta}\frac{\partial \zeta}{\partial P}\right]_t = \frac{\partial C_P}{\partial Q}$$

$$2\frac{\partial^2}{\partial P\,\partial Q}\int_T^t \bar{T}\,dt - t\frac{\partial^2 \alpha}{\partial P\,\partial Q} - \frac{\partial}{\partial P}\left[\dot{\xi}\frac{\partial \xi}{\partial Q} + \dot{\eta}\frac{\partial \eta}{\partial Q} + \dot{\zeta}\frac{\partial \zeta}{\partial Q}\right]_t = \frac{\partial C_Q}{\partial P}$$

that, when subtracted, yield the equation

$$[P, Q] = \frac{\partial C_Q}{\partial P} - \frac{\partial C_P}{\partial Q} \tag{8.2.10}$$

Since both C_P and C_Q are free of the time explicitly, we have shown that the Lagrange bracket of any two of the orbital elements, or any two independent functions of the orbital elements, does not contain the time explicitly; that is,

$$\frac{\partial}{\partial t} [P, Q] = 0 \tag{8.2.11}$$

We now wish to express C_P and C_Q in forms useful for calculation. Consequently, we introduce the notation

$$\Big[\sum_{\xi,\eta,\zeta} \dot{\xi} \frac{\partial \xi}{\partial P}\Big]_T = \Big[\dot{\xi} \frac{\partial \xi}{\partial P} + \dot{\eta} \frac{\partial \eta}{\partial P} + \dot{\zeta} \frac{\partial \zeta}{\partial P}\Big]_T$$

and note that, after the coordinates (ξ, η, ζ) and the velocity $(\dot{\xi}, \dot{\eta}, \dot{\zeta})$ are expressed in terms of the double-primed quantities ξ'', $\dot{\xi}''$, etc., and the properties of the matrix D given in (7.3.8), (7.3.9), and (7.3.11) are invoked, we can write

$$\Big[\sum_{\xi,\eta,\zeta} \dot{\xi} \frac{\partial \xi}{\partial P}\Big]_T = \Big[\dot{\eta}'' \frac{\partial \eta''}{\partial P} + h\mathbf{i}'' \cdot \frac{\partial \mathbf{j}''}{\partial P}\Big]_T \tag{8.2.12}$$

In forming (8.2.12), we have made use of the facts that $\dot{\xi}''(T) = 0$ and $h = \xi''\dot{\eta}'' - \eta''\dot{\xi}''$.

It is a straightforward matter to show that

$$\mathbf{i}'' \cdot \frac{\partial \mathbf{j}''}{\partial P} = -\frac{\partial \omega}{\partial P} - \cos i \frac{\partial \Omega}{\partial P} \tag{8.2.13}$$

where P is an orbital element or any function of the six orbital elements. Thus we can write

$$\Big[\sum_{\xi,\eta,\zeta} \dot{\xi} \frac{\partial \xi}{\partial P}\Big]_T = \Big[\dot{\eta}'' \frac{\partial \eta''}{\partial P} + h \cos i \frac{\partial \Omega}{\partial P} + h \frac{\partial \omega}{\partial P}\Big]_T \tag{8.2.14}$$

To evaluate the first term in the bracket on the right-hand side of (8.2.14), it is noted that

$$\dot{\eta}''(T) = \left(\frac{\mu}{p^*}\right)^{1/2} (1 + e)$$

$$\frac{\partial \eta''(T)}{\partial P} = \Big[\frac{\partial}{\partial P} (r \sin w)\Big]_T = \frac{p^*}{1 + e} \Big[\frac{\partial w}{\partial P}\Big]_T$$

The factor $[\partial w/\partial P]_T$ can be expressed in terms of $\partial T/\partial P$ by applying Leibnitz' rule to $\int_0^w r^2\, d\phi$. We have

$$r^2 \frac{\partial w}{\partial P} + 2\int_0^w r \frac{\partial r}{\partial P}\, d\phi = \frac{1}{2}\left(\frac{\mu}{p^*}\right)^{1/2} (t-T)\frac{\partial p^*}{\partial P} - (\mu p^*)^{1/2}\frac{\partial T}{\partial P}$$

whence, evaluating at time of pericentric passage,

$$\left[\frac{\partial w}{\partial P}\right]_T = -\frac{(1+e)^2\,\mu^{1/2}}{p^{*3/2}}\frac{\partial T}{\partial P}$$

Thus we have

$$\left[\dot{\eta}'' \frac{\partial \eta''}{\partial P}\right]_T = -\frac{\mu}{p^*}(1+e)^2\frac{\partial T}{\partial P}$$

When the kinetic energy per unit mass is evaluated at pericenter, where $w = 0$, we have

$$2\bar{T}(T) = \frac{\mu}{p^*}(1+e)^2$$

Hence (8.2.14) can be written as

$$\left[\sum_{\xi,\eta,\zeta} \dot{\xi}\frac{\partial \xi}{\partial P}\right]_T = -2\bar{T}(T)\frac{\partial T}{\partial P} + h\cos i\frac{\partial \Omega}{\partial P} + h\frac{\partial \omega}{\partial P} \tag{8.2.15}$$

In a similar fashion, we obtain the equation

$$\left[\sum_{\xi,\eta,\zeta} \dot{\xi}\frac{\partial \xi}{\partial Q}\right]_T = -2\bar{T}(T)\frac{\partial T}{\partial Q} + h\cos i\frac{\partial \Omega}{\partial Q} + h\frac{\partial \omega}{\partial Q} \tag{8.2.16}$$

When Equations (8.2.15) and (8.2.16) are substituted into the definitions of C_P and C_Q, we obtain the expressions

$$C_P = -T\frac{\partial \alpha}{\partial P} - h\cos i\frac{\partial \Omega}{\partial P} - h\frac{\partial \omega}{\partial P} \tag{8.2.17}$$

$$C_Q = -T\frac{\partial \alpha}{\partial Q} - h\cos i\frac{\partial \Omega}{\partial Q} - h\frac{\partial \omega}{\partial Q} \tag{8.2.18}$$

If (8.2.17) is differentiated partially with respect to Q and subtracted from the equation obtained by differentiating (8.2.18) partially with respect to P, we obtain the expression

$$[P,Q] = \frac{\partial(-T)}{\partial P}\frac{\partial \alpha}{\partial Q} - \frac{\partial(-T)}{\partial Q}\frac{\partial \alpha}{\partial P} + \frac{\partial \omega}{\partial P}\frac{\partial h}{\partial Q} - \frac{\partial \omega}{\partial Q}\frac{\partial h}{\partial P}$$
$$+ \frac{\partial \Omega}{\partial P}\frac{\partial}{\partial Q}(h\cos i) - \frac{\partial \Omega}{\partial Q}\frac{\partial}{\partial P}(h\cos i) \tag{8.2.19}$$

If (8.2.19) is expressed in Jacobian notation and (8.2.3) is used, we obtain the final formula

$$[P, Q] = \frac{\partial[-T, -\mu(1-e^2)/2p^*]}{\partial(P, Q)} + \frac{\partial(\omega, h)}{\partial(P, Q)} + \frac{\partial(\Omega, h\cos i)}{\partial(P, Q)} \tag{8.2.20}$$

8.3. THE LAGRANGE PLANETARY EQUATIONS FOR VARIOUS SETS OF ORBITAL ELEMENTS

In calculating a Lagrange bracket, using (8.2.20), it is important to note that if a parameter of one of the Jacobians is free of both of the elements with reference to which the bracket is taken, the contribution of that Jacobian to the bracket is zero. Thus, by inspection, it is clear that $[\omega, \Omega] = 0$. With respect to the set of orbital elements $\omega, \Omega, i, p^*, e, T$, the five nonvanishing brackets above the diagonal are

$$\begin{gathered}
[\omega, p^*] = \frac{1}{2}\left(\frac{\mu}{p^*}\right)^{1/2}, \qquad [\Omega, p^*] = \left[\frac{\mu}{p^*}\right]^{1/2}\frac{\cos i}{2} \\
[\Omega, i] = -h\sin i, \qquad [p^*, T] = \frac{\mu(1-e^2)}{2p^{*2}} \\
[e, T] = \frac{\mu e}{p^*}
\end{gathered} \tag{8.3.1}$$

Equations (8.1.9) can then be written in the form of the following array, where the 36 Lagrange brackets are arranged in the six columns to the left of the equal signs:

$$\begin{array}{ccccccc}
\dot{\omega} & \dot{\Omega} & (i)^{\cdot} & \dot{p}^* & \dot{e} & \dot{T} & \\
0 & 0 & 0 & \frac{1}{2}\left(\frac{\mu}{p^*}\right)^{1/2} & 0 & 0 & = \frac{\partial R}{\partial \omega} \\
0 & 0 & -h\sin i & \left(\frac{\mu}{p^*}\right)^{1/2}\frac{\cos i}{2} & 0 & 0 & = \frac{\partial R}{\partial \Omega} \\
0 & h\sin i & 0 & 0 & 0 & 0 & = \frac{\partial R}{\partial i} \\
-\frac{1}{2}\left(\frac{\mu}{p^*}\right)^{1/2} & -\left(\frac{\mu}{p^*}\right)^{1/2}\frac{\cos i}{2} & 0 & 0 & 0 & \frac{\mu(1-e^2)}{2p^{*2}} & = \frac{\partial R}{\partial p^*} \\
0 & 0 & 0 & 0 & 0 & \frac{\mu e}{p^*} & = \frac{\partial R}{\partial e} \\
0 & 0 & 0 & -\frac{\mu(1-e^2)}{2p^{*2}} & -\frac{\mu e}{p^*} & 0 & = \frac{\partial R}{\partial T}
\end{array} \tag{8.3.2}$$

To obtain a given equation, simply multiply the Lagrange bracket appearing in a given row and column by the corresponding time derivative appearing at the head of the column in which the bracket appears, sum such products over the entire row and set the sum of these products equal to the partial derivative appearing in the same row.

The Lagrange planetary equations corresponding to the array given in (8.3.2) for the elements ω, Ω, i, p^*, e, T are

$$\begin{aligned}
\dot{\omega} &= -2\left(\frac{p^*}{\mu}\right)^{1/2}\frac{\partial R}{\partial p^*} + \frac{(1-e^2)}{eh}\frac{\partial R}{\partial e} - (\cos i)\dot{\Omega} \\
\dot{p}^* &= 2\left(\frac{p^*}{\mu}\right)^{1/2}\frac{\partial R}{\partial \omega} \\
\dot{\Omega} &= \frac{1}{h\sin i}\frac{\partial R}{\partial i} \\
\dot{e} &= -\frac{(1-e^2)}{eh}\frac{\partial R}{\partial \omega} - \frac{p^*}{\mu e}\frac{\partial R}{\partial T} \\
(i)^{\cdot} &= -\frac{1}{h\sin i}\frac{\partial R}{\partial \Omega} + \frac{\cot i}{h}\frac{\partial R}{\partial \omega} \\
\dot{T} &= \frac{p^*}{\mu e}\frac{\partial R}{\partial e}
\end{aligned} \tag{8.3.3}$$

If the element p^* is replaced by the element a, we can introduce $p^* = a(1-e^2)$ into (8.2.20) to obtain the modified formula

$$[P, Q] = \frac{\partial(-T, -\mu/2a)}{\partial(P, Q)} + \frac{\partial(\Omega, h\cos i)}{\partial(P, Q)} + \frac{\partial(\omega, h)}{\partial(P, Q)} \tag{8.3.4}$$

It can be readily established that there are only six nonvanishing Lagrange brackets above the diagonal for the set of orbital elements a, e, T, ω, Ω, i. They are

$$\begin{aligned}
[\omega, a] &= \frac{h}{2a}, & [\Omega, e] &= -\left(\frac{\mu}{p^*}\right)^{1/2} ae\cos i \\
[\omega, e] &= -\left(\frac{\mu}{p^*}\right)^{1/2} ae, & [\Omega, i] &= -h\sin i \\
[\Omega, a] &= \frac{h\cos i}{2a}, & [a, T] &= \frac{\mu}{2a^2}
\end{aligned} \tag{8.3.5}$$

The corresponding array for these orbital elements takes the form

$$\begin{array}{cccccccc}
\dot{\omega} & \dot{\Omega} & (i)^{\cdot} & \dot{a} & \dot{e} & \dot{T} & & \\
0 & 0 & 0 & \dfrac{h}{2a} & -\left(\dfrac{\mu}{p^*}\right)^{1/2} ae & 0 & = & \dfrac{\partial R}{\partial \omega} \\
0 & 0 & -h \sin i & \dfrac{h \cos i}{2a} & -\left(\dfrac{\mu}{p^*}\right)^{1/2} ae \cos i & 0 & = & \dfrac{\partial R}{\partial \Omega} \\
0 & h \sin i & 0 & 0 & 0 & 0 & = & \dfrac{\partial R}{\partial i} \\
-\dfrac{h}{2a} & -\dfrac{h \cos i}{2a} & 0 & 0 & 0 & \dfrac{\mu}{2a^2} & = & \dfrac{\partial R}{\partial a} \\
\left(\dfrac{\mu}{p^*}\right)^{1/2} ae & \left(\dfrac{\mu}{p^*}\right)^{1/2} ae \cos i & 0 & 0 & 0 & 0 & = & \dfrac{\partial R}{\partial e} \\
0 & 0 & 0 & -\dfrac{\mu}{2a^2} & 0 & 0 & = & \dfrac{\partial R}{\partial T}
\end{array} \tag{8.3.6}$$

and the Lagrange planetary equations for ω, Ω, i, a, e, T are

$$\begin{aligned}
\dot{\omega} &= \frac{1}{ae}\left(\frac{p^*}{\mu}\right)^{1/2} \frac{\partial R}{\partial e} - (\cos i)\dot{\Omega}, & \dot{a} &= -\frac{2a^2}{\mu}\frac{\partial R}{\partial T} \\
\dot{\Omega} &= \frac{1}{h \sin i}\frac{\partial R}{\partial i}, & \dot{e} &= -\frac{1}{ae}\left(\frac{p^*}{\mu}\right)^{1/2} \frac{\partial R}{\partial \omega} - \frac{p^*}{\mu e}\frac{\partial R}{\partial T} \\
(i)^{\cdot} &= \frac{-1}{h \sin i}\frac{\partial R}{\partial \Omega} + \frac{\cot i}{h}\frac{\partial R}{\partial \omega}, & \dot{T} &= \frac{2a^2}{\mu}\frac{\partial R}{\partial a} + \frac{p^*}{\mu e}\frac{\partial R}{\partial e}
\end{aligned} \tag{8.3.7}$$

It is often convenient to replace the element T by one of the anomalies, say the mean anomaly M. Then, noting that the partial derivatives in Campbell's formula are to be taken at constant t and that $-T = (M/n) - t$, we have

$$P, Q: \qquad \frac{\partial(M/n - t)}{\partial} = \frac{\partial(M/n)}{\partial}$$

Thus the formula for use with the elements ω, Ω, i, a, e, M can be written

$$[P, Q] = \frac{\partial(M/n, -\mu/2a)}{\partial(P, Q)} + \frac{\partial(\Omega, h \cos i)}{\partial(P, Q)} + \frac{\partial(\omega, h)}{\partial(P, Q)} \tag{8.3.8}$$

The brackets can readily be computed using (8.3.8). It is found that the six nonvanishing brackets above the diagonal are

$$[\omega, a] = \frac{h}{2a}, \qquad [\omega, e] = -\left(\frac{\mu}{p^*}\right)^{1/2} ae$$

$$[\Omega, i] = -h \sin i, \qquad [\Omega, a] = \frac{h \cos i}{2a} \tag{8.3.9}$$

$$[\Omega, e] = -\left(\frac{\mu}{p^*}\right)^{1/2} ae \cos i, \qquad [a, M] = -\frac{\mu}{2na^2}$$

Before writing the array for the elements ω, Ω, i, a, e, M, we note that the time appears explicitly in M. Thus the functional dependence of $\mathbf{r}$ has the form

$$\mathbf{r} = \mathbf{r}(a, e, i, \omega, \Omega, M)$$

and t appears in $\mathbf{r}$ only through M. If the total time derivative of $\mathbf{r}$ is formed, we have

$$\dot{\mathbf{r}} = \frac{\partial \mathbf{r}}{\partial a}\dot{a} + \frac{\partial \mathbf{r}}{\partial e}\dot{e} + \frac{\partial \mathbf{r}}{\partial i}(i)^{\cdot} + \frac{\partial \mathbf{r}}{\partial \omega}\dot{\omega} + \frac{\partial \mathbf{r}}{\partial \Omega}\dot{\Omega} + \frac{\partial \mathbf{r}}{\partial M}\dot{M}$$

and the expression for $\dot{\mathbf{r}}$ is not in the proper general form of Lagrange's equations (7.2.9). It can be put in the proper form, however, by the addition of the well-chosen zero

$$\frac{\partial \mathbf{r}}{\partial t} - \frac{\partial \mathbf{r}}{\partial M}\frac{\partial M}{\partial t} = 0$$

which is simply a statement of the chain rule for partial derivatives. Since $\partial M/\partial t = n$, we can write

$$\dot{\mathbf{r}} = \frac{\partial \mathbf{r}}{\partial t} + \frac{\partial \mathbf{r}}{\partial a}\dot{a} + \frac{\partial \mathbf{r}}{\partial e}\dot{e} + \frac{\partial \mathbf{r}}{\partial i}(i)^{\cdot} + \frac{\partial \mathbf{r}}{\partial \omega}\dot{\omega} + \frac{\partial \mathbf{r}}{\partial \Omega}\dot{\Omega} + \frac{\partial \mathbf{r}}{\partial M}(\dot{M} - n)$$

We can now set the sum of the last six terms equal to zero, as before, to obtain three of the Lagrange equations. A similar procedure holds for getting the proper form of (7.2.13). Thus the array for these orbital elements has the form

$$
\begin{array}{cccccccc}
\dot{\omega} & \dot{\Omega} & (i)^{\cdot} & \dot{a} & \dot{e} & \dot{M}-n & & \\
0 & 0 & 0 & \dfrac{h}{2a} & -\left(\dfrac{\mu}{p^*}\right)^{1/2} ae & 0 & = & \dfrac{\partial R}{\partial \omega} \\
0 & 0 & -h\sin i & \dfrac{h\cos i}{2a} & -\left(\dfrac{\mu}{p^*}\right)^{1/2} ae\cos i & 0 & = & \dfrac{\partial R}{\partial \Omega} \\
0 & h\sin i & 0 & 0 & 0 & 0 & = & \dfrac{\partial R}{\partial i} \\
-\dfrac{h}{2a} & -\dfrac{h\cos i}{2a} & 0 & 0 & 0 & -\dfrac{\mu}{2na^2} & = & \left(\dfrac{\partial R}{\partial a}\right)_M \\
\left(\dfrac{\mu}{p^*}\right)^{1/2} ae & \left(\dfrac{\mu}{p^*}\right)^{1/2} ae\cos i & 0 & 0 & 0 & 0 & = & \dfrac{\partial R}{\partial e} \\
0 & 0 & 0 & \dfrac{\mu}{2na^2} & 0 & 0 & = & \dfrac{\partial R}{\partial M}
\end{array}
\tag{8.3.10}
$$

where the symbol $(\partial R/\partial a)_M$ means that the partial derivative is to be taken at constant M. The Lagrange planetary equations for the variables ω, Ω, i, a, e, M are

$$
\begin{aligned}
\dot{\omega} &= \frac{h}{\mu ae}\frac{\partial R}{\partial e} - (\cos i)\dot{\Omega}, &\qquad \dot{a} &= \frac{2na^2}{\mu}\frac{\partial R}{\partial M} \\
\dot{\Omega} &= \frac{1}{h\sin i}\frac{\partial R}{\partial i}, &\qquad \dot{e} &= -\frac{1}{ae}\left(\frac{p^*}{\mu}\right)^{1/2}\frac{\partial R}{\partial \omega} + \frac{np^*}{\mu e}\frac{\partial R}{\partial M} \\
(i)^{\cdot} &= -\frac{1}{h\sin i}\frac{\partial R}{\partial \Omega} + \frac{\cot i}{h}\frac{\partial R}{\partial \omega}, &\qquad \dot{M} &= n - \frac{np^*}{\mu e}\frac{\partial R}{\partial e} - \frac{2na^2}{\mu}\left(\frac{\partial R}{\partial a}\right)_M
\end{aligned}
\tag{8.3.11}
$$

8.4. RELATIONSHIPS BETWEEN PARTIAL DERIVATIVES OF *R* AND ACCELERATION COMPONENTS

It is clear that for a general perturbing acceleration $\mathbf{F}$, not necessarily conservative, Equations (8.1.9) can be written in the form

$$\sum_{k=1}^{6} \{[e_j, e_k]\}\, \dot{e}_k = \mathbf{F}\cdot\frac{\partial \mathbf{r}}{\partial e_j} \qquad (j = 1, 2, \ldots, 6)$$

Thus an alternative derivation of the Lagrange planetary equations obtained in Chapter 7 for general perturbing forces can be made if, instead of using the partial derivative notation $\partial R/\partial e_j$, $j = 1, 2, \ldots, 6$, we develop appropriate expressions for the dot products $\mathbf{F} \cdot \partial \mathbf{r}/\partial e_j$, in terms of the orbital elements and the acceleration components. A convenient way to do this is to relate the partial derivatives $\partial R/\partial e_j$ to the dot products $\mathbf{F} \cdot \partial \mathbf{r}/\partial e_j$ for the case of conservative perturbing forces. The resultant equations obviously apply whether or not the perturbing forces are conservative. To illustrate how this can be done, we consider the case for the orbital elements ω, Ω, i, a, e, T. We will express the partial derivatives in terms of the components R (observe that R is being used in two different senses) and S of the perturbing acceleration.

Thinking of the perturbing potential R as a function of the orbital elements, we write

$$\omega, \Omega, i, a, e, T\colon \qquad \frac{\partial R}{\partial} = \nabla R \cdot \frac{\partial \mathbf{r}}{\partial} = F_\xi \frac{\partial \xi}{\partial} + F_\eta \frac{\partial \eta}{\partial} + F_\zeta \frac{\partial \zeta}{\partial} \tag{8.4.1}$$

where

$$F_\xi = D_{11}F''_\xi + D_{21}F''_\eta + D_{31}F''_\zeta, \qquad \xi = D_{11}\xi'' + D_{21}\eta''$$

$$F_\eta = D_{12}F''_\xi + D_{22}F''_\eta + D_{32}F''_\zeta, \qquad \eta = D_{12}\xi'' + D_{22}\eta''$$

$$F_\zeta = D_{13}F''_\xi + D_{23}F''_\eta + D_{33}F''_\zeta, \qquad \zeta = D_{13}\xi'' + D_{23}\eta''$$

Since the partial derivatives of ξ, η, ζ appearing in (8.4.1) have the form

$$\omega, \Omega, i\colon \qquad \frac{\partial \xi}{\partial} = \xi'' \frac{\partial D_{11}}{\partial} + \eta'' \frac{\partial D_{21}}{\partial}$$

$$\frac{\partial \eta}{\partial} = \xi'' \frac{\partial D_{12}}{\partial} + \eta'' \frac{\partial D_{22}}{\partial}$$

$$\frac{\partial \zeta}{\partial} = \xi'' \frac{\partial D_{13}}{\partial} + \eta'' \frac{\partial D_{23}}{\partial}$$

$$a, e, T\colon \qquad \frac{\partial \xi}{\partial} = D_{11} \frac{\partial \xi''}{\partial} + D_{21} \frac{\partial \eta''}{\partial}$$

$$\frac{\partial \eta}{\partial} = D_{12} \frac{\partial \xi''}{\partial} + D_{22} \frac{\partial \eta''}{\partial}$$

$$\frac{\partial \zeta}{\partial} = D_{13} \frac{\partial \xi''}{\partial} + D_{23} \frac{\partial \eta''}{\partial}$$

we can, invoking the properties of the D matrix given in (7.3.8), (7.3.9), and (7.3.11), express Equations (8.4.1) as

$$a, e, T: \quad \frac{\partial R}{\partial} = \sum_{\xi,\eta,\zeta} \left[(D_{11}F''_\xi + D_{21}F''_\eta + D_{31}F''_\zeta)\left(D_{11}\frac{\partial \xi''}{\partial} + D_{21}\frac{\partial \eta''}{\partial}\right)\right]$$

$$= F''_\xi \frac{\partial \xi''}{\partial} + F''_\eta \frac{\partial \eta''}{\partial} \qquad (8.4.2)$$

$$\omega, \Omega, i: \quad \frac{\partial R}{\partial} = \sum_{\xi,\eta,\zeta} \left[(D_{11}F''_\xi + D_{21}F''_\eta + D_{31}F''_\zeta)\left(\xi''\frac{\partial D_{11}}{\partial} + \eta''\frac{\partial D_{21}}{\partial}\right)\right]$$

$$= (F''_\xi \eta'' - F''_\eta \xi'')\,\mathbf{i}'' \cdot \frac{\partial \mathbf{j}''}{\partial} + F''_\zeta \left(\xi''\mathbf{k}'' \cdot \frac{\partial \mathbf{i}''}{\partial} + \eta''\mathbf{k}'' \cdot \frac{\partial \mathbf{j}''}{\partial}\right)$$

$$= -rS\mathbf{i}'' \cdot \frac{\partial \mathbf{j}''}{\partial} + F''_\zeta \left(\xi''\mathbf{k}'' \cdot \frac{\partial \mathbf{i}''}{\partial} + \eta''\mathbf{k}'' \cdot \frac{\partial \mathbf{j}''}{\partial}\right) \qquad (8.4.3)$$

The partial derivatives of the double-primed variables in (8.4.2) have the form

$$a, e, T: \quad \frac{\partial \xi''}{\partial} = \frac{\partial}{\partial}(r \cos w) = -r \sin w \frac{\partial w}{\partial} + \cos w \frac{\partial r}{\partial}$$

$$\frac{\partial \eta''}{\partial} = \frac{\partial}{\partial}(r \sin w) = r \cos w \frac{\partial w}{\partial} + \sin w \frac{\partial r}{\partial}$$

and, from (7.5.8), we have

$$F''_\xi = R \cos w - S \sin w$$

$$F''_\eta = R \sin w + S \cos w$$

Substituting these results into (8.4.2), we obtain the expression

$$a, e, T: \quad \frac{\partial R}{\partial} = R\frac{\partial r}{\partial} + Sr\frac{\partial w}{\partial} \qquad (8.4.4)$$

Thus, using the expressions (see Exercise 11, Chapter 7) for $\partial r/\partial$ and $\partial w/\partial$ with respect to the elements a, T, e, we obtain the three equations

$$\frac{\partial R}{\partial a} = \left[\frac{r}{a} - \frac{3}{2}\frac{M}{(1-e^2)^{1/2}} e \sin w\right] R - \frac{p^*}{r}\frac{3M}{2(1-e^2)^{1/2}} S \qquad (8.4.5)$$

$$\frac{\partial R}{\partial T} = -\left(\frac{\mu}{p^*}\right)^{1/2} e(\sin w)\, R - \frac{(\mu p^*)^{1/2}}{r} S \qquad (8.4.6)$$

$$\frac{\partial R}{\partial e} = -a(\cos w)\, R + \frac{(r + p^*)}{(1-e^2)}(\sin w) S \qquad (8.4.7)$$

Expressions for the dot products appearing in (8.4.3) are given in (7.3.11) When these expressions are substituted into (8.4.3), we obtain

$$\frac{\partial R}{\partial \omega} = rS \tag{8.4.8}$$

$$\frac{\partial R}{\partial \Omega} = r\,(\cos i)S - r\cos(w + \omega)(\sin i)\,F''_{\zeta} \tag{8.4.9}$$

$$\frac{\partial R}{\partial i} = r[\sin(w + \omega)]\,F''_{\zeta} \tag{8.4.10}$$

If the expressions for the partial derivatives, given in (8.4.5)–(8.4.10), are substituted into (8.3.7), we obtain the Lagrange planetary equations derived in Chapter 7 for the orbital elements ω, Ω, i, a, e, T.

8.5. CANONICAL EQUATIONS

If we define new orbital elements by the relations

$$\begin{aligned} \alpha_1 &= -\mu/2a, \qquad & \beta_1 &= -T \\ \alpha_2 &= h, \qquad & \beta_2 &= \omega \\ \alpha_3 &= h\cos i, \qquad & \beta_3 &= \Omega \end{aligned} \tag{8.5.1}$$

and rewrite (8.2.20) in terms of these new variables, we obtain the formula

$$[P, Q] = \frac{\partial(\beta_1, \alpha_1)}{\partial(P, Q)} + \frac{\partial(\beta_2, \alpha_2)}{\partial(P, Q)} + \frac{\partial(\beta_3, \alpha_3)}{\partial(P, Q)} \tag{8.5.2}$$

If we now interpret $\mathbf{r}$ as a function of the new variables $\alpha_1, \alpha_2, \alpha_3$, $\beta_1, \beta_2, \beta_3$, t, Equations (8.1.2) and (8.1.3) take the form

$$\sum_{i=1}^{3} \frac{\partial \mathbf{r}}{\partial \alpha_i}\dot{\alpha}_i + \sum_{i=1}^{3} \frac{\partial \mathbf{r}}{\partial \beta_i}\dot{\beta}_i = \mathbf{0} \tag{8.5.3}$$

$$\sum_{i=1}^{3} \frac{\partial \dot{\mathbf{r}}}{\partial \alpha_i}\dot{\alpha}_i + \sum_{i=1}^{3} \frac{\partial \dot{\mathbf{r}}}{\partial \beta_i}\dot{\beta}_i = \nabla R \tag{8.5.4}$$

Repeating the procedure used to derive (8.1.4), we obtain the six equations

$$\sum_{i=1}^{3}\left(\frac{\partial \mathbf{r}}{\partial \alpha_j}\cdot\frac{\partial \dot{\mathbf{r}}}{\partial \alpha_i}-\frac{\partial \mathbf{r}}{\partial \alpha_i}\cdot\frac{\partial \dot{\mathbf{r}}}{\partial \alpha_j}\right)\dot{\alpha}_i+\sum_{i=1}^{3}\left(\frac{\partial \mathbf{r}}{\partial \alpha_j}\cdot\frac{\partial \dot{\mathbf{r}}}{\partial \beta_i}-\frac{\partial \mathbf{r}}{\partial \beta_i}\cdot\frac{\partial \dot{\mathbf{r}}}{\partial \alpha_j}\right)\dot{\beta}_i=\nabla R\cdot\frac{\partial \mathbf{r}}{\partial \alpha_j}\qquad (j=1,2,3)$$

$$\sum_{i=1}^{3}\left(\frac{\partial \mathbf{r}}{\partial \beta_j}\cdot\frac{\partial \dot{\mathbf{r}}}{\partial \alpha_i}-\frac{\partial \mathbf{r}}{\partial \alpha_i}\cdot\frac{\partial \dot{\mathbf{r}}}{\partial \beta_j}\right)\dot{\alpha}_i+\sum_{i=1}^{3}\left(\frac{\partial \mathbf{r}}{\partial \beta_j}\cdot\frac{\partial \dot{\mathbf{r}}}{\partial \beta_i}-\frac{\partial \mathbf{r}}{\partial \beta_i}\cdot\frac{\partial \dot{\mathbf{r}}}{\partial \beta_j}\right)\dot{\beta}_i=\nabla R\cdot\frac{\partial \mathbf{r}}{\partial \beta_j}\qquad (j=1,2,3)$$

which, when expressed in the notation of Lagrange brackets, can be written

$$\begin{aligned}\sum_{i=1}^{3}[\alpha_j,\alpha_i]\,\dot{\alpha}_i+\sum_{i=1}^{3}[\alpha_j,\beta_i]\,\dot{\beta}_i&=\frac{\partial R}{\partial \alpha_j}\qquad (j=1,2,3)\\ \sum_{i=1}^{3}[\beta_j,\alpha_i]\,\dot{\alpha}_i+\sum_{i=1}^{3}[\beta_j,\beta_i]\,\dot{\beta}_i&=\frac{\partial R}{\partial \beta_j}\qquad (j=1,2,3)\end{aligned}\tag{8.5.5}$$

From inspection of (8.5.2), it is clear that the only nonvanishing brackets are

$$\begin{aligned}[\alpha_1,\beta_1]&=-[\beta_1,\alpha_1]=-1\\ [\alpha_2,\beta_2]&=-[\beta_2,\alpha_2]=-1\\ [\alpha_3,\beta_3]&=-[\beta_3,\alpha_3]=-1\end{aligned}$$

Thus we are led to the array

$$\begin{array}{cccccccc}\dot{\alpha}_1 & \dot{\alpha}_2 & \dot{\alpha}_3 & \dot{\beta}_1 & \dot{\beta}_2 & \dot{\beta}_3 & & \\ 0 & 0 & 0 & -1 & 0 & 0 & = & \dfrac{\partial R}{\partial \alpha_1}\\ 0 & 0 & 0 & 0 & -1 & 0 & = & \dfrac{\partial R}{\partial \alpha_2}\\ 0 & 0 & 0 & 0 & 0 & -1 & = & \dfrac{\partial R}{\partial \alpha_3}\\ 1 & 0 & 0 & 0 & 0 & 0 & = & \dfrac{\partial R}{\partial \beta_1}\\ 0 & 1 & 0 & 0 & 0 & 0 & = & \dfrac{\partial R}{\partial \beta_2}\\ 0 & 0 & 1 & 0 & 0 & 0 & = & \dfrac{\partial R}{\partial \beta_3}\end{array}$$

The Lagrange planetary equations take the simple form

$$\begin{aligned}
\dot{\alpha}_1 &= \frac{\partial R}{\partial \beta_1}, & \dot{\beta}_1 &= -\frac{\partial R}{\partial \alpha_1} \\
\dot{\alpha}_2 &= \frac{\partial R}{\partial \beta_2}, & \dot{\beta}_2 &= -\frac{\partial R}{\partial \alpha_2} \\
\dot{\alpha}_3 &= \frac{\partial R}{\partial \beta_3}, & \dot{\beta}_3 &= -\frac{\partial R}{\partial \alpha_3}
\end{aligned} \tag{8.5.6}$$

Equations (8.5.6) are called *canonical* equations and the orbital elements α_j, β_j are said to be *canonically conjugate.*

Exercises

1. Show that

$$\mathbf{i}'' \cdot \frac{\partial \mathbf{j}''}{\partial P} = -\cos i \frac{\partial \Omega}{\partial P} - \frac{\partial \omega}{\partial P}$$

where P is an orbital element or any function of the six orbital elements.

2. (a) Obtain a formula for evaluating a Lagrange bracket for use with the orbital elements $\bar{\omega}$, Ω, i, a, e, ϵ, where

$$\bar{\omega} = \omega + \Omega, \qquad \epsilon = \bar{\omega} + n(t_0 - T)$$

(b) Evaluate the Lagrange brackets for this set and exhibit the 6×6 determinant of the coefficients of the $\dot{e}_k$, $k = 1, 2, \ldots, 6$.

(c) Write out the corresponding Lagrange planetary equations.

(d) Show that the determinant obtained in (b) is equal to $(\mu^3 a e^2 \sin^2 i)/4$. Thus the transformation breaks down for $e = 0$ or $i = 0$.

(e) The difficulties for small e appear in the equations for $\dot{e}$ and $\dot{\bar{\omega}}$. Show that if we set $a_1 = e \sin \bar{\omega}$, $b_1 = e \cos \bar{\omega}$, we can replace the equations for $\dot{e}$ and $\dot{\bar{\omega}}$ by the equations

$$\dot{a}_1 = \frac{(1 - e^2)^{1/2}}{na^2} \frac{\partial R}{\partial b_1} - \frac{a_1(1 - e^2)^{1/2}}{na^2[1 + (1 - e^2)^{1/2}]} \frac{\partial R}{\partial \epsilon} + \frac{b_1 \tan(i/2)}{na^2(1 - e^2)^{1/2}} \frac{\partial R}{\partial i}$$

$$\dot{b}_1 = \frac{-(1 - e^2)^{1/2}}{na^2} \frac{\partial R}{\partial a_1} - \frac{b_1(1 - e^2)^{1/2}}{na^2[1 + (1 - e^2)^{1/2}]} \frac{\partial R}{\partial \epsilon} - \frac{a_1 \tan(i/2)}{na^2(1 - e^2)^{1/2}} \frac{\partial R}{\partial i}$$

(f) The difficulties for small i appear in the equations for $(i)^{\cdot}$ and $\dot{\Omega}$. Show that if we set $a_2 = \tan i \sin \Omega$, $b_2 = \tan i \cos \Omega$, we can replace the equations for $(i)^{\cdot}$ and $\dot{\Omega}$ by the equations

$$\dot{a}_2 = \frac{1}{na^2(1-e^2)^{1/2}\cos^3 i}\frac{\partial R}{\partial b_2} - \frac{a_2}{2na^2(1-e^2)^{1/2}(\cos i)\cos^2(i/2)}\left[\frac{\partial R}{\partial \epsilon} + \frac{\partial R}{\partial \bar{\omega}}\right]$$

$$\dot{b}_2 = \frac{-1}{na^2(1-e^2)^{1/2}\cos^3 i}\frac{\partial R}{\partial a_2} - \frac{b_2}{2na^2(1-e^2)^{1/2}(\cos i)\cos^2(i/2)}\left[\frac{\partial R}{\partial \epsilon} + \frac{\partial R}{\partial \bar{\omega}}\right]$$

3. Neglect harmonics in the Earth's gravitational potential Φ higher than the second. Then we can approximate Φ by the expression

$$\Phi = -\frac{\mu}{\rho}\left[1 - \left(\frac{1}{\rho^2}\right)\frac{J_2(3\sin^2\delta - 1)}{2}\right] \qquad (\rho = \text{distance from origin})$$

in our usual notation. Here J_2 is a positive constant. The disturbing function R is then

$$R = -\frac{\mu}{2\rho^3}J_2(3\sin^2\delta - 1)$$

If we expand R in a Fourier series in terms of the mean anomaly M over the interval $0 < M < 2\pi$, the constant term in the expansion is that part of R, say R_s, which is responsible for the secular perturbations in the orbital elements of a close-Earth satellite.

(a) Show that

$$R_s = -\frac{\mu J_2(\frac{3}{2}\sin^2 i - 1)}{2a^3(1-e^2)^{3/2}}$$

(b) Show that, for the assumed perturbing function, there are no secular perturbations in a, e, and i. [See Y. Kozai. The Motion of a Close-Earth Satellite, *Astron. J.* **64**, 367 (1959).]

4. Verify Equation (8.2.12).
5. Verify the calculation of the Lagrange brackets given in (8.3.1).
6. Verify the calculation of the Lagrange brackets given in (8.3.5).
7. Verify the calculation of the Lagrange brackets given in (8.3.9).
8. Complete the verification begun in the text that, for the orbital elements ω, Ω, i, a, e, and M, $\dot{e}_6$ in Equation (7.2.13) must have the

form $\dot{e}_6 = \dot{M} - n$ when used in the Lagrange planetary equations.

9. Show that, where $R = R(\omega, \Omega, i, a, e, M)$,

$$\dot{w} = \frac{(\mu p^*)^{1/2}}{r^2}\Big[1 - \frac{2a^2}{\mu}\Big(\frac{\partial R}{\partial a}\Big)_M - \frac{p^*}{\mu e}\frac{\partial R}{\partial e} - \frac{r(r+p^*)\sin w}{\mu p^* e}\frac{\partial R}{\partial \omega} + \frac{r(r+p^*)\sin w}{\mu e(ap^*)^{1/2}}\frac{\partial R}{\partial M}\Big]$$

$$\dot{E} = \frac{an}{r}\Big[1 - \frac{p^*}{\mu e}\frac{\partial R}{\partial e} - \frac{2a^2}{\mu}\Big(\frac{\partial R}{\partial a}\Big)_M - \frac{(ap^*)^{1/2}}{\mu e}\sin E\frac{\partial R}{\partial \omega} + \frac{p^*\sin E}{\mu e}\frac{\partial R}{\partial M}\Big]$$

10. Show that, in terms of the components T and N of the perturbing acceleration,

$$\dot{w} = \frac{(\mu p^*)^{1/2}}{r^2} + \frac{2a\sin w}{r(\mu p^*)^{1/2}}\Big[\frac{(r+p^*)(\cos w + e)}{f} - \frac{p^*}{e}\Big(f - \frac{e\cos w}{f}\Big)\Big]T + \frac{1}{fr(\mu p^*)^{1/2}}\Big[\frac{p^{*2}\cos w}{e} + r(r+p^*)\sin^2 w\Big]N$$

$$\dot{E} = \frac{an}{r}\Big[1 - \frac{2\sin E}{nef}\Big(\frac{p^*}{\mu}\Big)^{1/2}T + \frac{(1-e^2)^{1/2}}{nf}\Big(\frac{p^*}{\mu}\Big)^{1/2}\Big(\sin^2 E + \frac{\cos E - e}{e}\Big)N\Big]$$

11. Show that, where $R = R(\omega, \Omega, i, a, e, E)$,

$$\dot{E} = \frac{na}{r} - \frac{2}{nr}\Big[\Big(\frac{\partial R}{\partial a}\Big)_E + \frac{h\sin E}{2na^3e}\frac{\partial R}{\partial \omega} + \frac{p^*}{2a^2e}\Big(\frac{\partial R}{\partial e}\Big)_E\Big]$$

Chapter 9

Lagrange and Hamilton Mechanics

9.1. CONSTRAINTS

We saw in Equation (2.2.10) that for constant masses the equations of motion for the *i*th of *n* particles, in motion with respect to an origin at one of them, namely, m_0, can be written

$$m_i\ddot{\mathbf{r}}_i = \sum_{j=0}^{n-1} \mathbf{f}_{ij} - \frac{m_i}{m_0}\sum_{j=1}^{n-1} \mathbf{f}_{0j} + \mathbf{F}_i^E - \frac{m_i}{m_0}\mathbf{F}_0^E \qquad (i \neq 0, \mathbf{f}_{ii} = 0)$$

where $\mathbf{f}_{ij}$ is the force on the *i*th particle produced by the *j*th particle along the line of centers and $\mathbf{F}_i^E$ includes all other forces acting upon the *i*th particle. If the symbol $\mathbf{F}_i$ represents the entire force on the *i*th particle, we have

$$\mathbf{F}_i = \sum_{j=0}^{n-1} \mathbf{f}_{ij} - \frac{m_i}{m_0}\sum_{j=1}^{n-1} \mathbf{f}_{0j} + \mathbf{F}_i^E - \frac{m_i}{m_0}\mathbf{F}_0^E \qquad (9.1.1)$$

whence we can write

$$m_i\ddot{\mathbf{r}}_i = \mathbf{F}_i \qquad (9.1.2)$$

Now, in addition to forces of a gravitational nature, the dynamical system may be subject to certain *constraints* that limit its freedom of motion. For example, if the system of n particles is confined to a closed container, the walls of the container impose the obvious constraint that the n particles must remain inside the container. Another example is found in the motion of a bead on a wire. Here the bead is constrained to follow the contour of the wire. A constraint is said to be *holonomic* if the constraint can be expressed as an equation between the coordinates and the time in the form

$$f(\mathbf{r}_1, \mathbf{r}_2, \mathbf{r}_3, \ldots, t) = 0$$

Otherwise the constraint is said to be *nonholonomic*. As an example of a holonomic constraint, we have the case of a particle confined to move on the surface of a smooth sphere of radius a. The equation of constraint is

$$r^2 = x^2 + y^2 + z^2 = a^2$$

The constraint would be nonholonomic if, in a gravitational field, the particle were placed at the top of the sphere and permitted to slide down and eventually fall off. Here the constraint is not expressed as an equation, but rather as an inequality

$$r^2 \geqslant a^2$$

If the constraints contain the time explicitly, they are said to be *rheonomous*; if not, they are said to be *scleronomous*. Thus we can speak of constraints as being holonomic-scleronomous, holonomic-rheonomous, nonholonomic-scleronomous, or nonholonomic-rheonomous. In the following discussion, we restrict ourselves to a consideration of holonomic constraints only.

When constraints are introduced into a dynamical problem, it means that the coordinates of the particles are no longer independent, nor are the equations of motion. We speak of a system of n particles, free of constraints, as having $3n$ degrees of freedom; that is, 3 degrees of freedom correspond to the 3 independent coordinates of each particle. If, however, there are k equations of holonomic constraint, we can eliminate k of the $3n$ coordinates and have $r = 3n - k$ independent coordinates remaining; the system is then said to have r degrees of freedom. We can thus introduce r new independent variables $q_1, q_2, \ldots, q_r$, into the problem and express the coordinates in terms of these new independent variables in the form

$$\begin{aligned}
\mathbf{r}_1 &= \mathbf{r}_1(q_1, q_2, \ldots, q_r, t) \\
\mathbf{r}_2 &= \mathbf{r}_2(q_1, q_2, \ldots, q_r, t) \\
&\vdots \\
\mathbf{r}_n &= \mathbf{r}_n(q_1, q_2, \ldots, q_r, t)
\end{aligned} \tag{9.1.3}$$

The q's introduced here are called *generalized coordinates*.

To illustrate, consider the two-dimensional motion of a particle of mass m constrained to move under gravity $\mathbf{g}$ on the surface of a smooth wire in the form of a semicircle of radius a (see Figure 9.1.1). The Newtonian form of the equations of motion is

$$m\ddot{x} = -R\frac{x}{a} \tag{9.1.4a}$$

$$m\ddot{y} = -mg - \frac{Ry}{a} \qquad (g = |\,\mathbf{g}\,|) \tag{9.1.4b}$$

where R is the magnitude of the force $\mathbf{R}$ exerted by the wire on the particle whose coordinates x and y satisfy the constraint relations

$$x = a\sin\theta, \qquad y = -a\cos\theta \tag{9.1.5}$$

and hence satisfy the equation

$$x^2 + y^2 = a^2 \tag{9.1.6}$$

The constraint is holonomic-scleronomous and the angle θ is the generalized coordinate. The reaction $\mathbf{R}$ can be eliminated between Equations (9.1.4a) and (9.1.4b). This procedure yields the equation

$$\ddot{y} = -g + \frac{y}{x}\ddot{x} \tag{9.1.7}$$

If Equations (9.1.5) are twice differentiated with respect to time, we obtain the expressions

$$\ddot{x} = (a\cos\theta)\ddot{\theta} - (a\sin\theta)\dot{\theta}^2, \qquad \ddot{y} = (a\sin\theta)\ddot{\theta} + (a\cos\theta)\dot{\theta}^2$$

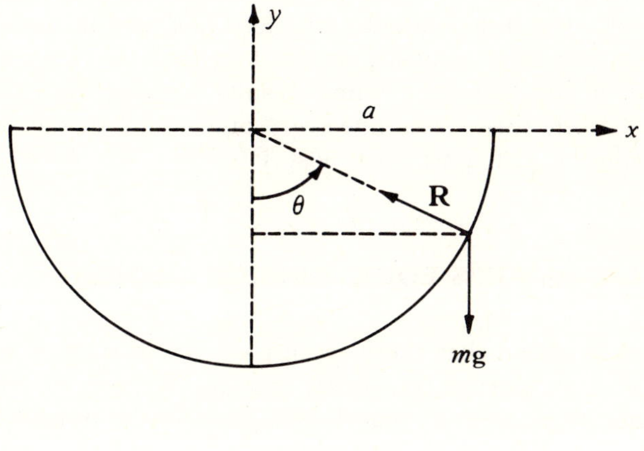

FIGURE 9.1.1

which, when substituted into (9.1.7) in conjunction with (9.1.5), lead to the equation

$$\ddot{\theta} = -\frac{g}{a}\sin\theta \tag{9.1.8}$$

Thus, starting with the two coordinates x and y, we introduce a new generalized coordinate θ through the equation of constraint. If (9.1.8) is solved for θ, we can then express x and y in terms of the time by substituting the solution $\theta(t)$ in (9.1.5).

We have, of course, seen another example of the use of generalized coordinates in the treatment of the unperturbed two-body orbit in which no constraints were involved. We were able to express the coordinates $\mathbf{r}$ in terms of the six independent orbital elements and the time. Thus the q's must not be thought of as coordinates only in the ordinary sense of Cartesian coordinates. They may, of course, be Cartesian coordinates, but many other parameters, such as energy and angular momentum, can also appear as generalized coordinates.

A further difficulty is introduced into a dynamical problem when constraints are involved. This difficulty arises from the fact that although the equations of constraint are known explicitly [Equations (9.1.5) in our example], the actual forces of constraint ($\mathbf{R}$ in our example) are not known in advance of the investigation. These unknown constraint forces have to be obtained after we obtain the solution. Thus, in the formulation of the dynamical problem, it is helpful to eliminate the forces of constraint so that we need deal only with the known applied forces. This can be accomplished in a formulation due to Lagrange, which we next discuss. The time derivatives of generalized coordinates are called *generalized velocities*, and it is a feature of the Lagrangian formulation that the generalized coordinates and generalized velocities are treated as independent variables. The justification for this procedure lies, in the final analysis, in the fact that the Lagrangian formulation leads to precisely the same results as the Newtonian formulation. Apart from this view, however, one could postulate the Lagrangian formulation just as we have postulated the Newtonian formulation.

9.2. LAGRANGE'S EQUATIONS OF MOTION

It is assumed that there are n particles in motion with respect to the origin of an inertial coordinate system. We assume further that $3n - r$ holonomic constraints may be involved in the dynamical system so that there are r degrees of freedom, $r \leqslant 3n$. Thus the coordinates $\mathbf{r}_i$ can be represented in the form

$$\mathbf{r}_i = \mathbf{r}_i(q_1, q_2, \ldots, q_r, t) \qquad (i = 1, 2, \ldots, n) \tag{9.2.1}$$

If the dot product of (1.2.1)—interpreted as Newton's second law for the ith particle—with $\partial \mathbf{r}_i/\partial q_j$ is formed, where q_j is one of the generalized coordinates $q_1, q_2, \ldots, q_r$, we can write

$$m_i \ddot{\mathbf{r}}_i \cdot \frac{\partial \mathbf{r}_i}{\partial q_j} = \mathbf{F}_i \cdot \frac{\partial \mathbf{r}_i}{\partial q_j} \tag{9.2.2}$$

The time derivative of (9.2.1) is

$$\dot{\mathbf{r}}_i = \frac{\partial \mathbf{r}_i}{\partial t} + \sum_{j=1}^{r} \frac{\partial \mathbf{r}_i}{\partial q_j} \dot{q}_j$$

Hence, treating the generalized coordinates and generalized velocities as independent and noting that neither $\partial \mathbf{r}_i/\partial t$ nor $\partial \mathbf{r}_i/\partial q_j$, $j = 1, 2, \ldots, r$, contains the velocities, we have

$$\frac{\partial \dot{\mathbf{r}}_i}{\partial \dot{q}_j} = \frac{\partial \mathbf{r}_i}{\partial q_j} \tag{9.2.3}$$

Substituting (9.2.3) into (9.2.2), we obtain

$$m_i \ddot{\mathbf{r}}_i \cdot \frac{\partial \dot{\mathbf{r}}_i}{\partial \dot{q}_j} = \mathbf{F}_i \cdot \frac{\partial \mathbf{r}_i}{\partial q_j} \tag{9.2.4}$$

In view of (9.2.3), we note that

$$\frac{d}{dt}\left(\dot{\mathbf{r}}_i \cdot \frac{\partial \dot{\mathbf{r}}_i}{\partial \dot{q}_j}\right) - \dot{\mathbf{r}}_i \cdot \frac{d}{dt}\left(\frac{\partial \mathbf{r}_i}{\partial q_j}\right) = \ddot{\mathbf{r}}_i \cdot \frac{\partial \dot{\mathbf{r}}_i}{\partial \dot{q}_j} \tag{9.2.5}$$

which can readily be verified. Thus we rewrite (9.2.4) as

$$m_i \frac{d}{dt}\left(\dot{\mathbf{r}}_i \cdot \frac{\partial \dot{\mathbf{r}}_i}{\partial \dot{q}_j}\right) - m_i \dot{\mathbf{r}}_i \cdot \frac{d}{dt}\left(\frac{\partial \mathbf{r}_i}{\partial q_j}\right) = \mathbf{F}_i \cdot \frac{\partial \mathbf{r}_i}{\partial q_j} \tag{9.2.6}$$

If $\mathbf{r}_i$ is such a function of the coordinates and time that the order of partial differentiation may be interchanged, we can write

$$\begin{aligned}
\frac{d}{dt}\left(\frac{\partial \mathbf{r}_i}{\partial q_j}\right) &= \frac{\partial}{\partial q_1}\left(\frac{\partial \mathbf{r}_i}{\partial q_j}\right)\dot{q}_1 + \frac{\partial}{\partial q_2}\left(\frac{\partial \mathbf{r}_i}{\partial q_j}\right)\dot{q}_2 + \cdots + \frac{\partial}{\partial q_r}\left(\frac{\partial \mathbf{r}_i}{\partial q_j}\right)\dot{q}_r + \frac{\partial}{\partial t}\left(\frac{\partial \mathbf{r}_i}{\partial q_j}\right) \\
&= \frac{\partial}{\partial q_j}\left(\frac{\partial \mathbf{r}_i}{\partial q_1}\dot{q}_1 + \frac{\partial \mathbf{r}_i}{\partial q_2}\dot{q}_2 + \cdots + \frac{\partial \mathbf{r}_i}{\partial q_r}\dot{q}_r + \frac{\partial \mathbf{r}_i}{\partial t}\right) \\
&= \frac{\partial \dot{\mathbf{r}}_i}{\partial q_j}
\end{aligned}$$

Thus (9.2.6) can be expressed in the form

$$\frac{d}{dt}\left(\frac{\partial}{\partial \dot{q}_j} \frac{m_i \dot{\mathbf{r}}_i \cdot \dot{\mathbf{r}}_i}{2}\right) - \frac{\partial}{\partial q_j}\left(\frac{m_i \dot{\mathbf{r}}_i \cdot \dot{\mathbf{r}}_i}{2}\right) = \mathbf{F}_i \cdot \frac{\partial \mathbf{r}_i}{\partial q_j} \tag{9.2.7}$$

Since

$$T_i = \tfrac{1}{2} m_i \dot{\mathbf{r}}_i \cdot \dot{\mathbf{r}}_i \tag{9.2.8}$$

where T_i is the kinetic energy of the ith particle referenced to the origin O of an inertial coordinate system, we have

$$\frac{d}{dt}\left(\frac{\partial T_i}{\partial \dot{q}_j}\right) - \frac{\partial T_i}{\partial q_j} = \mathbf{F}_i \cdot \frac{\partial \mathbf{r}_i}{\partial q_j} \tag{9.2.9}$$

If (9.2.9) is summed over the n particles, which are assumed to move about O, we have

$$\frac{d}{dt}\left(\frac{\partial T_1}{\partial \dot{q}_j}\right) + \frac{d}{dt}\left(\frac{\partial T_2}{\partial \dot{q}_j}\right) + \cdots + \frac{d}{dt}\left(\frac{\partial T_n}{\partial \dot{q}_j}\right)$$
$$- \left(\frac{\partial T_1}{\partial q_j} + \frac{\partial T_2}{\partial q_j} + \cdots + \frac{\partial T_n}{\partial q_j}\right) = \sum_{i=1}^{n} \mathbf{F}_i \cdot \frac{\partial \mathbf{r}_i}{\partial q_j}$$

or

$$\frac{d}{dt}\left(\frac{\partial T}{\partial \dot{q}_j}\right) - \frac{\partial T}{\partial q_j} = Q_j \qquad (j = 1, 2, \ldots, r) \tag{9.2.10}$$

where there is one such equation for each j and

$T =$ total kinetic energy of the system of n particles referenced to the origin O of the inertial system

$$Q_j \equiv \sum_{i=1}^{n} \mathbf{F}_i \cdot \frac{\partial \mathbf{r}_i}{\partial q_j}$$

The Q_j are called the components of the generalized force.

As pointed out earlier, the q_j do not have to have the dimensions of length, but it is to be noted that $Q_j\, dq_j$ must have the dimensions of work. This can readily be seen if we write

$$Q_j\, dq_j = \sum_{i=1}^{n} \mathbf{F}_i \cdot \frac{\partial \mathbf{r}_i}{\partial q_j}\, dq_j = \sum_{i=1}^{n} \mathbf{F}_i \cdot \left(d\mathbf{r}_i - \frac{\partial \mathbf{r}_i}{\partial t}\, dt - \cdots\right) \tag{9.2.11}$$

and note that $\mathbf{F}_i \cdot d\mathbf{r}_i$ has the dimensions of work.

When the forces are derivable from a scalar function of the coordinates, $V(\mathbf{r}_1, \mathbf{r}_2, \ldots, \mathbf{r}_n)$, where V is the potential energy of the system, we can write

$$Q_j = \sum_{i=1}^{n} (-\nabla_i V) \cdot \frac{\partial \mathbf{r}_i}{\partial q_j} \qquad (-\nabla_i V = \mathbf{F}_i) \quad (j = 1, 2, \ldots, r)$$

We can also write

$$Q_j = -\frac{\partial V}{\partial q_j} \qquad (j = 1, 2, \ldots, r) \tag{9.2.12}$$

since

$$\frac{\partial V}{\partial q_j} = \frac{\partial V}{\partial \xi_1}\frac{\partial \xi_1}{\partial q_j} + \frac{\partial V}{\partial \eta_1}\frac{\partial \eta_1}{\partial q_j} + \frac{\partial V}{\partial \zeta_1}\frac{\partial \zeta_1}{\partial q_j} + \cdots + \frac{\partial V}{\partial \zeta_n}\frac{\partial \zeta_n}{\partial q_j}$$

$$= \nabla_1 V \cdot \frac{\partial \mathbf{r}_1}{\partial q_j} + \cdots + \nabla_n V \cdot \frac{\partial \mathbf{r}_n}{\partial q_j} = \sum_{i=1}^{n} \nabla_i V \cdot \frac{\partial \mathbf{r}_i}{\partial q_j}$$

If (9.2.12) is substituted into (9.2.10), we obtain the equations

$$\frac{d}{dt}\left(\frac{\partial T}{\partial \dot{q}_j}\right) - \frac{\partial}{\partial q_j}(T - V) = 0 \qquad (j = 1, 2, \ldots, r)$$

Now if V is a function of the coordinates alone, $\partial V/\partial \dot{q}_j = 0$, and we have

$$\frac{d}{dt}\left[\frac{\partial}{\partial \dot{q}_j}(T - V)\right] - \frac{\partial}{\partial q_j}(T - V) = 0 \qquad (j = 1, 2, \ldots, r) \tag{9.2.13}$$

Equations (9.2.13) could still be written even if we relax our demand that V be a function of the coordinates alone. In fact, the only demand we need make of V is that it satisfy the conditions

$$\frac{d}{dt}\left(\frac{\partial V}{\partial \dot{q}_j}\right) - \frac{\partial V}{\partial q_j} = Q_j \qquad (j = 1, 2, \ldots, r) \tag{9.2.14}$$

and (9.2.13) will still be true. Of course, if V is a function of the coordinates alone, then (9.2.14) is clearly satisfied. But, more generally, we must allow the possibility that V may contain the velocities and the time. The form of (9.2.13) suggests that we define a new function L such that

$$\frac{d}{dt}\left(\frac{\partial L}{\partial \dot{q}_j}\right) - \frac{\partial L}{\partial q_j} = 0 \qquad (j = 1, 2, \ldots, r) \tag{9.2.15}$$

Thus L is whatever function is required to yield the correct equations of motion (i.e., Newton's second law) for the dynamical system under consideration. Equations (9.2.15) are called the Lagrange equations of motion and the function L is called the *Lagrangian function* or sometimes the ***kinetic potential***. For most dynamical systems, L will have the form

$$L(q_j\,;\dot{q}_j\,;t) = T(q_j\,;\dot{q}_j\,;t) - V(q_j\,;\dot{q}_j\,;t) \tag{9.2.16}$$

where, to abbreviate the notation, the convention is introduced that if f is to be thought of as a function of the variables q_j, $\dot{q}_j$, and t, $j = 1, 2, \ldots, r$, we write $f(q_j\,;\dot{q}_j\,;t)$. Also, we will use the notation $(q_j\,;\dot{q}_j\,;t)$ to refer to the set of variables $q_1, q_2, \ldots, q_r, \dot{q}_1, \dot{q}_2, \ldots, \dot{q}_r, t$.

The way in which the kinetic energy T depends on the generalized coordinates and velocities can be deduced by noting first that the rectangular coordinates of the n particles have the form

$$A_i = A_i(q_j\,;t) \qquad (A_i = \xi_i, \eta_i, \zeta_i\,;i = 1, 2, \ldots, n)$$

Thus the velocities are given by

$$\dot{A}_i = \frac{\partial A_i}{\partial t} + \sum_{j=1}^{r} \frac{\partial A_i}{\partial q_j}\dot{q}_j \qquad (A_i = \xi_i, \eta_i, \zeta_i)$$

Introducing the notation

$$\sum_{\xi,\eta,\zeta} \dot{\xi}_i^2 \equiv \dot{\xi}_i^2 + \dot{\eta}_i^2 + \dot{\zeta}_i^2$$

the kinetic energy T_i of the ith particle can be written

$$T_i = \frac{m_i}{2}\sum_{\xi,\eta,\zeta}\left[\left(\frac{\partial \xi_i}{\partial t}\right)^2 + 2\frac{\partial \xi_i}{\partial t}\sum_{j=1}^{r}\frac{\partial \xi_i}{\partial q_j}\dot{q}_j + \left(\sum_{j=1}^{r}\frac{\partial \xi_i}{\partial q_j}\dot{q}_j\right)^2\right]$$

If we designate by

$$T_{2i} = \frac{m_i}{2}\sum_{\xi,\eta,\zeta}\left(\sum_{j=1}^{r}\frac{\partial \xi_i}{\partial q_j}\dot{q}_j\right)^2$$ an homogeneous quadratic function of the $\dot{q}_j$ that also involves the q_j and t

$$T_{1i} = m_i\sum_{\xi,\eta,\zeta}\frac{\partial \xi_i}{\partial t}\sum_{j=1}^{r}\frac{\partial \xi_i}{\partial q_j}\dot{q}_j$$ an homogeneous linear function of the $\dot{q}_j$ that also involves the q_j and t

$$T_{0i} = \frac{m_i}{2}\sum_{\xi,\eta,\zeta}\left(\frac{\partial \xi_i}{\partial t}\right)^2$$ a term free of the $\dot{q}_j$ that contains the q_j and t

we can write the total kinetic energy T in the form

$$T = T_2 + T_1 + T_0 \tag{9.2.17}$$

where

$T_2 = \sum_{i=1}^{n} T_{2i}$ is an homogeneous quadratic function of the $\dot{q}_j$ that also involves the q_j and t

$T_1 = \sum_{i=1}^{n} T_{1i}$ is an homogeneous linear function of the $\dot{q}_j$ that also involves the q_j and t

$T_0 = \sum_{i=1}^{n} T_{0i}$ a term free of $\dot{q}_j$ that contains the q_j and t

In particular, it is clear that if the time does not appear explicitly in A_i, $i = 1, 2, \ldots, n$, then T_1 and T_0 are zero and T is an homogeneous quadratic function of the generalized velocities. Thus Euler's theorem on homogeneous functions permits us to write

$$2T(\dot{q}_j) = \sum_{j=1}^{r} \dot{q}_j \frac{\partial T}{\partial \dot{q}_j} \tag{9.2.18}$$

Example 9.2.1. A particle of mass m is moving in the $\xi''\eta''$–plane (see Figure 9.2.1) subject to an inverse square attractive force. It is desired to find the equations of motion by the Lagrangian method.

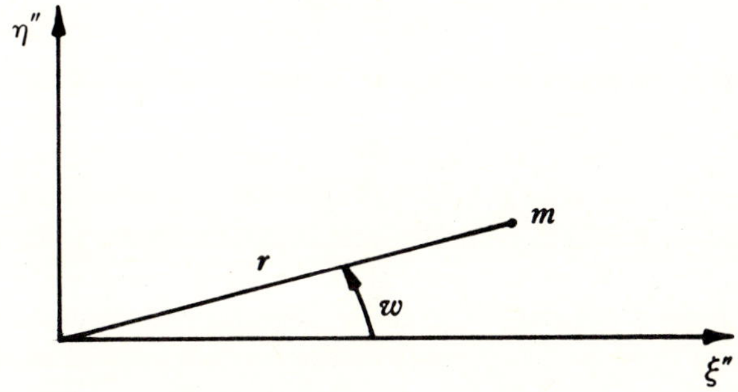

FIGURE 9.2.1

The kinetic and potential energies are given by the expressions

$$2T = m[(\dot{\xi}'')^2 + (\dot{\eta}'')^2]$$

$$V = -\frac{\mu m}{r} \qquad [r^2 = (\xi'')^2 + (\eta'')^2]$$

Since

$$\xi'' = r \cos w, \qquad \eta'' = r \sin w$$

and

$$\dot{\xi}'' = \dot{r} \cos w - (r \sin w)\dot{w}, \qquad \dot{\eta}'' = \dot{r} \sin w + (r \cos w)\dot{w}$$

it follows that

$$2T = m(\dot{r}^2 + r^2\dot{w}^2)$$

Hence the Lagrangian function is

$$L = T - V = \frac{m}{2}(\dot{r}^2 + r^2\dot{w}^2) + \frac{\mu m}{r}$$

Here the generalized coordinates are (r, w) and the generalized velocities are $(\dot{r}, \dot{w})$. The equations of motion for the generalized coordinates, given by (9.2.15), are readily found to be

$$\ddot{r} - r\dot{w}^2 = -\frac{\mu}{r^2}, \qquad r^2\dot{w} = \text{constant}$$

in agreement with the results already obtained for the two-body problem. The particle has only 2 deg of freedom due to the introduction of the holonomic-scleronomous constraint

$$\zeta'' = 0$$

9.3. THE HAMILTONIAN FUNCTION; HAMILTON'S CANONICAL EQUATIONS

We have seen that in the Lagrangian procedure for a system of n particles referenced to an inertial origin and subject to k constraints, we use $r = 3n - k$ generalized coordinates. For such a system, the $3n$ equations corresponding to Newton's second law are replaced by r second-order differential equations of motion for the r generalized coordinates. We also notice that, for conservative systems, the generalized velocities $\dot{q}$ appear in L only through the kinetic energy T. Now we observe that the magnitude of the linear momentum p_i of the ith particle may be obtained by differentiating its kinetic energy $T_i = \frac{1}{2}m_i v_i^2$ partially with respect to v_i. That is,

$$p_i = \frac{\partial T_i}{\partial v_i} = m_i v_i \tag{9.3.1}$$

By analogy with (9.3.1), we are led to define a quantity p_j, called the *generalized momentum conjugate to the generalized coordinate* q_j, by the equation

$$p_j = \frac{\partial L}{\partial \dot{q}_j} \tag{9.3.2}$$

It follows from (9.2.15) that if the Lagrangian function is free of a particular coordinate q_s (called an *ignorable coordinate*), then

$$\frac{d}{dt}\left(\frac{\partial L}{\partial \dot{q}_s}\right) = \dot{p}_s = 0$$

and the corresponding momentum p_s is a constant of the motion.

In the Lagrangian procedure, the generalized coordinates q_j and generalized velocities $\dot{q}_j$ are employed as a joint set of independent coordinates. We now introduce another procedure in which the generalized coordinates and generalized momenta form a joint set of independent coordinates.† In this new scheme, called the Hamiltonian procedure, the system of r second-order differential equations of motion arising in the Lagrangian procedure are replaced by $2r$ first-order differential equations of motion.

To introduce the Hamiltonian procedure, we assume that a Lagrangian function has been determined and write its total differential as

$$dL = \sum_{j=1}^{r}\left(\frac{\partial L}{\partial q_j}\,dq_j + \frac{\partial L}{\partial \dot{q}_j}\,d\dot{q}_j\right) + \frac{\partial L}{\partial t}\,dt \tag{9.3.3}$$

If the equations

$$\frac{d}{dt}\left(\frac{\partial L}{\partial \dot{q}_j}\right) = \frac{\partial L}{\partial q_j} = \dot{p}_j\,, \qquad p_j = \frac{\partial L}{\partial \dot{q}_j}$$

are used, then (9.3.3) can be written in the form

$$dL = \sum_{j=1}^{r}(\dot{p}_j\,dq_j + p_j\,d\dot{q}_j) + \frac{\partial L}{\partial t}\,dt \tag{9.3.4}$$

Since

$$d\left(\sum_{j=1}^{r} p_j\dot{q}_j\right) = \sum_{j=1}^{r} p_j\,d\dot{q}_j + \sum_{j=1}^{r} \dot{q}_j\,dp_j$$

the second sum on the right-hand side of (9.3.4) can be replaced, and we will then have the equation

$$dL = \sum_{j=1}^{r} \dot{p}_j\,dq_j - \sum_{j=1}^{r} \dot{q}_j\,dp_j + \frac{\partial L}{\partial t}\,dt + d\left(\sum_{j=1}^{r} p_j\dot{q}_j\right)$$

† There are two general approaches to the treatment of the time variable t. In the variational approach, one considers variations in the generalized coordinates and momenta at constant t. An alternative approach, and the one that will be used here, is to consider the time t as a variable independent of the q_j and p_j. Both treatments lead to the same equations of motion and this is the only criterion that we must satisfy.

which, when rearranged, can be written

$$d\left(\sum_{j=1}^{r} p_j\dot{q}_j - L\right) = \sum_{j=1}^{r} (\dot{q}_j\, dp_j - \dot{p}_j\, dq_j) - \frac{\partial L}{\partial t}\, dt \tag{9.3.5}$$

We now define a new function H, called the Hamiltonian function, by the equation

$$H(q_j\,;\, p_j\,;\, t) = \sum_{j=1}^{r} p_j\dot{q}_j - L(q_j\,;\,\dot{q}_j\,;\, t) \tag{9.3.6}$$

Hence

$$dH = \sum_{j=1}^{r} (\dot{q}_j\, dp_j - \dot{p}_j\, dq_j) - \frac{\partial L}{\partial t}\, dt \tag{9.3.7}$$

Although the right-hand side of (9.3.6) involves the $\dot{q}_j$, we consider H to be a function of the q_j, p_j, and t in the following sense. In nearly all physical problems, the Lagrangian function L is, at worst, quadratic in the $\dot{q}_j$. Thus, when we use the definition (9.3.2), the p_j are expressible as linear functions of the $\dot{q}_j$, which can be solved to express the $\dot{q}_j$ in terms of the generalized momenta. We can then write H explicitly in terms of the variables $(q_j\,;\, p_j\,;\, t)$. Accordingly, we restrict ourselves to dynamical systems in which it is so possible to write the Hamiltonian function. Since H is a function of the q_j, p_j, and t, its total differential has the form

$$dH = \sum_{j=1}^{r} \frac{\partial H}{\partial q_j}\, dq_j + \sum_{j=1}^{r} \frac{\partial H}{\partial p_j}\, dp_j + \frac{\partial H}{\partial t}\, dt \tag{9.3.8}$$

Equating (9.3.7) and (9.3.8) and noting that the q_j, p_j, and t are to be treated as independent coordinates, we obtain

$$\dot{q}_j = \frac{\partial H}{\partial p_j}, \qquad \dot{p}_j = -\frac{\partial H}{\partial q_j} \quad (j = 1, 2, \ldots, r) \tag{9.3.9}$$

$$\frac{\partial H}{\partial t} = -\frac{\partial L}{\partial t} \tag{9.3.10}$$

Equations (9.3.9) are called *Hamilton's canonical equations.*

Several important properties of the Hamiltonian function are stated in the following three theorems:

Theorem 9.3.1. If L is free of t explicitly, H is free of t explicitly, and conversely.

Proof. If $\partial L/\partial t = 0$,

$$\frac{\partial H}{\partial t} = -\frac{\partial L}{\partial t} = 0$$

and conversely.

Theorem 9.3.2. If H does not contain the time explicitly, it is a constant of the motion, and conversely.

Proof. If $\partial H/\partial t = 0$, we have

$$\dot{H} = \sum_{j=1}^{r} \frac{\partial H}{\partial q_j} \dot{q}_j + \sum_{j=1}^{r} \frac{\partial H}{\partial p_j} \dot{p}_j$$

$$= \sum_{j=1}^{r} (-\dot{p}_j \dot{q}_j + \dot{q}_j \dot{p}_j) = 0$$

and H is a constant. On the other hand, since $\dot{H} = \partial H/\partial t$, if $\dot{H} = 0$ it must be free of t explicitly.

Theorem 9.3.3. If the system contains the $\dot{q}_j$ only through T and T is a homogeneous quadratic function of the $\dot{q}_j$, then $H = 2T - L$. Moreover, if $L = T - V$ then $H = T + V$, the total energy of the system.

Proof. If the system contains the $\dot{q}_j$ only through T, then

$$\frac{\partial L}{\partial \dot{q}_j} = \frac{\partial T}{\partial \dot{q}_j} = p_j$$

If, in addition, T is a homogeneous quadratic function of the $\dot{q}_j$, then

$$H = \sum_{j=1}^{r} p_j \dot{q}_j - L = \sum_{j=1}^{r} \frac{\partial T}{\partial \dot{q}_j} \dot{q}_j - L = 2T - L$$

Example 9.3.1. Find the equations of motion (by the Hamiltonian method) of the particle of Example 9.2.1.

By definition, we write

$$H = p_r \dot{r} + p_w \dot{w} - L$$

where

$$L = \frac{m}{2}(\dot{r}^2 + r^2 \dot{w}^2) + \frac{\mu m}{r}$$

$$p_r = \frac{\partial L}{\partial \dot{r}} = m\dot{r}, \qquad p_w = \frac{\partial L}{\partial \dot{w}} = mr^2 \dot{w}$$

Replacing the generalized velocities in H by the generalized momenta, we have

$$H = \frac{1}{2m}\left(p_r^{\,2} + \frac{p_w^{\,2}}{r^2}\right) - \frac{\mu m}{r}$$

The canonical equations of motion are

$$\dot{r} = \frac{\partial H}{\partial p_r} = \frac{p_r}{m}, \qquad \dot{p}_r = -\frac{\partial H}{\partial r} = \frac{p_w^{\,2}}{mr^3} - \frac{\mu m}{r^2}$$

$$\dot{w} = \frac{\partial H}{\partial p_w} = \frac{p_w}{mr^2}, \qquad \dot{p}_w = -\frac{\partial H}{\partial w} = 0$$

These equations, clearly, are in agreement with the Lagrangian equations of motion, for we can replace p_r and p_w in the $\dot{p}_r$ and $\dot{p}_w$ equations by the generalized velocities and obtain

$$\ddot{r} - r\dot{w}^2 = -\frac{\mu}{r^2}, \qquad r^2\dot{w} = \text{constant}$$

as before. Observe that we could also have formed the Hamiltonian function ($H = T + V$) by using Theorem 9.3.3.

It is obvious from (9.3.9) that if H is free of the coordinate q_j, the corresponding conjugate momentum p_j is a constant of the motion. Such a coordinate is said to be *ignorable,* or *cyclic*. In fact, if all the position coordinates are cyclic (e.g., $H = p^2/2$ for the unbound particle of unit mass, moving with velocity $\mathbf{p}$), it follows from Hamilton's canonical equations that

$$\dot{q}_j = \frac{\partial H(p_j\,;\,t)}{\partial p_j} = f_j(p_j\,;\,t) \qquad (j = 1, 2,\ldots, r)$$

Thus each $\dot{q}_j$ is a function f_j of the constants p_j and the time t. If, in particular, H does not contain the time explicitly, then the $\dot{q}_j$ are constants and hence readily integrable to yield the coordinates q_j as linear functions of the time.

We now show that even if H contains the time explicitly, it can be replaced by a new Hamiltonian H^*, which does not contain t explicitly. Furthermore, this new Hamiltonian H^* yields exactly the same canonical equations for determining q_j and p_j as did the original H. The penalty we pay is that we must add one degree of freedom to the problem.

If H were free of time, its total derivative with respect to time would be

$$\frac{dH}{dt} = \sum_{j=1}^{r} \frac{\partial H}{\partial q_j}\dot{q}_j + \frac{\partial H}{\partial p_j}\dot{p}_j = 0$$

If, however, H contains the time explicitly, its total derivative with respect to time must be written

$$\frac{dH}{dt} = \sum_{j=1}^{r} \left(\frac{\partial H}{\partial q_j} \dot{q}_j + \frac{\partial H}{\partial p_j} \dot{p}_j \right) + \frac{\partial H}{\partial t} = \frac{\partial H}{\partial t}$$

Since $\dot{p}_j = -\partial H/\partial q_j$, the following scheme is suggested.

Interpret t wherever it appears in H as a generalized coordinate q_{r+1} and form a new Hamiltonian function H^*, defined by the equation

$$H^* = H(p_1, p_2, \ldots, p_r, q_1, q_2, \ldots, q_r, q_{r+1}) + p_{r+1} \tag{9.3.11}$$

where we interpret p_{r+1} as a momentum canonically conjugate to q_{r+1} and determined by the equation

$$\dot{p}_{r+1} = -\frac{\partial H}{\partial q_{r+1}} = -\frac{\partial H}{\partial t}$$

Then

$$\dot{H}^* = \dot{H} + \dot{p}_{r+1} = \frac{\partial H}{\partial t} - \frac{\partial H}{\partial t} = 0$$

and H^* is a constant of the motion.

Because of the way H^* is defined, the canonical equations for the q_j and p_j, $j = 1, 2, \ldots, r$, remain the same in terms of H^* as they are in terms of H. We have, in fact,

$$\begin{aligned} \dot{q}_j &= \frac{\partial H^*}{\partial p_j} = \frac{\partial H}{\partial p_j} \qquad (j = 1, 2, \ldots, r) \\ \dot{p}_j &= -\frac{\partial H^*}{\partial q_j} = -\frac{\partial H}{\partial q_j} \qquad (j = 1, 2, \ldots, r) \end{aligned} \tag{9.3.12}$$

and the two additional equations

$$\begin{aligned} \dot{q}_{r+1} &= \frac{\partial H^*}{\partial p_{r+1}} = 1 \\ \dot{p}_{r+1} &= -\frac{\partial H^*}{\partial q_{r+1}} = -\frac{\partial H}{\partial q_{r+1}} = -\frac{\partial H}{\partial t} \end{aligned} \tag{9.3.13}$$

Clearly (9.3.12) and (9.3.13) can be combined and written

$$\begin{aligned} \dot{q}_j &= \frac{\partial H^*}{\partial p_j} \qquad (j = 1, 2, \ldots, r+1) \\ \dot{p}_j &= -\frac{\partial H^*}{\partial q_j} \qquad (j = 1, 2, \ldots, r+1) \end{aligned} \tag{9.3.14}$$

Exercises

1. A mass m swings in a vertical plane on the end of an inextensible string of length l. Find its equation of motion.
2. A mass m swings in a vertical plane on the end of an inextensible string of length l. Show that the tension P of the string at any inclination θ to the vertical is

$$P = mg \cos \theta + ml\dot{\theta}^2$$

3. In a dynamical system with 2 degrees of freedom, the kinetic energy is

$$T = \frac{\dot{q}_1^2}{2(a + bq_2)} + \frac{1}{2} q_2^2 \dot{q}_2^2$$

and the potential energy is

$$V = c + dq_2$$

where a, b, c, d are constants. Show that the value of q_2 in terms of the time is given by an equation of the form

$$(q_2 - k)(q_2 + 2k)^2 = h(t - t_0)^2$$

where h, k, and t_0 are constants.†

4. The Lagrangian procedure is simplified considerably if the kinetic energy T can be expressed in the form

$$2T = Q_1\dot{q}_1^2 + Q_2\dot{q}_2^2 + Q_3\dot{q}_3^2$$

where the Q_i, $i = 1, 2, 3$, are functions of the coordinates. Show that this will be the case if orthogonal curvilinear coordinates are used to describe the motion.

5. Use the Lagrangian procedure to find the motion of a particle of unit mass descending on a helix under the influence of gravity, if the helix is constrained to turn about its axis, which is vertical, with a uniform angular velocity ω. Find the components of the reaction of the helix on the particle.
6. A particle of mass m moves on the inner surface of a cone of revolution, whose semivertical angle is α, under the action of a repulsive force $\mu m/r^3$ from the axis; the angular momentum of the particle about the

† Exercises 3 and 6 are reprinted from E. T. Whittaker, "A Treatise on the Analytical Dynamics of Particles and Rigid Bodies," 4th ed., pp. 69 and 116. Copyright 1944 (Am. ed.) by Cambridge Univ. Press, London and New York. Used with permission of Cambridge Univ. Press.

axis being $m(\mu \tan \alpha)^{1/2}$. Use the Lagrangian procedure to prove that the path of the particle is the arc of a hyperbola whose eccentricity is $\sec \alpha$.

7. A mass m is suspended by a spring of negligible mass, force constant k, and unstretched length b from a point having a constant upward acceleration a_0. Gravity g is acting vertically downward. Find L and from it the equation of motion in the vertical direction. Determine H and write Hamilton's equations. What is the period of the motion?†

8. A particle of mass m moves in a central field of attractive force of magnitude $(k/r^2)\, e^{-\alpha t}$, where k and α are constants, t is the time, and r is the distance of m from the force center. Find L and H. Is H the total energy? Is it constant?

9. In the nonlinear partial differential of first order

$$G(q_j\,;p_j\,;W) = 0 \tag{1}$$

let W be the dependent variable and let $(q_j\,;p_j \equiv \partial W/\partial q_j)$ be a set of $2r$ elements composed of r independent variables, q_j, $j \equiv 1, 2,\ldots, r$, and r partial derivatives, $\partial W/\partial q_j$, $j \equiv 1, 2,\ldots, r$.
Show that if

$$u = g(q_j\,;W) = 0 \tag{2}$$

is a solution of (1), we have

$$p_j = -\frac{\partial u/\partial q_j}{\partial u/\partial W} \qquad (j = 1, 2,\ldots, r)$$

Thus show that if we introduce the transformation

$$W = q_{r+1}, \qquad P_j = \frac{\partial u}{\partial q_j} \qquad (j = 1, 2,\ldots, r+1) \tag{3}$$

Equation (1) can be transformed into a nonlinear partial differential equation of first order

$$F(q_j\,;P_j) = 0 \qquad (j = 1, 2,\ldots, r+1) \tag{4}$$

which does not contain the dependent variable but does contain one more independent variable. If we can find a solution $u = 0$ of (4), replacement of q_{r+1} by W yields the solution (2) of the original equation (1).

† Exercises 7 and 8 are reprinted from R. A. Becker, "Introduction to Theoretical Mechanics," pp. 339 and 340. Copyright 1954 by McGraw-Hill, New York. Used with permission of McGraw-Hill Book Company.

Chapter 10

Canonical Transformations

10.1. GENERALIZED DEFINITION OF A LAGRANGE BRACKET

In view of the developments in Chapter 9, it is tempting to speculate on the possibility of finding a simultaneous transformation of coordinates and momenta that preserves the canonical form of the equations of motion and yet leads to cyclic variables or, at least, a new set of variables in which the original problem may be handled more simply. To obtain insight into this general question, we consider the properties of transformations of coordinates and momenta that preserve the canonical form. Such transformations are said to be canonical.

We begin by generalizing our definition of the Lagrange bracket. In the definition of the Lagrange bracket given in (8.1.5), we were considering the position vector $\mathbf{r}$ and the velocity vector $\dot{\mathbf{r}}$ for only a single particle without constraints. If we are studying a dynamical system of n particles, subject to k constraints, we have $r = 3n - k$ degrees of freedom. In the Lagrangian formulation, we associate with the system a set of r independent generalized coordinates q_j, $j = 1, 2, \ldots, r$. We then define a Lagrange bracket of P and Q by the equation

$$[P, Q] = \sum_{j=1}^{r} \left(\frac{\partial q_j}{\partial P} \frac{\partial \dot{q}_j}{\partial Q} - \frac{\partial q_j}{\partial Q} \frac{\partial \dot{q}_j}{\partial P} \right) \tag{10.1.1}$$

where P and Q are any two of $2r$ distinct functions of $(q_j\,;\,\dot{q}_j)$. Clearly (10.1.1) embraces our earlier definition for a single particle with coordinates (ξ, η, ζ) and velocity $(\dot{\xi}, \dot{\eta}, \dot{\zeta})$. For this case, (10.1.1) becomes

$$[P, Q] = \sum_{\xi,\eta,\zeta} \left(\frac{\partial \xi}{\partial P} \frac{\partial \dot{\xi}}{\partial Q} - \frac{\partial \xi}{\partial Q} \frac{\partial \dot{\xi}}{\partial P} \right)$$

When the Hamiltonian formulation is employed, we associate a set of generalized coordinates q_j and generalized momenta p_j, $j = 1, 2, \ldots, r$, with the dynamical system and, in this notation, the Lagrange bracket is defined by the equation

$$[P, Q] = \sum_{j=1}^{r} \left(\frac{\partial q_j}{\partial P} \frac{\partial p_j}{\partial Q} - \frac{\partial q_j}{\partial Q} \frac{\partial p_j}{\partial P} \right) \tag{10.1.2}$$

10.2. SUFFICIENCY CRITERIA FOR A CANONICAL TRANSFORMATION

In order to develop criteria that can be used to test to see if a transformation is canonical, we consider a set $(q_j\,;\,p_j)$ of generalized coordinates and momenta satisfying the equations

$$\dot{q}_j = \frac{\partial H(q_j\,;\,p_j\,;\,t)}{\partial p_j}, \qquad \dot{p}_j = -\frac{\partial H(q_j\,;\,p_j\,;\,t)}{\partial q_j} \qquad (j = 1, 2, \ldots, r) \tag{10.2.1}$$

We want to transform to a new set of variables $(Q_k\,;\,P_k)$, which satisfy the equations

$$\dot{Q}_k = \frac{\partial H(Q_k\,;\,P_k\,;\,t)}{\partial P_k}, \qquad \dot{P}_k = -\frac{\partial H(Q_k\,;\,P_k\,;\,t)}{\partial Q_k} \qquad (k = 1, 2, \ldots, r) \tag{10.2.2}$$

It is understood that $H(Q_k\,;\,P_k\,;\,t)$ is to be formed by replacing the variables q_j and p_j in $H(q_j\,;\,p_j\,;\,t)$, using the transformation equations

$$\begin{aligned} q_j &= q_j(Q_k\,;\,P_k\,;\,t) \\ p_j &= p_j(Q_k\,;\,P_k\,;\,t) \end{aligned} \qquad (k = 1, 2, \ldots, r;\ j = 1, 2, \ldots, r) \tag{10.2.3}$$

If Equations (10.2.3) are differentiated with respect to t, we obtain

$$\dot{q}_j = \frac{\partial q_j}{\partial t} + \sum_{k=1}^{r} \frac{\partial q_j}{\partial Q_k} \dot{Q}_k + \sum_{k=1}^{r} \frac{\partial q_j}{\partial P_k} \dot{P}_k = \frac{\partial H(q_j\,;\,p_j\,;\,t)}{\partial p_j} \tag{10.2.4}$$

$$\dot{p}_j = \frac{\partial p_j}{\partial t} + \sum_{k=1}^{r} \frac{\partial p_j}{\partial Q_k} \dot{Q}_k + \sum_{k=1}^{r} \frac{\partial p_j}{\partial P_k} \dot{P}_k = -\frac{\partial H(q_j\,;\,p_j\,;\,t)}{\partial q_j} \tag{10.2.5}$$

Let α represent any element in the set $(Q_k\,;\,P_k)$. Multiply (10.2.4), successively, by $\partial p_j/\partial\alpha$, $j = 1, 2,\ldots, r$, and multiply (10.2.5), successively, by $-\partial q_j/\partial\alpha$, $j = 1, 2,\ldots, r$. Then add the $2r$ products to obtain the equation

$$\sum_{j=1}^{r}\left(\frac{\partial q_j}{\partial t}\frac{\partial p_j}{\partial\alpha} - \frac{\partial p_j}{\partial t}\frac{\partial q_j}{\partial\alpha}\right) + \sum_{k=1}^{r}\sum_{j=1}^{r}\left(\frac{\partial q_j}{\partial Q_k}\frac{\partial p_j}{\partial\alpha} - \frac{\partial p_j}{\partial Q_k}\frac{\partial q_j}{\partial\alpha}\right)\dot{Q}_k$$

$$+ \sum_{k=1}^{r}\sum_{j=1}^{r}\left(\frac{\partial q_j}{\partial P_k}\frac{\partial p_j}{\partial\alpha} - \frac{\partial p_j}{\partial P_k}\frac{\partial q_j}{\partial\alpha}\right)\dot{P}_k = \sum_{j=1}^{r}\left(\frac{\partial H}{\partial p_j}\frac{\partial p_j}{\partial\alpha} + \frac{\partial H}{\partial q_j}\frac{\partial q_j}{\partial\alpha}\right)$$

which can be expressed in Lagrange bracket notation. Noting that t and α are independent variables, we have

$$[t, \alpha] + \sum_{k=1}^{r}[Q_k\,, \alpha]\,\dot{Q}_k + \sum_{k=1}^{r}[P_k\,, \alpha]\,\dot{P}_k = \frac{\partial H}{\partial\alpha} \tag{10.2.6}$$

The transformation will be canonical if, for every k and j, we satisfy the sufficient conditions

$$[t, \alpha] = 0 \tag{10.2.7a}$$

$$[Q_k\,, Q_j] = 0 \tag{10.2.7b}$$

$$[P_k\,, P_j] = 0 \tag{10.2.7c}$$

$$[Q_k\,, P_j] = -[P_k\,, Q_j] = \delta_{kj} \tag{10.2.7d}$$

where δ_{kj} is the Kronecker delta. The canonical equations in the new variables $(Q_k\,;\,P_k)$ are given by (10.2.2).

10.3. DIFFERENTIAL FORMS OF CANONICAL TRANSFORMATIONS

We now wish to show that a transformation $q_j = q_j(Q_k\,;\,P_k\,;\,t)$, $p_j = p_j(Q_k\,;\,P_k\,;\,t)$, such that

$$\sum_{j=1}^{r}(p_j\,dq_j - P_j\,dQ_j) = dW_1(q_j\,;\,Q_j\,;\,t) \tag{10.3.1}$$

where W_1 is an arbitrary function of the $2r + 1$ independent coordinates $(q_j\,;\,Q_j\,;\,t)$ is canonical. Note that even though the q_j are related to the Q_k through the transformation equations, as long as the q_j are also related to the independent quantities P_k we can, for fixed values of the Q_k, vary the q_j by varying the P_k. Hence the q_j and the Q_j can be treated as independent. For example, although $p^* = a(1 - e^2)$, we may consider p^* and e as independent variables. On the other hand, it is quite possible that all, or any one, of the q_j may be dependent upon the Q_j as is illustrated by the extended point transformation discussed later in this chapter.

Using the transformation equations for (10.2.3), we can write

$$dq_j = \sum_{k=1}^{r} \left(\frac{\partial q_j}{\partial Q_k} dQ_k + \frac{\partial q_j}{\partial P_k} dP_k \right) + \frac{\partial q_j}{\partial t} dt \tag{10.3.2}$$

If (10.3.2) is substituted for dq_j in (10.3.1), we can write

$$-\sum_{k=1}^{r} P_k \, dQ_k + \sum_{j=1}^{r} p_j \left[\sum_{k=1}^{r} \left(\frac{\partial q_j}{\partial Q_k} dQ_k + \frac{\partial q_j}{\partial P_k} dP_k \right) + \frac{\partial q_j}{\partial t} dt \right]$$

$$= \sum_{k=1}^{r} \frac{\partial W_1}{\partial Q_k} dQ_k + \sum_{k=1}^{r} \frac{\partial W_1}{\partial P_k} dP_k + \frac{\partial W_1}{\partial t} dt \tag{10.3.3}$$

where, in the right-hand side of (10.3.3), we have used the fact that W_1 is a function of P_j through the equations of transformation. Equating the coefficients in (10.3.3), we obtain the $2r + 1$ equations

$$-P_k + \sum_{l=1}^{r} p_l \frac{\partial q_l}{\partial Q_k} = \frac{\partial W_1}{\partial Q_k} \qquad (k = 1, 2, \ldots, r) \tag{10.3.4a}$$

$$\sum_{l=1}^{r} p_l \frac{\partial q_l}{\partial P_k} = \frac{\partial W_1}{\partial P_k} \qquad (k = 1, 2, \ldots, r) \tag{10.3.4b}$$

$$\sum_{l=1}^{r} p_l \frac{\partial q_l}{\partial t} = \frac{\partial W_1}{\partial t} \tag{10.3.4c}$$

To avoid ambiguity in notation in the following development, the summation index l is used to replace j.

If we form the partial derivative of (10.3.4b) with respect to Q_k and the partial derivative of (10.3.4a) with respect to P_k, we obtain the equations

$$\sum_{l=1}^{r} \frac{\partial p_l}{\partial Q_k} \frac{\partial q_l}{\partial P_k} + \sum_{l=1}^{r} p_l \frac{\partial^2 q_l}{\partial Q_k \, \partial P_k} = \frac{\partial^2 W_1}{\partial Q_k \, \partial P_k}$$

$$-1 + \sum_{l=1}^{r} \frac{\partial p_l}{\partial P_k} \frac{\partial q_l}{\partial Q_k} + \sum_{l=1}^{r} p_l \frac{\partial^2 q_l}{\partial P_k \, \partial Q_k} = \frac{\partial^2 W_1}{\partial Q_k \, \partial P_k}$$

which can be subtracted to yield

$$\sum_{l=1}^{r} \left(\frac{\partial p_l}{\partial P_k} \frac{\partial q_l}{\partial Q_k} - \frac{\partial p_l}{\partial Q_k} \frac{\partial q_l}{\partial P_k} \right) = [Q_k, P_k] = 1 \qquad (k = 1, 2, ..., r) \tag{10.3.5}$$

If the partial derivative of (10.3.4a) with respect to Q_j is formed and if in (10.3.4a) k is changed to j and the partial derivative with respect to Q_k is formed, we obtain the equations

$$\sum_{l=1}^{r} \frac{\partial p_l}{\partial Q_j} \frac{\partial q_l}{\partial Q_k} + \sum_{l=1}^{r} p_l \frac{\partial^2 q_l}{\partial Q_j \, \partial Q_k} = \frac{\partial^2 W_1}{\partial Q_j \, \partial Q_k}$$

$$\sum_{l=1}^{r} \frac{\partial p_l}{\partial Q_k} \frac{\partial q_l}{\partial Q_j} + \sum_{l=1}^{r} p_l \frac{\partial^2 q_l}{\partial Q_k \, \partial Q_j} = \frac{\partial^2 W_1}{\partial Q_k \, \partial Q_j}$$

which, when subtracted, yield

$$\sum_{l=1}^{r} \left(\frac{\partial q_l}{\partial Q_k} \frac{\partial p_l}{\partial Q_j} - \frac{\partial q_l}{\partial Q_j} \frac{\partial p_l}{\partial Q_k} \right) = [Q_k, Q_j] = 0 \qquad (k = 1, 2, ..., r; \; j = 1, 2, ..., r) \tag{10.3.6}$$

Similarly, it can readily be shown that

$$[P_k, P_j] = 0 \qquad (k = 1, 2, ..., r; \; j = 1, 2, ..., r) \tag{10.3.7}$$

$$[Q_k, P_j] = 0 \qquad (k = 1, 2, ..., r; \; j = 1, 2, ..., r; \; k \neq j) \tag{10.3.8}$$

$$[t, Q_k] = [t, P_k] = 0 \qquad (k = 1, 2, ..., r) \tag{10.3.9}$$

Thus we have verified that a transformation of variables

$$q_j = q_j(Q_k \, ; P_k \, ; t), \qquad p_j = p_j(Q_k \, ; P_k \, ; t)$$

such that (10.3.1) is satisfied, is canonical. In similar fashion, it can be shown that a transformation of variables from $(q_j \, ; p_j)$ to $(Q_j \, ; P_j)$ satisfying any one of the three alternative conditions

$$\sum_{j=1}^{r} (p_j \, dq_j + Q_j \, dP_j) = dW_2(q_j \, ; P_j \, ; t) \tag{10.3.10}$$

$$\sum_{j=1}^{r} (q_j \, dp_j + P_j \, dQ_j) = -dW_3(p_j \, ; Q_j \, ; t) \tag{10.3.11}$$

$$\sum_{j=1}^{r} (-q_j \, dp_j + Q_j \, dP_j) = dW_4(p_j \, ; P_j \, ; t) \tag{10.3.12}$$

is also canonical. The left-hand side of each equation is to be an exact differential of some arbitrary function W_i, $i = 2, 3, 4$, of the independent variables which are indicated by the notation.

It is clear that if the transformation equations are free of t, the criterion $[t, \alpha] = 0$ is automatically satisfied. In the proof that the transformation satisfying (10.3.1) is canonical, Equation (10.3.4c) is used only to establish that $[t, \alpha] = 0$. Thus it is clear that the transformations satisfying (10.3.1) and (10.3.10)–(10.3.12), where W_i, $i = 1, 2, 3, 4$, is free of t, are canonical as long as the variables appearing in W_i can be considered independent.

The question now arises as to how a transformation should be chosen to ensure that it is canonical. We discuss this point in connection with choosing a transformation that satisfies the differential condition (10.3.1).

First we suppose that W_1 does not contain the time t explicitly. Then (10.3.1) will be satisfied identically if the transformation equations are

$$p_j = \frac{\partial W_1}{\partial q_j}, \qquad P_j = -\frac{\partial W_1}{\partial Q_j} \qquad (j = 1, 2, \ldots, r) \tag{10.3.13a}$$

since

$$dW_1 = \sum_{j=1}^{r} \frac{\partial W_1}{\partial q_j} dq_j + \sum_{j=1}^{r} \frac{\partial W_1}{\partial Q_j} dQ_j$$

and we can equate coefficients in (10.3.1) to obtain (10.3.13a).

If W_1 contains t explicitly, we cannot equate coefficients in (10.3.1), for the total differential

$$dW_1 = \sum_{j=1}^{r} \frac{\partial W_1}{\partial q_j} dq_j + \sum_{j=1}^{r} \frac{\partial W_1}{\partial Q_j} dQ_j + \frac{\partial W_1}{\partial t} dt$$

contains the extra term $(\partial W_1/\partial t)\, dt$. Suppose, however, we require that the transformation satisfy the modified differential condition

$$\sum_{j=1}^{r} (p_j\, dq_j - P_j\, dQ_j) + \frac{\partial W_1}{\partial t} dt = dW_1(q_j, Q_j, t) \tag{10.3.13b}$$

Then the transformation equations will still have the form given in (10.3.13a), but is the transformation now canonical?

The development proceeds as before in Equations (10.3.2) through (10.3.8) with the exception that (10.3.4c) is replaced by

$$\sum_{j=1}^{r} - p_j \frac{\partial q_j}{\partial t} = 0$$

Thus all of conditions (10.2.7) hold except that for $[t, \alpha]$. If the details are carried out, it becomes clear that the condition $[t, \alpha] = 0$ is replaced by the modified condition

$$[t, \alpha] = -\frac{\partial^2 W_1}{\partial \alpha \, \partial t} \tag{10.3.14}$$

Hence conditions (10.2.7) are not satisfied and we can not conclude that (10.2.2) holds. Returning to (10.2.6), however, we note that if the condition (10.3.14) replaces the condition that $[t, \alpha] = 0$, then (10.2.2) will still hold except that H must be replaced by $K = H + (\partial W_1/\partial t)$. To see this, set $\alpha = P_j$ and Q_j, successively, and use (10.3.14) in (10.2.6). For each j, we can then write the two equations

$$\dot{Q}_j = \frac{\partial H}{\partial P_j} + \frac{\partial^2 W_1}{\partial P_j \, \partial t} = \frac{\partial}{\partial P_j}\left(H + \frac{\partial W_1}{\partial t}\right)$$

$$\dot{P}_j = -\frac{\partial H}{\partial Q_j} - \frac{\partial^2 W_1}{\partial Q_j \, \partial t} = -\frac{\partial}{\partial Q_j}\left(H + \frac{\partial W_1}{\partial t}\right)$$

Thus the transformation given by (10.3.13a), with W_1 a function of the variables $(q_j\,; Q_j\,; t)$, is again canonical if the Hamiltonian function transforms to a new Hamiltonian function K given by

$$K = H + \frac{\partial W_1}{\partial t} \tag{10.3.15}$$

and the new variables satisfy the canonical equations

$$\dot{Q}_j = \frac{\partial K}{\partial P_j}, \qquad \dot{P}_j = -\frac{\partial K}{\partial Q_j} \qquad (j = 1, 2, \ldots, r) \tag{10.3.16}$$

Equations of transformation corresponding to the differential conditions (10.3.10)–(10.3.12) are readily deduced in the same way as were Equations (10.3.13a). In each case, if the generating function W_i, $i = 1, 2, 3, 4$, contains t explicitly, the new Hamiltonian K is given by

$$K = H + \frac{\partial W_i}{\partial t}$$

where K is understood to be expressed in terms of Q_j, P_j, and t. If W_i does not contain t explicitly, the new Hamiltonian function is obtained from the old Hamiltonian function by replacing the old variables, using the equations of transformation. The equations of transformation, for the generating functions W_2, W_3, and W_4, are

$$p_j = \frac{\partial W_2(q_j\,;\,P_j\,;\,t)}{\partial q_j}, \qquad Q_j = \frac{\partial W_2(q_j\,;\,P_j\,;\,t)}{\partial P_j} \tag{10.3.17}$$

$$q_j = -\frac{\partial W_3(p_j\,;\,Q_j\,;\,t)}{\partial p_j}, \qquad P_j = -\frac{\partial W_3(p_j\,;\,Q_j\,;\,t)}{\partial Q_j} \tag{10.3.18}$$

$$q_j = -\frac{\partial W_4(p_j\,;\,P_j\,;\,t)}{\partial p_j}, \qquad Q_j = \frac{\partial W_4(p_j\,;\,P_j\,;\,t)}{\partial P_j} \tag{10.3.19}$$

with $j = 1, 2, \ldots, r$.

10.4. EXAMPLES OF CANONICAL TRANSFORMATIONS

If we take the generating function to be

$$W_2 = \sum_{j=1}^{r} q_j P_j$$

and use (10.3.17), we obtain the identity transformation

$$p_j = \frac{\partial W_2}{\partial q_j} = P_j, \qquad Q_j = \frac{\partial W_2}{\partial P_j} = q_j \qquad (j = 1, 2, \ldots, r)$$

If the generating function is taken to be

$$W_1 = \sum_{j=1}^{r} q_j Q_j$$

and (10.3.13a) is used, we obtain the transformation equations

$$q_j = -P_j, \qquad p_j = Q_j \qquad (j = 1, 2, \ldots, r)$$

Transformations in which the new and old variables satisfy either (10.3.1), (10.3.10), (10.3.11), or (10.3.12), for which no demand is made that the variables appearing in the W_i be independent, are called *contact transformations*. For the special case where each q_j is expressible in terms of the Q_k alone, by relations not involving p_k or P_k, the transformation is called an *extended point transformation*.

Consider an extended point transformation

$$q_j = f_j(Q_k\,;\,t) \qquad (j = 1, 2, \ldots, r) \tag{10.4.1}$$

where the f_j are functions of the new coordinates and possibly the time. Let the generating function W_3 be represented by

$$W_3(Q_j\,;\,p_j\,;\,t) = -\sum_{j=1}^{r} f_j p_j \tag{10.4.2}$$

Using (10.3.18), we obtain as the equations of transformation

$$q_j = f_j(Q_k\,;\,t), \qquad P_j = \sum_{k=1}^{r} \frac{\partial f_k(Q_k\,;\,t)}{\partial Q_j}\,p_k \qquad (j = 1, 2,\ldots, r) \tag{10.4.3}$$

Since the $f_j(Q_k\,;\,t)$ have not been specified, we have thus shown that all extended point transformations are canonical. It is understood, of course, that having chosen the transformation equations for q_j, the corresponding transformation equations for the p_j must be obtained from the second set of equations in (10.4.3).

If we return to the example in Chapter 9 in which we considered the Hamiltonian formulation of the motion of a particle in an inverse square field, and write the Hamilton equations in rectangular coordinates, we have

$$\dot{\xi}'' = \frac{\partial H}{\partial p_{\xi''}}, \qquad \dot{p}_{\xi''} = -\frac{\partial H}{\partial \xi''}$$

$$\dot{\eta}'' = \frac{\partial H}{\partial p_{\eta''}}, \qquad \dot{p}_{\eta''} = -\frac{\partial H}{\partial \eta''}$$

where

$$H = \frac{1}{2m}\,(p_{\xi''}^2 + p_{\eta''}^2) - \mu m[(\xi'')^2 + (\eta'')^2]^{-1/2}$$

If we use the extended point transformation

$$\xi'' = r\cos w, \qquad p_{\xi''} = p_r \cos w - \frac{\sin w}{r}\,p_w$$

$$\eta'' = r\sin w, \qquad p_{\eta''} = p_r \sin w + \frac{\cos w}{r}\,p_w$$

we transform to a new set of variables (r, w, p_r, p_w). Since the extended point transformation is canonical and free of t, we know that the new variables satisfy the canonical equations

$$\dot{r} = \frac{\partial H}{\partial p_r}, \qquad \dot{p}_r = -\frac{\partial H}{\partial r}, \qquad \dot{w} = \frac{\partial H}{\partial p_w}, \qquad \dot{p}_w = -\frac{\partial H}{\partial w}$$

where

$$H = \frac{1}{2m}\left(p_r^{\,2} + \frac{1}{r^2}\,p_w^{\,2}\right) - \frac{\mu m}{r}$$

It should also be noted that if the criteria given in (10.2.7) are met, and hence the transformation is canonical, the Jacobian determinant of the transformation does not vanish (see Exercise 4 of this chapter).

Exercises

1. Verify Equations (10.3.7)–(10.3.9) in the text.
2. Show that the transformation

$$q_j = (2Q_j)^{1/2} \cos P_j, \qquad p_j = (2Q_j)^{1/2} \sin P_j \quad (j = 1, 2, \ldots, r)$$

(from modified Delaunay variables to Poincaré variables) is canonical by showing that Equations (10.2.7) in the text are satisfied.

3. Show that the transformation in Exercise 2 is canonical by exhibiting a function $W_2(q_j\,;\,P_j)$ which satisfies the differential condition

$$\sum_{j=1}^{r} (p_j\, dq_j + Q_j\, dP_j) = dW_2(q_j\,;\,P_j)$$

4. Show that

$$J \equiv \begin{vmatrix} \dfrac{\partial q_1}{\partial Q_1} & \dfrac{\partial q_1}{\partial Q_2} & \cdots & \dfrac{\partial q_1}{\partial Q_r} & \dfrac{\partial q_1}{\partial P_1} & \dfrac{\partial q_1}{\partial P_2} & \cdots & \dfrac{\partial q_1}{\partial P_r} \\ \vdots & \vdots & & \vdots & \vdots & \vdots & & \vdots \\ \dfrac{\partial q_r}{\partial Q_1} & \dfrac{\partial q_r}{\partial Q_2} & \cdots & \dfrac{\partial q_r}{\partial Q_r} & \dfrac{\partial q_r}{\partial P_1} & \dfrac{\partial q_r}{\partial P_2} & \cdots & \dfrac{\partial q_r}{\partial P_r} \\ \dfrac{\partial p_1}{\partial Q_1} & \dfrac{\partial p_1}{\partial Q_2} & \cdots & \dfrac{\partial p_1}{\partial Q_r} & \dfrac{\partial p_1}{\partial P_1} & \dfrac{\partial p_1}{\partial P_2} & \cdots & \dfrac{\partial p_1}{\partial P_r} \\ \vdots & \vdots & & \vdots & \vdots & \vdots & & \vdots \\ \dfrac{\partial p_r}{\partial Q_1} & \dfrac{\partial p_r}{\partial Q_2} & \cdots & \dfrac{\partial p_r}{\partial Q_r} & \dfrac{\partial p_r}{\partial P_1} & \dfrac{\partial p_r}{\partial P_2} & \cdots & \dfrac{\partial p_r}{\partial P_r} \end{vmatrix} = \pm 1$$

where J is the Jacobian determinant for a transformation of coordinates

$$q_j = q_j(Q_k\,;\,P_k\,;\,t), \qquad p_j = p_j(Q_k\,;\,P_k\,;\,t) \qquad (j = 1, 2, \ldots, r\,;\,k = 1, 2, \ldots, r)$$

which satisfies the criteria (10.2.7).

5. Verify that the extended point transformation

$$\xi'' = r \cos w, \qquad \eta'' = r \sin w$$

is canonical by exhibiting a function W_3 which satisfies the differential condition

$$\sum_{j=1}^{r} (q_j\, dp_j + P_j\, dQ_j) = -dW_3(p_j\,;\,Q_j\,;\,t)$$

6. Verify or disprove that Equations (10.2.7) of the text are also necessary conditions that the transformation be canonical.
7. A class of operations is said to have the group property if (1) it contains the identity operator, (2) the inverse to each operator is also a member of the class, and (3) the product of any two operators also belongs to the class. Show that the canonical transformations for a system of r degrees of freedom has the group property.†
8. Show that the variables $(q; p)$ which satisfy the canonical equations

$$\dot{q} = \frac{\partial H}{\partial p}, \qquad \dot{p} = -\frac{\partial H}{\partial q}, \qquad H = \frac{1}{2}(p^2 + k^2q^2)$$

arc transformed by the equations

$$q = \left(\frac{2Q}{k}\right)^{1/2} \cos P, \qquad p = (2Qk)^{1/2} \sin P \qquad (k = \text{positive real constant})$$

to new variables which satisfy the canonical system

$$\dot{Q} = \frac{\partial K}{\partial P}, \qquad \dot{P} = -\frac{\partial K}{\partial Q}, \qquad K = kQ^{\ddagger}$$

9. Show that the transformation defined by the equations

$$Q_1 = q_1^2 + \lambda^2 p_1^2, \qquad Q_2 = \frac{1}{2\lambda^2}(q_1^2 + q_2^2 + \lambda^2 p_1^2 + \lambda^2 p_2^2)$$

$$2\lambda P_1 = \arctan\left(\frac{q_1}{\lambda p_1}\right) - \arctan\left(\frac{q_2}{\lambda p_2}\right), \qquad P_2 = \lambda \arctan\left(\frac{q_2}{\lambda p_2}\right)$$

where λ is a constant, is a canonical transformation that reduces the dynamical system whose Hamiltonian function is

$$\tfrac{1}{2}(p_1^2 + p_2^2 + \lambda^{-2}q_1^2 + \lambda^{-2}q_2^2)$$

to the dynamical system whose Hamiltonian function is Q_2.

† From H. Goldstein, "Classical Mechanics," p. 272. Copyright 1950 by Addison-Wesley, Reading, Massachusetts. Used with permission of Addison-Wesley.

‡ Exercises 8 and 9 are reprinted from E. T. Whittaker, "A Treatise on the Analytical Dynamics of Particles and Rigid Bodies," 4th ed., pp. 307 and 311. Copyright 1944 (Am. ed.) by Cambridge Univ. Press, London and New York. Used with permission of Cambridge Univ. Press.

Chapter 11

Hamilton–Jacobi Theory with Examples

11.1. THE HAMILTON–JACOBI DIFFERENTIAL EQUATION

In essence, we have seen that for an arbitrary function $W(q_j; Q_j; t)$ of the old and new coordinates and the time, if the variables $(q_j; p_j)$ satisfying the Hamilton equations of motion

$$\dot{q}_j = \frac{\partial H}{\partial p_j}, \qquad \dot{p}_j = -\frac{\partial H}{\partial q_j} \qquad (j = 1, 2,..., r) \tag{11.1.1}$$

are transformed by the $2r$ equations

$$P_j = -\frac{\partial W}{\partial Q_j}(q_j\,; Q_j\,; t), \qquad p_j = \frac{\partial W}{\partial q_j}(q_j\,; Q_j\,; t) \tag{11.1.2}$$

the new variables $(Q_j; P_j)$ will satisfy the Hamilton equations

$$\dot{Q}_j = \frac{\partial K}{\partial P_j}, \qquad \dot{P}_j = -\frac{\partial K}{\partial Q_j} \qquad (j = 1, 2,..., r) \tag{11.1.3}$$

where

$$K = H + \frac{\partial W}{\partial t}$$

Similar statements can be made for the other three generating functions W_2, W_3, and W_4.

We now examine the question of choosing $W(q_j; Q_j; t)$ so that an advantage is gained by the transformation. Suppose, for example, that we are able to transform from the $(q_j; p_j)$ to a new set of variables $(Q_j; P_j)$, where Q_j and P_j are constants for every j. Then we can solve equations (11.1.2) for the q_j and p_j and thus obtain a solution to the dynamical problem. Certainly if K is zero, from (11.1.3), the Q_j and P_j are constants for every j, and we obtain the condition on W that

$$H(q_j\,;p_j\,;t) + \frac{\partial W}{\partial t} = 0 \tag{11.1.4}$$

If we replace the p_j in (11.1.4) using (11.1.2), we obtain the Hamilton–Jacobi partial differential equation for determining a generating function W. This equation may be written in the form

$$H\left(q_j\,;\frac{\partial W}{\partial q_j};t\right) + \frac{\partial W}{\partial t} = 0 \tag{11.1.5}$$

Since the Hamiltonian function will generally contain terms that are quadratic in the momenta variables, the partial derivatives $\partial W/\partial q_j$ will appear in (11.1.5) to the second degree. Thus it is a nonlinear partial differential equation of order one, in the $r + 1$ independent variables $(q_j; t)$, which can be represented in the functional form

$$f\left(q_j\,;t;\frac{\partial W}{\partial q_j};\frac{\partial W}{\partial t}\right) = 0 \tag{11.1.6}$$

where W does not appear explicitly in (11.1.6). Since there are $r + 1$ independent variables in (11.1.6), a complete solution must contain exactly $r + 1$ independent constants of integration. The fact, however, that W does not appear explicitly in (11.1.6) permits us to deduce that one of the constants of integration must appear in W as an additive constant, since it must be automatically eliminated from each of the equations obtained by taking partial derivatives of W with respect to the $r + 1$ independent variables. Thus in determining W we need only concern ourselves with determining r independent constants of integration. The additive constant plays no role in the transformation because only partial derivatives of the generating function are involved. An amplification of these points, with examples, is contained in the next several paragraphs.

A solution of (11.1.6) is assumed, in the form

$$W = W(q_j\,;\alpha_k\,;t) \tag{11.1.7}$$

where the α_k, $k = 1, 2,..., r, r + 1$, are $r + 1$ independent constants of integration. To deduce the differential equation that gave rise to this solution, we can write

$$g(q_j; \alpha_k; W; t) = W - W(q_j; \alpha_k; t) = 0 \tag{11.1.8}$$

and differentiate g totally with respect to each of the $r + 1$ independent variables. In this way we obtain a system of $r + 1$ equations

$$\begin{aligned}
\frac{dg}{dq_1} &= \frac{\partial g}{\partial q_1} + \frac{\partial g}{\partial W}\frac{\partial W}{\partial q_1} = 0 \\
\frac{dg}{dq_2} &= \frac{\partial g}{\partial q_2} + \frac{\partial g}{\partial W}\frac{\partial W}{\partial q_2} = 0 \\
&\vdots \\
\frac{dg}{dq_r} &= \frac{\partial g}{\partial q_r} + \frac{\partial g}{\partial W}\frac{\partial W}{\partial q_r} = 0 \\
\frac{dg}{dt} &= \frac{\partial g}{\partial t} + \frac{\partial g}{\partial W}\frac{\partial W}{\partial t} = 0
\end{aligned} \tag{11.1.9}$$

which are to be used to eliminate r of the independent constants α_k in forming the differential equation (11.1.6). Equation (11.1.7) cannot be used to eliminate the remaining independent constant, for this procedure would cause W to appear explicitly in (11.1.6). Thus this remaining constant must appear in such a way that it is not present in any of the equations of (11.1.9); that is, it must appear as an additive constant in W.

Example 11.1.1. Let W be the dependent variable, q_1 and q_2 the two independent variables, and consider the primitive

$$g(q_1, q_2, \alpha_1, \alpha_2, W) = W - \alpha_1 q_1^2 - \alpha_2 q_2 = 0 \tag{11.1.10}$$

where α_1 and α_2 are two independent constants. Taking partial derivatives of g with respect to q_1 and q_2, we obtain the two equations

$$\frac{\partial W}{\partial q_1} - 2\alpha_1 q_1 = 0, \qquad \frac{\partial W}{\partial q_2} - \alpha_2 = 0 \tag{11.1.11}$$

hence it follows that

$$\alpha_1 = \frac{1}{2q_1}\frac{\partial W}{\partial q_1}, \qquad \alpha_2 = \frac{\partial W}{\partial q_2} \tag{11.1.12}$$

Eliminating α_1 and α_2 from (11.1.10), we obtain the differential equation

$$W = \frac{q_1}{2}\frac{\partial W}{\partial q_1} + q_2\frac{\partial W}{\partial q_2} \tag{11.1.13}$$

It is clear that if the differential equation is to be free of W, then (11.1.10) cannot be used to eliminate the constants. They must be eliminated and the differential equation formed only through the use of (11.1.11). This is, of course, not possible if α_1 and α_2 are independent constants. The only way a second independent constant could appear in W is as an additive constant, which would, therefore, not appear in (11.1.11). This point is illustrated in the next example.

Example 11.1.2. Consider the primitive

$$g(q_1, q_2, \alpha_1, \alpha_2, W) = W - \alpha_1 q_1 - \alpha_1^2 q_2 - \alpha_2 = 0 \qquad (11.1.14)$$

where again α_1 and α_2 are independent constants. The counterpart equations of (11.1.11) are

$$\frac{\partial W}{\partial q_1} - \alpha_1 = 0, \qquad \frac{\partial W}{\partial q_2} - \alpha_1^2 = 0 \qquad (11.1.15)$$

from which we obtain the differential equation

$$\frac{\partial W}{\partial q_2} - \left(\frac{\partial W}{\partial q_1}\right)^2 = 0 \qquad (11.1.16)$$

which does not contain W explicitly.

If we can find a complete solution of the Hamilton–Jacobi equation (11.1.5)—that is, (neglecting the additive constant) one with r arbitrary independent constants α_j—we have solved the dynamical problem. To see this, assume that we have determined a complete solution $W(q_j; \alpha_j; t)$ of (11.1.5), where we neglect the additive constant. Since W has been so determined that the new Hamiltonian K is zero, the new generalized coordinates Q_j and generalized momenta P_j are constants. The set of constants α_j will serve to play the role of the new generalized coordinates, that is, the Q_j. We designate the constant momenta, conjugate to α_j, by the notation β_j. Then the β_j play the role of the new momenta, that is, the P_j.

The transformation equations (10.3.13a) can then be written

$$\begin{aligned} \beta_j &= -\frac{\partial W(q_j; \alpha_j; t)}{\partial \alpha_j} \\ p_j &= \frac{\partial W(q_j; \alpha_j; t)}{\partial q_j} \qquad (j = 1, 2, \ldots, r) \end{aligned} \qquad (11.1.17)$$

If the equations for the β_j can be solved to express the q_j as functions of the constants α_j and β_j and the time t, the expressions for the q_j can be substituted into the p_j equations to obtain the p_j as functions of α_j, β_j, and t. The constants α_j and β_j are called *canonical constants*.

If H is free of t, we have seen that it is a constant of the motion, say α. For such a dynamical system, we have from (11.1.5)

$$\frac{\partial W}{\partial t} = -\alpha \tag{11.1.18}$$

Integrating,

$$W(q_j; \alpha, \alpha_2, \alpha_3, \ldots, \alpha_r, t) = -\alpha t + W^*(q_j; \alpha, \alpha_2, \alpha_3, \ldots, \alpha_r) \tag{11.1.19}$$

where W^* does not contain t explicitly. Thus one of the constants appearing in W is equal to the constant value of H. Designate this constant α_1. Frequently α_1 will be the total energy but it does not have to be. Since $\partial t/\partial q_j$ and $\partial \alpha_1/\partial q_j$ are both zero for every j, we have

$$\frac{\partial W}{\partial q_j} = \frac{\partial W^*}{\partial q_j}$$

and we can write (11.1.5) in the form

$$H\left(q_j; \frac{\partial W^*}{\partial q_j}\right) = \alpha_1 \tag{11.1.20}$$

Thus if the Hamiltonian of the dynamical system does not contain the time explicitly, we will have solved the problem of describing the motion of the system if we are able to determine a complete solution W^* of (11.1.20). If W^* has been obtained, the solution proceeds as follows:

$$\begin{aligned}
\beta_1 &= -\frac{\partial W}{\partial \alpha_1} = t - \frac{\partial W^*(q_j; \alpha_j)}{\partial \alpha_1} \\
\beta_j &= -\frac{\partial W}{\partial \alpha_j} = -\frac{\partial W^*(q_j; \alpha_j)}{\partial \alpha_j} \qquad (j = 2, 3, \ldots, r) \\
p_j &= \frac{\partial W^*(q_j; \alpha_j)}{\partial q_j} \qquad (j = 1, 2, \ldots, r)
\end{aligned} \tag{11.1.21}$$

Example 11.1.3. Discuss the dynamical problem of the unbound particle of unit mass and velocity $\mathbf{v}$ in the framework of Hamilton–Jacobi theory.

Since no forces act upon the particle, the Hamiltonian function H is simply the kinetic energy referenced to an inertial origin. Thus we write

$$H = \frac{p^2}{2} \qquad (p = |\mathbf{v}|) \tag{11.1.22}$$

The symbol q will be used to express the position coordinate.

We will discuss the problem in the following order:

Step 1. We write the Hamilton equations and solve for p and q.

Step 2. We find W^* and hence the generator W of a transformation to canonical constants.

Step 3. We write the equations of transformation and hence the solution for p and q.

Step 4. As a check, we show that the new Hamiltonian function is equal to zero, thus verifying that the new parameters are constants.

Step 5. We compare the solutions obtained in steps 1 and 3.

Step 6. We verify that the transformation in step 3 is a canonical transformation.

Step 1. The Hamilton equations are

$$\dot{p} = -\frac{\partial H}{\partial q} = 0, \qquad \dot{q} = \frac{\partial H}{\partial p} = p \tag{11.1.23}$$

Thus the solution is given by

$$p = p_0 \qquad (p_0 = \text{constant}), \qquad q = p_0 t + q_0 \tag{11.1.24}$$

where q_0 is the initial value of q.

Step 2. Since H is free of t, it is a constant of the motion, say α. Thus we write

$$H = \frac{p^2}{2} = \alpha \qquad (\alpha > 0) \tag{11.1.25}$$

If the expression

$$p = \frac{\partial W}{\partial q} = \frac{\partial W^*}{\partial q} = \frac{dW^*}{dq}$$

is substituted into (11.1.25), we obtain the ordinary differential equations

$$\frac{dW^*}{dq} = \pm(2\alpha)^{1/2} \tag{11.1.26}$$

Since p is positive, we consider only the positive root and integrate (11.1.26) to obtain

$$W^* = (2\alpha)^{1/2} q \tag{11.1.27}$$

The arbitrary additive constant appearing in the integration has been dropped, for only partial derivatives of W^* are of interest in forming the transformation. From (11.1.19), we can write immediately

$$W = -\alpha t + (2\alpha)^{1/2} q \tag{11.1.28}$$

Step 3. The equations

$$p = \frac{\partial W}{\partial q} = (2\alpha)^{1/2}, \qquad \beta = -\frac{\partial W}{\partial \alpha} = t - \frac{q}{(2\alpha)^{1/2}} \tag{11.1.29}$$

define a canonical transformation from $(q; p)$ to the constants $(\alpha; \beta)$, where β is the generalized momentum conjugate to α. Expressing q in terms of α, β, and t, the solution of the dynamical problem can be written as

$$p = (2\alpha)^{1/2}, \qquad q = (2\alpha)^{1/2}(t - \beta) \tag{11.1.30}$$

Step 4. Simply as a check on our work, we use (11.1.28) and find that

$$K = H(\alpha; \beta) + \frac{\partial W}{\partial t} = \alpha + (-\alpha) = 0 \tag{11.1.31}$$

Since

$$\dot{\alpha} = \frac{\partial K}{\partial \beta} = 0, \qquad \dot{\beta} = -\frac{\partial K}{\partial \alpha} = 0 \tag{11.1.32}$$

it follows that α and β are indeed constants of the motion and (11.1.30) is the solution of the dynamical problem.

Step 5. It is a simple matter to relate the constants α and β to the initial conditions. From (11.1.30), we note that $(2\alpha)^{1/2} = p_0$. When $t = 0$, $q = q_0$, and $\beta = -q_0/(2\alpha)^{1/2} = -q_0/p_0$. Thus Equations (11.1.24) and (11.1.30) are identical.

Step 6. To prove that the transformation is canonical, we show that W satisfies the modified differential condition (10.3.13b). We have

$$\begin{aligned} dW(q; \alpha; t) &= -\alpha\, dt - t\, d\alpha + (2\alpha)^{1/2}\, dq + \frac{q}{(2\alpha)^{1/2}}\, d\alpha \\ &= p\, dq - \beta\, d\alpha - \alpha\, dt \end{aligned}$$

11.2. TREATMENT OF PERTURBATIONS

Suppose we have a dynamical system described by variables $(q_j; p_j)$ which satisfy the canonical equations

$$\dot{q}_j = \frac{\partial H}{\partial p_j}, \qquad \dot{p}_j = -\frac{\partial H}{\partial q_j} \qquad (j = 1, 2, \ldots, r) \tag{11.2.1}$$

where H is a function of the variables $(q_j; p_j; t)$. If we are confronted with the problem of solving (11.2.1), an attempt might be made to obtain a generating function W of a canonical transformation to new parameters $(\alpha_j; \beta_j)$, which are constants of the motion. Thus the Hamilton–Jacobi differential equation

$$H\left(q_j\,;\frac{\partial W}{\partial q_j};t\right)+\frac{\partial W}{\partial t}=0$$

would be formed and an attempt made to solve this partial differential equation for W by the method of separation of variables or some other suitable method. If we are successful in doing this, we can proceed, as described in the preceding section of this chapter, to obtain the solution for the q_j and p_j through the use of either (11.1.17) or (11.1.21), depending on whether or not H contained the time explicitly. In either case, we would have obtained, in principle, the complete solution to the dynamical problem described by (11.2.1).

In practice, the matter is not so simply handled. Even for problems of moderate complexity, the Hamiltonian function H is generally of such a form that it is not possible to find a coordinate system in which the partial differential equation (11.1.5) will separate. In such a case, we are usually not able to obtain an exact generating function W to use in writing the solutions from either (11.1.17) or (11.1.21). To illustrate, the Hamiltonian function, for the problem of a particle moving in the gravitational field of the Earth, expressed in spherical coordinates, leads to a nonseparable Hamilton–Jacobi differential equation. For problems such as this, special perturbative procedures must be invoked.

It may be possible to make a convenient decomposition of H into the sum of two terms, H_0 and H_1, in the form

$$H(q_j\,;p_j\,;t)=H_0(q_j\,;p_j\,;t)+H_1(q_j\,;p_j\,;t) \tag{11.2.2}$$

where a complete integral of the Hamilton–Jacobi equation for H_0 is known. Now any arbitrary function $W(q_j; Q_j; t)$ will serve as a generator of a canonical transformation from the generalized variables $(q_j; p_j)$ to a new set of variables $(Q_j\,; P_j)$. Suppose that we choose as our generator the function W_0 obtained in the following way. In (11.2.1), if the Hamiltonian function H is expressible as in (11.2.2), replace H by its first term H_0 and consider the canonical equations

$$\dot{q}_j=\frac{\partial H_0}{\partial p_j},\qquad \dot{p}_j=-\frac{\partial H_0}{\partial q_j}\qquad (j=1,2,\ldots,r) \tag{11.2.3}$$

To solve (11.2.3), we seek a generating function $W_0(q_j; \alpha_j; t)$ of a canonical transformation to new parameters $(\alpha_j; \beta_j)$, which are constants of the motion. The generating function W_0 is to be obtained by solving the Hamilton–Jacobi equation

$$H_0\left(q_j\,;\frac{\partial W_0}{\partial q_j};\,t\right)+\frac{\partial W_0}{\partial t}=0 \tag{11.2.4}$$

or, alternatively, if H_0 does not contain t explicitly, we can write

$$H_0\left(q_j\,;\frac{\partial W_0{}^*}{\partial q_j}\right)=\alpha_1 \qquad [W_0{}^*=W_0{}^*(q_j\,;\alpha_j)] \tag{11.2.5}$$

Assume that we are able to solve either (11.2.4) or (11.2.5) for W_0, or $W_0{}^*$, as required. Returning to the original problem (11.2.1), we choose the generating function W to be either W_0 or $W_0{}^*$ depending on whether H_0 does or does not contain t explicitly. (In the following exposition, we will assume that H_0 contains t explicitly.) If $W_0(q_j; \alpha_j; t)$ is the generator of the transformation, the old generalized variables $(q_j; p_j)$ that satisfy the canonical equations (11.2.1) will be transformed to new generalized variables $(\alpha_j; \beta_j)$ satisfying the canonical equations

$$\dot{\alpha}_j=\frac{\partial K}{\partial \beta_j},\qquad \dot{\beta}_j=-\frac{\partial K}{\partial \alpha_j} \tag{11.2.6}$$

where

$$K=H_0+H_1+\frac{\partial W_0}{\partial t}=H_1$$

since W_0 is a solution of (11.2.4). As usual, it is understood that K is to be expressed in terms of the new variables $(\alpha_j; \beta_j)$. Since $K(\alpha_j; \beta_j; t) = H_1(\alpha_j; \beta_j; t)$, we can write (11.2.6) in the form

$$\dot{\alpha}_j=\frac{\partial H_1(\alpha_j\,;\beta_j\,;t)}{\partial \beta_j},\qquad \dot{\beta}_j=-\frac{\partial H_1(\alpha_j\,;\beta_j\,;t)}{\partial \alpha_j} \tag{11.2.7}$$

If H_0 is so chosen that we are able to obtain the generating function W_0 from (11.2.4), in the process of doing so, we automatically obtain the solution to the dynamical problem for which H_0 is the entire Hamiltonian. Such a solution, in which the α_j and β_j, $j = 1, 2,\ldots, r$, are constants, is called an *unperturbed solution* and H_0 is referred to as the *unperturbed* Hamiltonian. The second term H_1, in H, is called the *perturbing* Hamiltonian and the solution of the H–problem is called the *perturbed solution.*

For the unperturbed problem, the $\dot{\alpha}_j$ and $\dot{\beta}_j$ are zero, since H_1 is zero. If $H_1 \neq 0$ and we are able to integrate Equations (11.2.7) either exactly or approximately, we can express the α_j and β_j as functions of time.

In the process of obtaining the unperturbed solution, we have, in principle at least, obtained ordinary algebraic equations (not differential equations) which express the q_j and p_j of the H_0-problem in terms of the α_j and β_j. In these equations, we now replace α_j and β_j as functions of time obtained by integrating (11.2.7). We then have the q_j and p_j of the H–problem expressed as functions of t. Hence we have the complete solution to the perturbed dynamical problem.

We examine the motion of the simple harmonic oscillator as an example of the theory. Because of the uncomplicated nature of this particular dynamical problem, we will be able to demonstrate three different approaches to its solution, each of which illustrates a separate facet of the Hamiltonian procedure. In one part of the example, it will be noticed that we choose the potential energy as H_1 without regard to its magnitude relative to the kinetic energy H_0. We can do this here because the perturbation equations (11.2.7) are simple enough that we can integrate them exactly. Generally speaking, however, in any dynamical problem of interest, H_1 must be small in some sense with respect to H_0 so that we can obtain meaningful approximate solutions to (11.2.7).

Example 11.2.1. Discuss the dynamical problem of a simple harmonic oscillator of unit mass and unit force constant in the framework of Hamilton–Jacobi theory.

If we use the symbol q to represent the position coordinate, the kinetic energy T, referenced to an inertial origin, is $\dot{q}^2/2$ and the potential energy V is $q^2/2$. A Lagrangian function L is given by

$$L = T - V = \frac{\dot{q}^2}{2} - \frac{q^2}{2}$$

From its definition, $p = \partial L/\partial \dot{q} = \dot{q}$, so that the Hamiltonian function H is written

$$H = p\dot{q} - L = H_0 + H_1 \qquad \left(H_0 = \frac{p^2}{2};\ H_1 = \frac{q^2}{2}\right)$$

We will discuss the problem in the following three ways:

1. We will obtain a direct solution of Hamilton's equations without transforming to new parameters.

2. We will solve the H-problem in the spirit of Section 11.1, by introducing a transformation to canonical constants.

3. We will solve the problem in the spirit of this section, by considering H_0 as the unperturbed Hamiltonian and H_1 as the perturbing Hamiltonian.

1. Direct Solution without Transformation

We want to obtain a direct solution to the differential system given by the canonical equations

$$\dot{q} = \frac{\partial H}{\partial p}, \qquad \dot{p} = -\frac{\partial H}{\partial q} \qquad \left(H = \frac{p^2}{2} + \frac{q^2}{2}\right)$$

and the initial conditions that at $t = 0$, $q = q_0$, $p = p_0$. Explicitly, we have

$$\dot{q} = p, \qquad \dot{p} = -q \tag{11.2.8}$$

The equation

$$\ddot{q} + q = 0 \tag{11.2.9}$$

follows directly from (11.2.8). Its general solution is $q = A \sin t + B \cos t$, where A and B are constants to be determined from the initial conditions. Thus the two equations

$$q = p_0 \sin t + q_0 \cos t, \qquad p = p_0 \cos t - q_0 \sin t \tag{11.2.10}$$

give the solution to the dynamical problem.

2. Solution by Transformation to Canonical Constants

Considering again the canonical equations

$$\dot{q} = \frac{\partial H}{\partial p}, \qquad \dot{p} = -\frac{\partial H}{\partial q} \qquad \left(H = \frac{p^2}{2} + \frac{q^2}{2}\right)$$

we seek a generator of a transformation to canonical constants. Since H is free of time explicitly, we write the Hamilton–Jacobi equation in the form

$$q^2 + \left[\frac{\partial W^*(q; \alpha)}{\partial q}\right]^2 = 2\alpha \tag{11.2.11}$$

But α is a constant of the motion, equal to the total energy. Hence $\partial W^*/\partial q = dW^*/dq$ and we may write

$$\frac{dW^*}{dq} = \pm(2\alpha - q^2)^{1/2} \tag{11.2.12}$$

where we retain both signs, for $p = dW^*/dq$ will alternate sign during the motion. Since p must remain real, we see that q^2 is bounded by the condition

$$q^2 \leqslant 2\alpha$$

The integrals of (11.2.12) are readily found to be

$$W^* = \pm \left[\frac{q}{2}(2\alpha - q^2)^{1/2} + \alpha \arcsin \frac{q}{\sqrt{2\alpha}}\right] \tag{11.2.13}$$

neglecting the arbitrary constant of integration. The transformation equations (11.1.21) are

$$\begin{aligned} \beta &= t - \frac{\partial W^*}{\partial \alpha} = t \mp \arcsin \frac{q}{2\alpha} \\ p &= \frac{\partial W^*}{\partial q} = \pm(2\alpha - q^2)^{1/2} \end{aligned} \tag{11.2.14}$$

Expressing q and p in terms of the canonical constants $(\alpha; \beta)$ and the time t, we have

$$q = \mp(2\alpha)^{1/2} \sin(t - \beta), \qquad p = \mp(2\alpha)^{1/2} \cos(t - \beta) \tag{11.2.15}$$

The canonical constants are related to the initial conditions through the equations

$$\sin \beta = \pm \frac{q_0}{(2\alpha)^{1/2}}, \qquad \cos \beta = \mp \frac{p_0}{(2\alpha)^{1/2}}, \qquad 2\alpha = q_0^2 + p_0^2$$

When these expressions are introduced into (11.2.15), Equations (11.2.10) are obtained.

3. Solution by Means of a Perturbation Technique

We return again to the Hamiltonian function $H = H_0 + H_1$ and interpret $H_0 = p^2/2$ as the unperturbed Hamiltonian. We then seek a solution W_0^* of the differential equation

$$\left(\frac{\partial W_0^*}{\partial q}\right)^2 = \left(\frac{dW_0^*}{dq}\right)^2 = 2\alpha \tag{11.2.16}$$

Integrating (11.2.16), we obtain

$$W_0^* = \pm(2\alpha)^{1/2} q \tag{11.2.17}$$

as the generator of a canonical transformation to the canonical variables $(\alpha; \beta)$ through the transformation

$$\begin{aligned} \beta &= t - \frac{\partial W_0^*}{\partial \alpha} = t \pm \frac{q}{(2\alpha)^{1/2}} \\ p &= \frac{\partial W_0^*}{\partial q} = \pm (2\alpha)^{1/2} \end{aligned} \tag{11.2.18}$$

Solving for q and p, we write

$$q = \pm(2\alpha)^{1/2}(t - \beta), \qquad p = \pm(2\alpha)^{1/2} \tag{11.2.19}$$

To determine the time dependence of α and β, we first express H_1 in terms of $(\alpha; \beta)$, using (11.2.19). Since

$$H_1 = \alpha(t - \beta)^2$$

the time derivatives of α and β are given by the equations

$$\begin{aligned} \dot{\alpha} &= \frac{\partial H_1}{\partial \beta} = -2\alpha(t - \beta) \\ \dot{\beta} &= -\frac{\partial H_1}{\partial \alpha} = -(t - \beta)^2 \end{aligned} \tag{11.2.20}$$

To integrate (11.2.20), set $\tan z = t - \beta$. Then $\dot{\beta} = 1 - (\sec^2 z)\dot{z}$ and it follows directly that $\dot{z} = 1$. Thus we have

$$z = t - b \qquad (b = \text{constant})$$

and

$$\beta = t - \tan z = t - \tan(t - b) \tag{11.2.21}$$

Using this equation for β in the first equation of (11.2.20), we have

$$\dot{\alpha} = -2\alpha \tan(t - b)$$

from which it follows immediately that

$$\alpha = k \cos^2(t - b) \tag{11.2.22}$$

where k is a constant of integration.

When the time dependencies of α and β are substituted into (11.2.19) the original variables $(q; p)$ can be expressed as

$$\begin{aligned} q &= \pm[2k\cos^2(t - b)]^{1/2}\{t - [t - \tan(t - b)]\} \\ p &= \pm[2k\cos^2(t - b)]^{1/2} \end{aligned} \tag{11.2.23}$$

Clearly Equations (11.2.23) are equivalent to

$$q = \pm(2k)^{1/2}\sin(t - b), \qquad p = \pm(2k)^{1/2}\cos(t - b) \tag{11.2.24}$$

Since Equations (11.2.24) are of the same form as (11.2.15), it is obvious that when the initial conditions are introduced, Equations (11.2.10) are obtained. Note carefully that in the second treatment of the problem, we executed a canonical transformation in which the α and β appearing in (11.2.15) are *canonical constants* and must not be identified with the time-dependent α and β which were introduced in the third treatment of the problem.

11.3. THE PLANETARY n-BODY PROBLEM

In an earlier chapter, the equations of motion were developed for a particle of mass m, under the attraction of a mass m_0 and acted upon by other point masses m_j, $j = 1, 2, \ldots, n - 2$. For the special case where the masses are constant, where the mutual attractions are of inverse-square form, and where no external fields are present, the equations of motion for m may be written (see Section 7.2):

$$\ddot{\mathbf{r}} = -\frac{\mu \mathbf{r}}{r^3} + \nabla R$$

$$R = k^2 \sum_{j=1}^{n-2} m_j \left(\frac{1}{\rho_j} - \frac{\xi\xi_j + \eta\eta_j + \zeta\zeta_j}{r_j^3} \right)$$

In the expression for R, ρ_j is the distance between m and m_j and r_j is the distance between m_0 and m_j.

For the particular case of the solar system, we refer to the n-body problem as the *planetary n-body problem*. We then interpret the particle m_0 to be the Sun and the particle m to be one of the planets perturbed by the influences of the remaining planets. In the subsequent discussion, it is not necessary to select a fundamental plane of reference. Hence we will use the notation $OXYZ$ to designate an inertial reference system, with origin at heliocenter. The coordinates of m in this reference frame are then (X, Y, Z) and the coordinates of the other planets referenced to the heliocenter are designated by (X_j, Y_j, Z_j), $j = 1, 2, \ldots, n - 2$. In this notation, the equations of motion for m are

$$\ddot{\mathbf{r}} = -\frac{\mu \mathbf{r}}{r^3} + \nabla R \tag{11.3.1}$$

where

$$R = k^2 \sum_{j=1}^{n-2} m_j \left(\frac{1}{\rho_j} - \frac{XX_j + YY_j + ZZ_j}{r_j^3} \right)$$

$$\rho_j^2 = (X - X_j)^2 + (Y - Y_j)^2 + (Z - Z_j)^2$$

$$r_j^2 = X_j^2 + Y_j^2 + Z_j^2$$

$$r^2 = X^2 + Y^2 + Z^2$$

Since the negative of the gradient of

$$\phi = -\frac{\mu}{r} - R \tag{11.3.2}$$

is

$$-\nabla\phi = -\nabla\left(-\frac{\mu}{r}\right) + \nabla R = -\frac{\mu\mathbf{r}}{r^3} + \nabla R$$

we can rewrite (11.3.1) in the form

$$\ddot{\mathbf{r}} = -\nabla\phi = -\frac{\partial\phi}{\partial X}\mathbf{i} - \frac{\partial\phi}{\partial Y}\mathbf{j} - \frac{\partial\phi}{\partial Z}\mathbf{k}$$

or, in component form,

$$\ddot{X} = -\frac{\partial\phi}{\partial X}, \qquad \ddot{Y} = -\frac{\partial\phi}{\partial Y}, \qquad \ddot{Z} = -\frac{\partial\phi}{\partial Z} \tag{11.3.3}$$

Equations (11.3.3) are the correct equations of motion for the mass m. The same equations are readily deducible from the Lagrangian procedure if L is taken to be

$$L = \frac{T}{m} - \phi = \frac{1}{2}(\dot{X}^2 + \dot{Y}^2 + \dot{Z}^2) - \phi(X, Y, Z, t) \tag{11.3.4}$$

for then

$$\frac{d}{dt}\left(\frac{\partial L}{\partial \dot{X}}\right) - \frac{\partial L}{\partial X} = \frac{d}{dt}\dot{X} + \frac{\partial\phi}{\partial X} = 0$$

and hence

$$\ddot{X} = -\frac{\partial\phi}{\partial X}$$

Similar results are obtained for the Y and Z coordinates.

In applying the Lagrangian procedure here, it should be noted that we consider ϕ to be a function of the three generalized coordinates X, Y, Z and the time t. This is because the coordinates (X_j, Y_j, Z_j) of the perturbing planets which appear in R are considered to be time-dependent parameters.

Since L contains the velocities only through T, T is homogeneous quadratic in the velocities, and the Lagrangian is of the form $T/m - \phi$, the Hamiltonian function per unit mass H is $T/m + \phi$. The momentum conjugate to X is given by

$$p_X = \frac{\partial L}{\partial \dot{X}} = \dot{X}$$

with similar expressions for p_Y and p_Z. Thus we have

$$H = \frac{1}{2}(p_X^2 + p_Y^2 + p_Z^2) - \frac{\mu}{r} - R(X, Y, Z; X_j, Y_j, Z_j) \qquad (j = 1, 2, \ldots, n-2) \tag{11.3.5}$$

which can be written in the form

$$H = H_0(X, Y, Z, p_X, p_Y, p_Z) + H_1(X, Y, Z; X_j, Y_j, Z_j) \tag{11.3.6}$$

where

$$H_0 = \frac{1}{2}(p_X^2 + p_Y^2 + p_Z^2) - \frac{\mu}{r}$$

$$H_1 = -R(X, Y, Z; X_j, Y_j, Z_j) \qquad (j = 1, 2, \ldots, n-2)$$

The canonical system is

$$\begin{aligned} \dot{X} &= \frac{\partial H}{\partial p_X}, & \dot{p}_X &= -\frac{\partial H}{\partial X} \\ \dot{Y} &= \frac{\partial H}{\partial p_Y}, & \dot{p}_Y &= -\frac{\partial H}{\partial Y} \\ \dot{Z} &= \frac{\partial H}{\partial p_Z}, & \dot{p}_Z &= -\frac{\partial H}{\partial Z} \end{aligned} \tag{11.3.7}$$

The motion of a planet is dominated by the H_0 term in its Hamiltonian H. The contribution of H_1 to the motion is small. This is because of the appearance of the mass of a perturbing planet as a multiplicative factor in each term of R. Since the gravitational effect of the perturbing masses is to be compared to the gravitational effect of the Sun and since the ratio of the mass of a planet to the solar mass is never larger than 0.001, the contribution of ∇R to the force field is much smaller than the contribution of the $\mu\mathbf{r}/r^3$ term in (11.3.1). This fact is brought out quite clearly if we write an explicit expression for ∇R in the form

$$\nabla R = k^2 \sum_{j=1}^{n-2} m_j \left(\frac{\mathbf{u}_j}{\rho_j^2} - \frac{\mathbf{u}_{0j}}{r_j^2} \right) \tag{11.3.8}$$

where

$$\mathbf{u}_j = \frac{(X_j - X)}{\rho_j}\mathbf{i} + \frac{(Y_j - Y)}{\rho_j}\mathbf{j} + \frac{(Z_j - Z)}{\rho_j}\mathbf{k}$$

$$\mathbf{u}_{0j} = \frac{X_j}{r_j}\mathbf{i} + \frac{Y_j}{r_j}\mathbf{j} + \frac{Z_j}{r_j}\mathbf{k}$$

are unit vectors directed from m to m_j and m_0 to m_j respectively.

Explicitly, then, Equation (11.3.1) can be written

$$\ddot{\mathbf{r}} = -\frac{\mu \mathbf{u}_0}{r^2} + k^2 \sum_{j=1}^{n-2} m_j \left(\frac{\mathbf{u}_j}{\rho_j^2} - \frac{\mathbf{u}_{0j}}{r_j^2} \right) \tag{11.3.9}$$

where

$$\mathbf{u}_0 = \frac{X\mathbf{i} + Y\mathbf{j} + Z\mathbf{k}}{r}$$

is a unit vector directed from heliocenter to the planet m. The distances ρ_j and r_j will usually be at least of the same order of magnitude as r; $\mu = k^2(m_0 + m) \approx k^2$ solar masses and $k^2 m_j \leqslant 0.001\, k^2$ solar masses. Thus the contribution of ∇R, and hence $H_1 = -R$, to the character of the motion is of the nature of a small perturbing acceleration.

The perturbation technique discussed in Section 11.2 can be applied to obtain an approximate solution to the canonical equations (11.3.7). Since H_0 does not contain the time explicitly, we can search for a generating function

$$W_0(X, Y, Z, \alpha_1, \alpha_2, \alpha_3, t) = -\alpha_1 t + W_0^*(X, Y, Z, \alpha_1, \alpha_2, \alpha_3)$$

where W_0^* is a solution of the Hamilton–Jacobi equation

$$H_0\left(X, Y, Z, \frac{\partial W_0^*}{\partial X}, \frac{\partial W_0^*}{\partial Y}, \frac{\partial W_0^*}{\partial Z}\right) = \alpha_1$$

As we shall see, the Hamilton–Jacobi equation for W_0^* is readily separable when expressed in spherical coordinates, and we will be able to obtain the generating function W_0^* as a function of the spherical coordinates and the canonical variables α_1, α_2, and α_3. We can thus express the spherical coordinates of a given planet m in terms of the canonical variables $(\alpha_j; \beta_j)$, $j = 1, 2, 3$. The time dependence of these canonical variables is then given by the equations

$$\begin{aligned} \dot{\alpha}_j &= \frac{\partial H_1(\alpha_j\,;\beta_j\,;t)}{\partial \beta_j} \\ \dot{\beta}_j &= -\frac{\partial H_1}{\partial \alpha_j}(\alpha_j\,;\beta_j\,;t) \qquad (j = 1, 2, 3) \end{aligned} \tag{11.3.10}$$

where

$$H_1 = -R(\alpha_1, \alpha_2, \alpha_3, \beta_1, \beta_2, \beta_3, t)$$

and the dependence of H_1 on the time t is through the coordinates (X_j, Y_j, Z_j) of the disturbing planets. In principle, we could obtain approximate expressions for the time dependence of the coordinates

(X_j, Y_j, Z_j) by solving, for each planet, the corresponding unperturbed H_0–problem. These approximate solutions could be substituted into H_1 as it appears in (11.3.10) and approximate values of the time dependence of the α_j and β_j obtained. This could be done for each planet and hence improved values of the $(\alpha_j; \beta_j)$ obtained for each planet. These improved values could then be used to obtain a higher-order approximation by repeating the procedure. Each successive iteration would introduce a higher power of the masses and a higher order of approximation.

11.4. THE UNPERTURBED SOLUTION

We consider the problem of solving the canonical equations (11.3.7) for the special case that $H_1 = 0$. It will be convenient to use the spherical coordinates (r, Λ, B) in the heliocentric ecliptic system depicted in Figure 11.4.1. Here Λ is the celestial longitude, B is the celestial latitude, and we have the extended point transformation

$$\begin{aligned} X &= r \cos B \cos \Lambda \\ Y &= r \cos B \sin \Lambda \\ Z &= r \sin B \end{aligned} \tag{11.4.1}$$

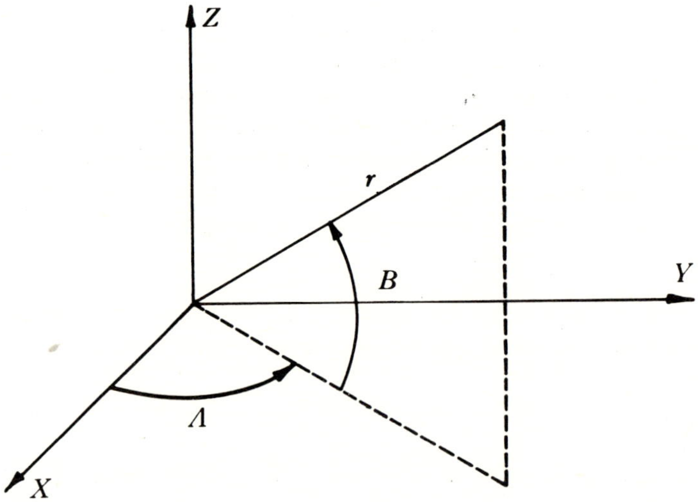

FIGURE 11.4.1

The Lagrangian function L_0, for the unperturbed problem, is

$$L_0 = \frac{1}{2}(\dot{X}^2 + \dot{Y}^2 + \dot{Z}^2) + \frac{\mu}{r} \tag{11.4.2}$$

If Equations (11.4.1) are used to express L_0 in terms of the spherical coordinates, it is found that

$$L_0 = \frac{1}{2}[\dot{r}^2 + (r\cos B)^2\dot{\Lambda}^2 + r^2\dot{B}^2] + \frac{\mu}{r} \tag{11.4.3}$$

The momenta conjugate to the coordinates (r, Λ, B) are given by

$$\begin{aligned} p_r &= \frac{\partial L}{\partial \dot{r}} = \dot{r} \\ p_\Lambda &= \frac{\partial L}{\partial \dot{\Lambda}} = (r^2\cos^2 B)\dot{\Lambda} \\ p_B &= \frac{\partial L}{\partial \dot{B}} = r^2\dot{B} \end{aligned} \tag{11.4.4}$$

Using the definition of the Hamiltonian function, we write

$$\begin{aligned} H_0 &= p_r\dot{r} + p_\Lambda\dot{\Lambda} + p_B\dot{B} - L_0 \\ &= \frac{1}{2}\left(p_r^2 + \frac{p_\Lambda^2}{r^2\cos^2 B} + \frac{p_B^2}{r^2}\right) - \frac{\mu}{r} = \alpha_1 \end{aligned} \tag{11.4.5}$$

where α_1 is a constant. Thus the differential equation from which $W_0^*(r, \Lambda, B, \alpha_1, \alpha_2, \alpha_3)$ is to be determined is

$$\left(\frac{\partial W_0^*}{\partial r}\right)^2 + \frac{1}{r^2\cos^2 B}\left(\frac{\partial W_0^*}{\partial \Lambda}\right)^2 + \frac{1}{r^2}\left(\frac{\partial W_0^*}{\partial B}\right)^2 = 2\alpha_1 + \frac{2\mu}{r} \tag{11.4.6}$$

Equation (11.4.6) can be solved by the method of separation of variables, if we assume a solution W_0^* of the form

$$W_0^* = W_{01}^*(r) + W_{02}^*(\Lambda) + W_{03}^*(B) \tag{11.4.7}$$

Substituting the assumed solution into (11.4.6), we obtain

$$\left(\frac{dW_{01}^*}{dr}\right)^2 + \frac{1}{r^2\cos^2 B}\left(\frac{dW_{02}^*}{d\Lambda}\right)^2 + \frac{1}{r^2}\left(\frac{dW_{03}^*}{dB}\right)^2 = 2\alpha_1 + 2\frac{\mu}{r}$$

which, after multiplying each side by r^2, leads to the equations

$$\sec^2 B \left(\frac{dW_{02}^*}{d\Lambda}\right)^2 + \left(\frac{dW_{03}^*}{dB}\right)^2 = 2\alpha_1 r^2 + 2\mu r - r^2 \left(\frac{dW_{01}^*}{dr}\right)^2 = \alpha_2{}^2$$

where $\alpha_2{}^2$ is a separation constant. Rewriting these equations, we have

$$\left(\frac{dW_{01}^*}{dr}\right)^2 = 2\alpha_1 + \frac{2\mu}{r} - \frac{\alpha_2{}^2}{r^2} \tag{11.4.8}$$

$$\left(\frac{dW_{03}^*}{dB}\right)^2 + \sec^2 B \left(\frac{dW_{02}^*}{d\Lambda}\right)^2 = \alpha_2{}^2 \tag{11.4.9}$$

Since (11.4.9) can be further separated into

$$\left(\frac{dW_{02}^*}{d\Lambda}\right)^2 = \frac{\alpha_2{}^2}{\sec^2 B} - \frac{1}{\sec^2 B}\left(\frac{dW_{03}^*}{dB}\right)^2 = \alpha_3{}^2 \tag{11.4.10}$$

where $\alpha_3{}^2$ is a separation constant, we thus obtain the following three ordinary differential equations

$$p_r = \frac{dW_{01}^*}{dr} = \left(2\alpha_1 + \frac{2\mu}{r} - \frac{\alpha_2{}^2}{r^2}\right)^{1/2} \tag{11.4.11a}$$

$$p_\Lambda = \frac{dW_{02}^*}{d\Lambda} = \alpha_3 \tag{11.4.11b}$$

$$p_B = \frac{dW_{03}^*}{dB} = (\alpha_2{}^2 - \alpha_3{}^2 \sec^2 B)^{1/2} \tag{11.4.11c}$$

for the regime in which $\dot{r}$, $\dot{\Lambda}$, and $\dot{B}$ are nonnegative. Other regimes can be handled in a similar fashion by an appropriate choice of sign. Integrating (11.4.11), we can write

$$W_0{}^* = \int_{r_1}^{r} \left(2\alpha_1 + \frac{2\mu}{r} - \frac{\alpha_2{}^2}{r^2}\right)^{1/2} dr + \int_0^B (\alpha_2{}^2 - \alpha_3{}^2 \sec^2 B)^{1/2}\, dB + \alpha_3 \Lambda \tag{11.4.12}$$

As will be seen shortly, the arbitrary lower limit r_1 which appears on the first integral in (11.4.12) is permissible because it will be chosen so that no new independent constant (that is, in addition to α_1, α_2, α_3) is introduced in $W_0{}^*$. Since p_B must be real, we must have $\alpha_2{}^2 \geqslant \alpha_3{}^2 \sec^2 B$ This implies that

$$|\alpha_2| \geqslant |\alpha_3|\,|\sec B| \geqslant |\alpha_3|$$

If we set

$$J = (2\alpha_1 r^2 + 2\mu r - \alpha_2^2)^{1/2}$$
$$N = (\alpha_2^2 - \alpha_3^2 \sec^2 B)^{1/2}$$

we can write the generating function $W_0 = W_0^* - \alpha_1 t$ in the form

$$W_0 = \int_{r_1}^{r} \frac{J}{r}\, dr + \int_0^B N\, dB + \alpha_3 \Lambda - \alpha_1 t \tag{11.4.13}$$

The transformation equations that involve the new variables β_1, β_2, and β_3 are

$$-\beta_1 = \frac{\partial W_0}{\partial \alpha_1} = \int_{r_1}^{r} \frac{\partial}{\partial \alpha_1}\left(\frac{J}{r}\right) dr + \frac{\partial r_1}{\partial \alpha_1}\left[\frac{J}{r}\right]_{r=r_1} - t$$

$$-\beta_2 = \frac{\partial W_0}{\partial \alpha_2} = \int_{r_1}^{r} \frac{\partial}{\partial \alpha_2}\left(\frac{J}{r}\right) dr + \frac{\partial r_1}{\partial \alpha_2}\left[\frac{J}{r}\right]_{r=r_1} + \int_0^B \frac{\partial N}{\partial \alpha_2}\, dB \tag{11.4.14}$$

$$-\beta_3 = \frac{\partial W_0}{\partial \alpha_3} = \frac{\partial r_1}{\partial \alpha_3}\left[\frac{J}{r}\right]_{r=r_1} + \int_0^B \frac{\partial N}{\partial \alpha_3}\, dB + \Lambda$$

where, in writing (11.4.14), we have made use of the fact that the old coordinates are to be treated as being independent of the new coordinates. If we choose r_1 to be a zero of J/r, then

$$\left[\frac{J}{r}\right]_{r=r_1} = 0 \tag{11.4.15}$$

and we can use (11.4.15) to express r_1 in terms of the canonical constants α_1 and α_2. With this restriction on r_1, after computing the partial derivatives of the integrands appearing in (11.4.14), we can write

$$t - \beta_1 = I_0, \qquad -\beta_2 = -\alpha_2(I_1 - I_2), \qquad -\beta_3 = \Lambda - \alpha_3 I_3 \tag{11.4.16}$$

where

$$I_0 = \int_{r_1}^{r} \frac{r}{J}\, dr, \qquad I_1 = \int_{r_1}^{r} \frac{1}{rJ}\, dr$$
$$I_2 = \int_0^B \frac{1}{N}\, dB, \qquad I_3 = \int_0^B \frac{\sec^2 B}{N}\, dB \tag{11.4.17}$$

The integrals I_0, I_1, I_2, and I_3 defined above are elementary and can readily be evaluated (see Exercise 4 at end of chapter). In what follows, we consider the motion when α_1 is negative. For this case, the explicit equations of canonical transformation are

$$\begin{aligned}
t - \beta_1 &= \left[\frac{J}{2\alpha_1} + \frac{\mu}{2\alpha_1(-2\alpha_1)^{1/2}} \sin^{-1}\left\{\frac{2\alpha_1 r + \mu}{(\mu^2 + 2\alpha_1\alpha_2{}^2)^{1/2}}\right\}\right]_{r_1}^{r} \\
\beta_2 &= \left[\sin^{-1}\left\{\frac{\mu r - \alpha_2{}^2}{r(\mu^2 + 2\alpha_1\alpha_2{}^2)^{1/2}}\right\}\right]_{r_1}^{r} - \sin^{-1}\left\{\frac{\sin B}{[1 - (\alpha_3{}^2/\alpha_2{}^2)]^{1/2}}\right\} \\
A &- \sin^{-1}\left\{\frac{\tan B}{[(\alpha_2{}^2/\alpha_3{}^2) - 1]^{1/2}}\right\} = -\beta_3 \\
p_r &= \frac{J}{r}, \qquad p_A - \alpha_3, \qquad p_B = N
\end{aligned} \tag{11.4.18}$$

The formal solution to the unperturbed problem is contained in (11.4.18), although the character of the motion is not immediately evident from the form of these equations. The primary interest here, however, is simply to identify the canonical variables with the constant orbital elements (a, e, i, ω, Ω, T) so that we can relate the perturbations of these canonical variables to the perturbations of the more easily interpreted orbital elements. Thus we assume a priori knowledge that the solutions of (11.4.18) are conic sections. Furthermore, we restrict our considerations to the case of elliptic orbits, for which the total energy α_1 is negative. The identification is developed in the next section.

11.5. IDENTIFICATION OF CANONICAL VARIABLES WITH THE ORBITAL ELEMENTS FOR $\alpha_1 < 0$

Return to (11.4.11) and observe that the radial momentum component

$$p_r = \frac{J}{r} = \frac{1}{r}(2\alpha_1 r^2 + 2\mu r - \alpha_2{}^2)^{1/2}$$

must be real. We seek to determine what restrictions this places on r if $\alpha_1 < 0$. We can write

$$p_r = \frac{1}{r}[-2\alpha_1 A(r)]^{1/2} \tag{11.5.1}$$

and p_r will be real whenever the function $A(r)$, defined by the equation

$$A(r) = -r^2 - \frac{\mu}{\alpha_1} r + \frac{\alpha_2{}^2}{2\alpha_1} \tag{11.5.2}$$

is nonnegative. The expression defining $A(r)$ can be written

$$A(r) = -\left(r^2 + \frac{\mu}{\alpha_1} r - \frac{\alpha_2{}^2}{2\alpha_1}\right)$$
$$= -\left[\left(r + \frac{\mu}{2\alpha_1}\right)^2 - \frac{\mu^2 + 2\alpha_1\alpha_2{}^2}{4\alpha_1{}^2}\right] \tag{11.5.3}$$

Since

$$\left(r + \frac{\mu}{2\alpha_1}\right)^2 \geqslant 0$$

the expression in the square bracket has its least value when $r = -\mu/2\alpha_1$. A negative sign precedes the square bracket in (11.5.3) so that the parabola $A(r)$ opens downward and will have its maximum value

$$A_{\max} = \frac{\mu^2 + 2\alpha_1\alpha_2{}^2}{4\alpha_1{}^2} \tag{11.5.4}$$

when $r = -\mu/2\alpha_1$.

The zeros of $A(r)$ are the two solutions

$$r = -\frac{\mu}{2\alpha_1} \mp \frac{1}{2}\left(\frac{\mu^2 + 2\alpha_1\alpha_2{}^2}{\alpha_1{}^2}\right)^{1/2}$$

of the quadratic equation

$$r^2 + \frac{\mu}{\alpha_1} r - \frac{\alpha_2{}^2}{2\alpha_1} = 0$$

The zeros r_1 and r_2, corresponding to the minus and plus signs respectively, will be real and distinct with $r_1 < r_2$ if $A_{\max}$ is positive, that is, if

$$\mu^2 + 2\alpha_1\alpha_2{}^2 > 0$$

With this restriction on $A_{\max}$, p_r will be real for all values of r in the interval $0 < r_1 \leqslant r \leqslant r_2$. Since α_1 is negative, it is clear that r_1 is positive. The sum and product of the two roots are given by the equations

$$r_1 + r_2 = -\frac{\mu}{\alpha_1}, \qquad r_1 r_2 = -\frac{\alpha_2{}^2}{2\alpha_1} \tag{11.5.5}$$

Note that r_1 is expressed in terms of α_1 and α_2 and does not introduce any new independent constants. The minimum and maximum values of r correspond to the pericentric and apocentric points of an elliptic orbit. We have, therefore, in the usual notation

$$r_1 = a(1 - e), \qquad r_2 = a(1 + e)$$

Substituting for r_1 and r_2, expressed in terms of the canonical constants, we find

$$\alpha_1 = -\frac{\mu}{2a}, \qquad \alpha_2 = [\mu a(1 - e^2)]^{1/2} = (\mu p^*)^{1/2} = h \tag{11.5.6}$$

The third equation of (11.4.18) can be written

$$\sin(\Lambda + \beta_3) \cos B = \left(\frac{\alpha_2^2}{\alpha_3^2} - 1\right)^{-1/2} \sin B \tag{11.5.7}$$

or, alternatively,

$$\cos \beta_3 \sin \Lambda \cos B + \sin \beta_3 \cos \Lambda \cos B - \left(\frac{\alpha_2^2}{\alpha_3^2} - 1\right)^{-1/2} \sin B = 0$$

If the equations of transformation (11.4.1) are introduced, we have

$$X \sin \beta_3 + Y \cos \beta_3 - \left(\frac{\alpha_2^2}{\alpha_3^2} - 1\right)^{-1/2} Z = 0 \tag{11.5.8}$$

Since the coefficients of the rectangular coordinates are constants, it is clear from (11.5.8) that the motion takes place in a plane that contains the Sun and is inclined at some angle i to the ecliptic plane. From spherical trigonometry, we have

$$\cos B \sin(\Lambda - \Omega) = \sin u \cos i$$
$$\sin B = \sin i \sin u$$

whence, dividing one equation by the other, we have

$$\sin(\Lambda - \Omega) = \cot i \tan B \tag{11.5.9}$$

Let us set $\alpha_3/\alpha_2 = \cos i$, where i is a constant, $0 < i < \pi/2$. Then

$$\left(\frac{\alpha_2^2}{\alpha_3^2} - 1\right)^{1/2} = \tan i$$

and (11.5.7) can be written

$$\sin(\Lambda + \beta_3) = \cot i \tan B \tag{11.5.10}$$

If we compare (11.5.9) and (11.5.10), we can identify

$$\beta_3 = -\Omega$$
$$i = \text{angle of inclination} \tag{11.5.11}$$

The constant β_1 can readily be identified with a time of perihelion passage T. With reference to the first equation in (11.4.18), when the particle passes through pericenter, that is, $r = r_1$,

$$T - \beta_1 = 0$$

or

$$\beta_1 = T \tag{11.5.12}$$

where T is a time of pericentric passage.

One canonical constant remains to be identified, namely, β_2. To relate β_2 to the orbital elements, we proceed as follows. Return to the second equation in (11.4.18) and designate the first term on the right-hand side of the equation by $f(r)$. We can then write

$$\beta_2 = f(r) - \sin^{-1}\left(\frac{\sin B}{\sin i}\right)$$

or, since $\sin B = \sin i \sin u$,

$$u + \beta_2 = f(r)$$

When $r = r_1$, then $u = \omega$, the argument of perihelion, whence we have

$$\beta_2 = -\omega \tag{11.5.13}$$

In summary, we write the following relationships:

$$\begin{aligned} \alpha_1 &= -\frac{\mu}{2a}, & \beta_1 &= T \\ \alpha_2 &= [\mu a(1 - e^2)]^{1/2} = h, & \beta_2 &= -\omega \\ \alpha_3 &= h \cos i, & \beta_3 &= -\Omega \end{aligned} \tag{11.5.14}$$

The inverse relations are

$$\begin{aligned} a &= -\frac{\mu}{2\alpha_1}, & \omega &= -\beta_2 \\ e &= (1 + 2\alpha_1\alpha_2^2\mu^{-2})^{1/2}, & \Omega &= -\beta_3 \\ i &= \cos^{-1}\left(\frac{\alpha_3}{\alpha_2}\right), & T &= \beta_1 \\ \bar{\omega} &= \Omega + \omega = -\beta_2 - \beta_3, & n &= \frac{(-2\alpha_1)^{3/2}}{\mu} \\ \epsilon &= \bar{\omega} + n(t_0 - T) = -\beta_2 - \beta_3 + \frac{(-2\alpha_1)^{3/2}(t_0 - \beta_1)}{\mu} \end{aligned} \tag{11.5.15}$$

11.6. FURTHER COMMENTS ON THE PERTURBED SOLUTION

In Section 11.3 we began the study of the motion of a planet about the Sun for which the Hamiltonian H is of the form

$$H = H_0(X, Y, Z, p_X, p_Y, p_Z) + H_1(X, Y, Z; X_j, Y_j, Z_j)$$

where

$$H_0 = \frac{1}{2}(p_X^2 + p_Y^2 + p_Z^2) - \frac{\mu}{r}$$

$$H_1 = -R(X, Y, Z; X_j, Y_j, Z_j) \qquad (j = 1, 2, \ldots, n-2)$$

An exact solution has been obtained to the unperturbed problem in which we took the Hamiltonian to be simply H_0. We expressed H_0 in spherical coordinates and then separated the associated Hamilton–Jacobi equation to obtain a generator W_0 of a transformation from the old coordinates and momenta $(r, \Lambda, B, p_r, p_\Lambda, p_B)$ to new coordinates and momenta $(\alpha_1, \alpha_2, \alpha_3, \beta_1, \beta_2, \beta_3)$ which are canonical constants. The transformation equations (11.4.18) represent the solution to the unperturbed problem because they permit us to determine the spherical coordinates (r, Λ, B) and their conjugate momenta (p_r, p_Λ, p_B) as functions of the canonical constants and the time t.

To obtain the complete solution to the perturbed problem, we must obtain the time variations of the canonical variables $(\alpha_1, \alpha_2, \alpha_3, \beta_1, \beta_2, \beta_3)$ from the differential equations (11.3.10), where it is understood that R is to be considered a function of the canonical variables through the transformation equations. The explicit dependence of R upon t enters through the coordinates of the perturbing planets.

Equations (11.3.10) are to be integrated to determine explicit expressions for the time dependence of $(\alpha_j; \beta_j)$, $j = 1, 2, 3$. These expressions are to be substituted for the canonical "constants" in (11.4.18). We will then have the complete solution to the perturbed problem.

In earlier chapters we obtained a solution to the unperturbed problem, which, for elliptic motion, was expressed by the equations

$$\begin{aligned}
\mu &= k^2(m_0 + m) = n^2a^3 \\
M &= \epsilon + n(t - t_0) - \bar{\omega} = n(t - T) \\
M &= E - e \sin E, \qquad r = a(1 - e \cos E) \qquad\qquad (11.6.1) \\
\tan\frac{w}{2} &= \left(\frac{1+e}{1-e}\right)^{1/2} \tan\frac{E}{2}, \qquad u = w + \bar{\omega} - \Omega \\
\sin B &= \sin i \sin u \qquad \tan(\Lambda - \Omega) = \cos i \tan u
\end{aligned}$$

Equations (11.6.1) are, of course, entirely equivalent to (11.4.18). Since we have relations between the canonical variables and the orbital elements in (11.5.15), we may, if we wish, use (11.6.1) to express the solution to the perturbed problem if it is understood that osculating elements (obtained from a knowledge of the time dependence of the canonical variables) are to be used in these equations.

The Lagrangian L for the perturbed problem is

$$L = L_0 + R(r, \Lambda, B;\ X_j\,, Y_j\,, Z_j) \tag{11.6.2}$$

Since R is free of the velocities, then $\partial L/\partial \dot{r} = \partial L_0/\partial \dot{r}$, etc. Hence the momenta equations for the perturbed and unperturbed problem are of the same form. Thus the equations

$$p_r = \frac{J}{r} = \dot{r}, \qquad p_\Lambda = (r^2 \cos^2 B)\dot{\Lambda} = \alpha_3\,, \qquad p_B = r^2\dot{B} = N$$

can be used to calculate the instantaneous velocities $\dot{r}$, $\dot{\Lambda}$, and $\dot{B}$ for the perturbed problem provided that instantaneous values of both the spherical coordinates and the $(\alpha_j; \beta_j)$ are used. This, of course, corresponds to the use of osculating elements in our earlier development of the Lagrange planetary equations for the orbital elements.

The Lagrange planetary equations for a specific set of orbital elements are readily derivable from the results thus far obtained in our study of Hamilton mechanics. To illustrate, we will obtain the Lagrange planetary equations for the orbital elements a, e, i, Ω, $\bar{\omega}$, ϵ.

Denoting the orbital elements by e_k, $k = 1, 2, \ldots, 6$, and using the fact that the e_k depend on the canonical variables $(\alpha_j\,; \beta_j)$, $j = 1, 2, 3$, through (11.5.15), we can write, for each k,

$$e_k = e_k(\alpha_1\,, \alpha_2\,, \alpha_3\,, \beta_1\,, \beta_2\,, \beta_3) \tag{11.6.3}$$

The time derivative of e_k is given by

$$\dot{e}_k = \sum_{j=1}^{3} \left(\frac{\partial e_k}{\partial \alpha_j}\,\dot{\alpha}_j + \frac{\partial e_k}{\partial \beta_j}\,\dot{\beta}_j\right) \tag{11.6.4}$$

If the perturbing function R is introduced, we can write

$$\dot{e}_k = \sum_{j=1}^{3} \left(-\frac{\partial R}{\partial \beta_j}\,\frac{\partial e_k}{\partial \alpha_j} + \frac{\partial e_k}{\partial \beta_j}\,\frac{\partial R}{\partial \alpha_j}\right) \tag{11.6.5}$$

The perturbing function can also be considered as a function of the e_k and the time t. In which case we can write

$$\frac{\partial R}{\partial \alpha_j} = \sum_{l=1}^{6} \frac{\partial R}{\partial e_l} \frac{\partial e_l}{\partial \alpha_j}$$
$$\frac{\partial R}{\partial \beta_j} = \sum_{l=1}^{6} \frac{\partial R}{\partial e_l} \frac{\partial e_l}{\partial \beta_j} \qquad (j = 1, 2, 3) \tag{11.6.6}$$

If (11.6.6) in substituted into (11.6.5), we find that

$$\dot{e}_k = \sum_{j=1}^{3} \frac{\partial e_k}{\partial \beta_j} \Big(\sum_{l=1}^{6} \frac{\partial R}{\partial e_l} \frac{\partial e_l}{\partial \alpha_j} \Big) - \sum_{j=1}^{3} \frac{\partial e_k}{\partial \alpha_j} \Big(\sum_{l-1}^{6} \frac{\partial R}{\partial e_l} \frac{\partial e_l}{\partial \beta_j} \Big)$$

When the order of summation is interchanged, we have

$$\dot{e}_k = \sum_{l=1}^{6} \frac{\partial R}{\partial e_l} \Big[\sum_{j=1}^{3} \Big(\frac{\partial e_l}{\partial \alpha_j} \frac{\partial e_k}{\partial \beta_j} - \frac{\partial e_l}{\partial \beta_j} \frac{\partial e_k}{\partial \alpha_j} \Big) \Big]$$

or

$$\dot{e}_k = \sum_{l=1}^{6} \{e_l , e_k\} \frac{\partial R}{\partial e_l} \qquad (k = 1, 2, \ldots, 6) \tag{11.6.7}$$

where

$$\{e_l , e_k\} = \sum_{j=1}^{3} \Big(\frac{\partial e_l}{\partial \alpha_j} \frac{\partial e_k}{\partial \beta_j} - \frac{\partial e_l}{\partial \beta_j} \frac{\partial e_k}{\partial \alpha_j} \Big) \tag{11.6.8}$$

is called a Poisson bracket for the elements e_l and e_k. As with the Lagrange bracket, it is clear from the definition that

$$\{e_l , e_l\} = 0 \tag{11.6.9}$$

$$\{e_l , e_k\} = -\{e_k , e_l\} \tag{11.6.10}$$

Equations (11.6.7) represent the Lagrange planetary equations. Thirty-six Poisson brackets appear in these equations, but, in view of (11.6.9) and (11.6.10), we have to compute only fifteen brackets. The procedure is straightforward. To illustrate, we determine $\{a, \epsilon\}$, using the identification

$$\begin{aligned} e_1 &= a, & e_4 &= \Omega \\ e_2 &= e, & e_5 &= \bar{\omega} \\ e_3 &= i, & e_6 &= \epsilon \end{aligned} \tag{11.6.11}$$

From the definition, we write

$$\{a, \epsilon\} = \sum_{j=1}^{3} \left(\frac{\partial a}{\alpha_j} \frac{\partial \epsilon}{\partial \beta_j} - \frac{\partial a}{\partial \beta_j} \frac{\partial \epsilon}{\partial \alpha_j} \right) = \frac{\partial a}{\partial \alpha_1} \frac{\partial \epsilon}{\partial \beta_1} = \frac{-2}{na} \tag{11.6.12}$$

Similarly, the remaining nonzero Poisson brackets above the diagonal in the 6×6 array of brackets appearing in (11.6.7) are found to be

$$\{e, \bar{\omega}\} = \frac{\cot \phi}{na^2}, \qquad \{i, \bar{\omega}\} = \frac{\tan(i/2)}{na^2 \cos \phi} = \{i, \epsilon\}$$

$$\{e, \epsilon\} = \frac{\tan(\phi/2) \cos \phi}{na^2}, \qquad \{i, \Omega\} = \frac{1}{na^2 \sin i \cos \phi}$$

where $\sin \phi = e$.

The Lagrange planetary equations (11.6.7) now take the form

$$\begin{aligned}
\dot{a} &= \frac{2}{na} \frac{\partial R}{\partial \epsilon} \\
\dot{e} &= -\frac{\cot \phi}{na^2} \frac{\partial R}{\partial \bar{\omega}} - \frac{\tan(\phi/2) \cos \phi}{na^2} \frac{\partial R}{\partial \epsilon} \\
(i)^{\boldsymbol{\cdot}} &= -\frac{1}{na^2 \sin i \cos \phi} \frac{\partial R}{\partial \Omega} - \frac{\tan(i/2)}{na^2 \cos \phi} \left(\frac{\partial R}{\partial \bar{\omega}} + \frac{\partial R}{\partial \epsilon} \right) \\
\dot{\Omega} &= \frac{1}{na^2 \sin i \cos \phi} \frac{\partial R}{\partial i} \\
\dot{\bar{\omega}} &= \frac{\cot \phi}{na^2} \frac{\partial R}{\partial e} + \frac{\tan(i/2)}{na^2 \cos \phi} \frac{\partial R}{\partial i} \\
\dot{\epsilon} &= -\frac{2}{na} \frac{\partial R}{\partial a} + \frac{\tan(\phi/2) \cos \phi}{na^2} \frac{\partial R}{\partial e} + \frac{\tan(i/2)}{na^2 \cos \phi} \frac{\partial R}{\partial i}
\end{aligned} \tag{11.6.13}$$

for elements $a, e, i, \Omega, \bar{\omega}, \epsilon$.

There is a difficulty involved in using (11.6.13) to study the planetary n-body problem. The nature of this difficulty and a way of circumventing it will be described in the remaining part of this section. To reduce the problem to its essential terms, we consider the perturbing function R of (11.3.1) for the case of only one perturbing planet. If we represent the perturbing planet by m_1, we can write

$$R = k^2 m_1 \left(\frac{1}{\rho_1} - \frac{XX_1 + YY_1 + ZZ_1}{r_1^3} \right) \tag{11.6.14}$$

where ρ_1 is the distance from m to m_1 and r_1 is the distance from m_0 to

m_1. The unsubscripted and subscripted coordinates refer to the masses m and m_1 respectively.

It can be shown that the disturbing function R, given in (11.6.14), can be developed as a sum (see Plummer [7]):

$$R = \sum F(a, a_1, e, e_1, i, i_1) \cos\{h[\epsilon + n(t - t_0)] + h_1[\epsilon_1 + n_1(t - t_0)] + g\bar{\omega} + g_1\bar{\omega}_1 + f\Omega + f_1\Omega_1\} \tag{11.6.15}$$

where, in the coefficients F of the trigonometric terms, a, e, i, Ω, $\bar{\omega}$, ϵ and a_1, e_1, i_1, Ω_1, $\bar{\omega}_1$, ϵ_1 are the osculating elements of the perturbed and perturbing orbits respectively. The h, h_1, g, g_1, f, f_1 are integers positive, negative, or zero, and the summation extends over all possible values of these integers; t_0 is the epoch time of the mean longitude L. Since $L = M + \bar{\omega} = \epsilon + n(t - t_0)$ when $t = t_0$, we have $L_0 = \epsilon$.

It is to be noted in (11.6.15) that the element a appears explicitly in the coefficients of the series and implicitly through n in the arguments of the trigonometric terms. If we refer to the Lagrange planetary equations (11.6.13), we observe that $\partial R/\partial a$ appears in only one equation, namely, the equation for ϵ. The appearance of a in both the coefficients and trigonometric arguments of the series in R is awkward because the formation of $\partial R/\partial a$ will introduce Poisson terms in which we have the products of t and trigonometric terms appearing in the $\dot{\epsilon}$ equation. These Poisson terms introduce inconveniently large amplitudes in the subsequent analysis, and hence it is desirable to eliminate the difficulty if possible.

If we take the partial derivative of R with respect to n, considering a to be constant, we obtain

$$\left(\frac{\partial R}{\partial n}\right)_a = -(t - t_0) \sum hF(a, a_1, e, e_1, i, i_1) \sin A \tag{11.6.16}$$

where we have set

$$A = h[\epsilon + n(t - t_0)] + h_1[\epsilon_1 + n_1(t - t_0)] + g\bar{\omega} + g_1\bar{\omega}_1 + f\Omega + f_1\Omega_1$$

If we form $\partial R/\partial \epsilon$ we obtain

$$\frac{\partial R}{\partial \epsilon} = -\sum hF(a, a_1, e, e_1, i, i_1) \sin A \tag{11.6.17}$$

Comparing (11.6.16) and (11.6.17), we see that

$$\left(\frac{\partial R}{\partial n}\right)_a = (t - t_0) \frac{\partial R}{\partial \epsilon} \tag{11.6.18}$$

On the other hand, we note that

$$\frac{\partial R}{\partial a} = \left(\frac{\partial R}{\partial a}\right)_n + \left(\frac{\partial R}{\partial n}\right)_a \frac{dn}{da} \tag{11.6.19}$$

If we substitute from (11.6.18) into (11.6.19), we obtain

$$\frac{\partial R}{\partial a} = \left(\frac{\partial R}{\partial a}\right)_n + (t - t_0)\frac{dn}{da}\frac{\partial R}{\partial \epsilon} \tag{11.6.20}$$

Now substituting for $\partial R/\partial\epsilon$ from the first equation of (11.6.13), we find

$$\frac{\partial R}{\partial a} = \left(\frac{\partial R}{\partial a}\right)_n + \frac{na(t - t_0)}{2}\frac{dn}{dt} \tag{11.6.21}$$

The last equation of (11.6.13) may now be written

$$\dot{\epsilon} + (t - t_0)\dot{n} = \frac{-2}{na}\left(\frac{\partial R}{\partial a}\right)_n + \frac{\tan(\phi/2)\cos\phi}{na^2}\frac{\partial R}{\partial e} + \frac{\tan(i/2)}{na^2\cos\phi}\frac{\partial R}{\partial i} \tag{11.6.22}$$

If we define a new element ϵ' by the equation

$$\epsilon' = \epsilon + n(t - t_0) - \int_{t_0}^{t} n\,dt \tag{11.6.23}$$

then

$$\dot{\epsilon}' = \dot{\epsilon} + n + \dot{n}(t - t_0) - n = \dot{\epsilon} + \dot{n}(t - t_0) \tag{11.6.24}$$

Since the mean longitude $L = \epsilon + n(t - t_0)$, in terms of the element ϵ', we have

$$L = \epsilon' + \rho \tag{11.6.25}$$

where we have set

$$\rho = \int_{t_0}^{t} n\,dt \tag{11.6.26}$$

If $\epsilon + n(t - t_0)$ is replaced by $\epsilon' + \rho$ in (11.6.15), we see that

$$\frac{\partial R}{\partial \epsilon'} = \frac{\partial R}{\partial \epsilon}$$

Thus the Lagrange planetary equations (11.6.13) will have exactly the same form for the elements a e, i, Ω, $\bar{\omega}$, ϵ' as for the elements a, e, i, Ω, $\bar{\omega}$, ϵ. Explicitly, we have for the elements a, e, i, Ω, $\bar{\omega}$, ϵ'

$$\begin{aligned}
\dot{a} &= \frac{2}{na}\frac{\partial R}{\partial \epsilon'} \\
\dot{e} &= -\frac{\cot\phi}{na^2}\frac{\partial R}{\partial \bar{\omega}} - \frac{\tan(\phi/2)\cos\phi}{na^2}\frac{\partial R}{\partial \epsilon'} \\
(i)^{\cdot} &= -\frac{1}{na^2 \sin i \cos\phi}\frac{\partial R}{\partial \Omega} - \frac{\tan(i/2)}{na^2\cos\phi}\left(\frac{\partial R}{\partial \bar{\omega}} + \frac{\partial R}{\partial \epsilon'}\right) \\
\dot{\Omega} &= \frac{1}{na^2 \sin i \cos\phi}\frac{\partial R}{\partial i} \\
\dot{\bar{\omega}} &= \frac{\cot\phi}{na^2}\frac{\partial R}{\partial \epsilon'} + \frac{\tan(i/2)}{na^2\cos\phi}\frac{\partial R}{\partial i} \\
\dot{\epsilon}' &= \frac{-2}{na}\left(\frac{\partial R}{\partial a}\right)_n + \frac{\tan(\phi/2)\cos\phi}{na^2}\frac{\partial R}{\partial e} + \frac{\tan(i/2)}{na^2\cos\phi}\frac{\partial R}{\partial i}
\end{aligned} \tag{11.6.27}$$

and

$$L = \epsilon' + \int_{t_0}^{t} n\,dt \tag{11.6.28}$$

The penalty we must pay for replacing ϵ by ϵ' appears through (11.6.28). In order to determine the mean longitude L, two integrations are necessary. One, to determine n, and then the integration over n from t_0 to t. Equation (11.6.28) is not in a very convenient form for calculating L. It is desirable to express $\rho = \int_{t_0}^{t} n\,dt$ in terms of the unperturbed contribution, plus the contributions due to the introduction of the perturbing function R. To this end, we write

$$\rho = \int_{t_0}^{t} n\,dt, \qquad \dot{\rho} = n = \mu^{1/2}a^{-3/2}$$

$$\ddot{\rho} = \dot{n} = \frac{-3}{2}\mu^{1/2}a^{-5/2}\dot{a} = \frac{-3}{2}\mu^{1/2}a^{-5/2}\frac{2}{na}\frac{\partial R}{\partial \epsilon'} = \frac{-3}{a^2}\frac{\partial R}{\partial \epsilon'}$$

for ρ and its first two time derivatives. To integrate this last equation between t and t_0, we first write

$$d\dot{\rho} = \frac{-3}{a^2}\frac{\partial R}{\partial \epsilon'}\,dt$$

If we let zero subscripts indicate evaluation at $t = t_0$, we can write

$$\int_{\dot{\rho}_0}^{\dot{\rho}} d\dot{\rho} = -3 \int_{t_0}^{t} \frac{1}{a^2} \frac{\partial R}{\partial \epsilon'} \, dt$$

Since $\dot{\rho} = n$, it follows that $\dot{\rho} - \dot{\rho}_0 = n - n_0$. Thus $n = n_0 - 3 \int_{t_0}^{t} (1/a^2)(\partial R/\partial \epsilon') \, dt$. Substituting for n in the definition of ρ, we have

$$\begin{aligned} \rho &= \int_{t_0}^{t} \left(n_0 - 3 \int_{t_0}^{t} \frac{1}{a^2} \frac{\partial R}{\partial \epsilon'} \, dt \right) dt \\ &= n_0(t - t_0) - 3 \int_{t_0}^{t} \left(\int_{t_0}^{t} \frac{1}{a^2} \frac{\partial R}{\partial \epsilon'} \, dt \right) dt \qquad (11.6.29) \end{aligned}$$

A number of authors have been active in the development of more modern perturbation theories. A new perturbation method introduced by Arenstorf, which is based upon the foundations laid by several researchers [8–12], may be cited as an example. The Arenstorf approach is applicable to a perturbed analytic system of first-order differential equations that is sufficiently close to an associated (integrable) unperturbed autonomous system.

The method consists of transforming a perturbed system of first-order differential equations, by a suitable time-dependent coordinate transformation, to another system of first-order differential equations in such a way that the second system of differential equations is less perturbed in some sense than the original system. This leads to an iterative procedure that reduces the original system to an associated unperturbed autonomous system. This procedure is then reversed, and a method is obtained through which the solution of the perturbed system of equations can be expressed in terms of the solution of the associated unperturbed system of equations and the time. In order to use this method for a preassigned (finite) time interval, one must restrict the variables to a region that will be determined in the iterative procedure. The convergence of the successive perturbation functions to zero and the sequence of combined coordinate transformations to a transformation that expresses the perturbed solution in terms of the time and of the solution to the unperturbed problem should be very fast, whenever the method is applicable.

New perturbation theories developed by Hori [12a] and, more recently, by Deprit [12b] may also be cited as examples. Both of these theories are based upon the use of Lie series to generate successive canonical transformations which reduce the order of the equations of motion for a dynamical system (see also Section 13.2 and Section 14.9).

11.7. GENERALIZED POISSON BRACKETS AND THEIR RELATION TO GENERALIZED LAGRANGE BRACKETS

In Equation (10.1.2) we introduced a generalized definition of the Lagrange bracket for the quantities P and Q. Specifically, we have

$$[P, Q] = \sum_{j=1}^{r} \left(\frac{\partial q_j}{\partial P}\frac{\partial p_j}{\partial Q} - \frac{\partial q_j}{\partial Q}\frac{\partial p_j}{\partial P}\right)$$

where r is the number of degrees of freedom. There are $2r$ independent variables, hence $2r$ orbital elements. The quantities P and Q are any two of these orbital elements or any two of $2r$ independent functions of these orbital elements. Replace the symbols P and Q by the symbols C_l and C_i, respectively, $l = 1, 2,..., 2r$; $i = 1, 2,..., 2r$. We designate the Lagrange bracket for C_l and C_i by $L_{l,i}$ and write

$$L_{l,i} = [C_l, C_i] = \sum_{j=1}^{r} \left(\frac{\partial q_j}{\partial C_l}\frac{\partial p_j}{\partial C_i} - \frac{\partial q_j}{\partial C_i}\frac{\partial p_j}{\partial C_l}\right) \tag{11.7.1}$$

The generalized Poisson bracket for C_l and C_k, $k = 1, 2,..., 2r$, we designate by $P_{l,k} = \{C_l, C_k\}$ and we define $P_{l,k}$ by the equation

$$P_{l,k} = \{C_l, C_k\} = \sum_{j=1}^{r} \left(\frac{\partial C_l}{\partial q_j}\frac{\partial C_k}{\partial p_j} - \frac{\partial C_l}{\partial p_j}\frac{\partial C_k}{\partial q_j}\right) \tag{11.7.2}$$

There is a close connection between the Lagrange and Poisson brackets. In fact, it can be shown that (see Exercise 6 at end of chapter)

$$\sum_{l=1}^{2r} L_{l,i}P_{l,k} = \begin{cases} 1 & (i = k) \\ 0 & (i \neq k) \end{cases} \tag{11.7.3}$$

Equations (11.7.3) can be represented in the form

$$L_{1,i}P_{1,k} + L_{2,i}P_{2,k} + \cdots + L_{l,i}P_{l,k} + \cdots + L_{2r,i}P_{2r,k} = \delta_{i,k} \tag{11.7.4}$$

where $\delta_{i,k}$ is the Kronecker delta and $i = 1, 2,..., 2r$; $k = 1,.2,..., 2r$. If we fix k and let i range over all values from 1 to $2r$, we obtain the array

$$\begin{array}{lllllll}
L_{1,1}P_{1,k} & + L_{2,1}P_{2,k} & + \cdots & + L_{l,1}P_{l,k} & + \cdots & + L_{2r,1}P_{2r,k} & = \delta_{1,k} \\
L_{1,2}P_{1,k} & + L_{2,2}P_{2,k} & + \cdots & + L_{l,2}P_{l,k} & + \cdots & + L_{2r,2}P_{2r,k} & = \delta_{2,k} \\
\vdots & \vdots & & \vdots & & \vdots & \vdots \\
L_{1,2r}P_{1,k} & + L_{2,2r}P_{2,k} & + \cdots & + L_{l,2r}P_{l,k} & + \cdots & + L_{2r,2r}P_{2r,k} & = \delta_{2r,k}
\end{array} \tag{11.7.5}$$

If we define Δ to be the determinant

$$\Delta = \begin{vmatrix} L_{1,1} & L_{2,1} & \cdots & L_{l,1} & \cdots & L_{2r,1} \\ L_{1,2} & L_{2,2} & & L_{l,2} & & L_{2r,2} \\ \vdots & \vdots & & \vdots & & \vdots \\ L_{1,2r} & L_{2,2r} & & L_{l,2r} & & L_{2r,2r} \end{vmatrix} \tag{11.7.6}$$

and assume that Δ is not zero, we can solve (11.7.5) for $P_{l,k}$ to obtain

$$P_{l,k} = \frac{1}{\Delta} \begin{vmatrix} L_{1,1} & L_{2,1} & \cdots & \delta_{1,k} & \cdots & L_{2r,1} \\ L_{1,2} & L_{2,2} & & \delta_{2,k} & & L_{2r,2} \\ \vdots & \vdots & & \vdots & & \vdots \\ L_{1,2r} & L_{2,2r} & & \delta_{2r,k} & & L_{2r,2r} \end{vmatrix} \tag{11.7.7}$$

Consider the determinant in the numerator of (11.7.7). Assume that l has been chosen. This fixes the column in which the $L_{l,i}$ are replaced by the $\delta_{i,k}$. Now assume that k has been chosen. We then select $i = k$ and this determines the only nonzero entry in the $\delta_{i,k}$ column. This nonzero entry, which has the value 1, is the element that replaces $L_{l,k}$. Expanding about the entries in the $\delta_{i,k}$ column, we see that the value of the determinant is clearly equal to the value of the cofactor of $L_{l,k}$. Designate this cofactor by $L'_{l,k}$. Then we have

$$P_{l,k} = \frac{1}{\Delta} L'_{l,k} \tag{11.7.8}$$

Similarly, it can be shown that the Lagrange bracket $L_{l,k}$ can be obtained from the formula

$$L_{l,k} = \frac{1}{\Delta^*} P'_{l,k} \tag{11.7.9}$$

where

$$\Delta^* = \begin{vmatrix} P_{1,1} & P_{2,1} & \cdots & P_{2r,1} \\ P_{1,2} & P_{2,2} & & P_{2r,2} \\ \vdots & \vdots & & \vdots \\ P_{1,2r} & P_{2,2r} & & P_{2r,2r} \end{vmatrix}$$

and $P'_{l,k}$ is the cofactor of $P_{l,k}$ in Δ^*. It is clear that for $r = 3$, the determinant Δ^* is simply the determinant of the coefficients of the $\partial R/\partial e_l$, $l = 1, 2, \ldots, 6$, which appear in the Lagrange planetary equations given in (11.6.7). The Poisson bracket elements of the Δ^* determinant can be obtained from a knowledge of the corresponding Lagrange bracket elements of the Δ determinant, as will be made clear in the subsequent discussion.

The determinant $\varDelta$ can be written in the form

$$\varDelta = L_{1,1}L'_{1,1} + L_{2,1}L'_{2,1} + \cdots + L_{2r,1}L'_{2r,1} = \sum_{j=1}^{2r} L_{j,1}L'_{j,1}$$

so that

$$L'_{j,1} = \frac{\partial \varDelta}{\partial L_{j,1}} \qquad (j = 1, 2, \ldots, 2r)$$

More generally, for any element $L_{l,k}$ of $\varDelta$, we can write $L'_{l,k} = \partial\varDelta/\partial L_{l,k}$ so that (11.7.8) becomes

$$P_{l,k} = \frac{1}{\varDelta}\frac{\partial \varDelta}{\partial L_{l,k}} \tag{11.7.10}$$

If we interchange an even number of rows or columns of $\varDelta$, the value of the determinant is unchanged. If we interchange an odd number of rows or columns, the value of the determinant only changes sign. Call the determinant rearranged in any way $\varDelta_1$. Then $\varDelta_1 = \pm\varDelta$, and (11.7.10) can be written

$$P_{l,k} = \frac{1}{\varDelta_1}\frac{\partial \varDelta_1}{\partial L_{l,k}} = \frac{1}{\varDelta_1} L'_{l,k} \tag{11.7.11}$$

where $L'_{l,k}$ is the cofactor of $L_{l,k}$ in the matrix $\varDelta_1$.

Campbell's formula for computing a Lagrange bracket, in a form convenient for use with the orbital elements $e_1 = a$, $e_2 = e$, $e_3 = i$, $e_4 = \Omega$, $e_5 = \bar{\omega}$, $e_6 = \epsilon$, is

$$[P, Q] = \frac{\partial[(\epsilon - \bar{\omega})/n, -\mu/2a]}{\partial(P, Q)} + \frac{\partial(\bar{\omega} - \Omega, h)}{\partial(P, Q)} + \frac{\partial(\Omega, h\cos i)}{\partial(P, Q)} \tag{11.7.12}$$

Using this formula, and setting $e = \sin\phi$, we obtain the following nonzero Lagrange brackets:

$$\begin{aligned}
L_{1,6} &= \frac{-na}{2} = -L_{6,1} \\
L_{1,4} &= \frac{na\cos\phi(1 - \cos i)}{2} = -L_{4,1} \\
L_{2,4} &= -na^2\tan\phi(1 - \cos i) = -L_{4,2} \\
L_{1,5} &= \frac{na(1 - \cos\phi)}{2} = -L_{5,1} \\
L_{2,5} &= na^2\tan\phi = -L_{5,2} \\
L_{3,4} &= na^2\cos\phi\sin i = -L_{4,3}
\end{aligned} \tag{11.7.13}$$

The corresponding determinant Δ can be written in the form

$$\Delta = \begin{vmatrix} 0 & 0 & 0 & L_{4,1} & L_{5,1} & L_{6,1} \\ 0 & 0 & 0 & L_{4,2} & L_{5,2} & 0 \\ 0 & 0 & 0 & L_{4,3} & 0 & 0 \\ L_{1,4} & L_{2,4} & L_{3,4} & 0 & 0 & 0 \\ L_{1,5} & L_{2,5} & 0 & 0 & 0 & 0 \\ L_{1,6} & 0 & 0 & 0 & 0 & 0 \end{vmatrix}$$

We can rearrange this determinant by interchanging columns 1 and 6, 2 and 5, 3 and 4, to obtain

$$\Delta_1 = \begin{vmatrix} L_{6,1} & L_{5,1} & L_{4,1} & 0 & 0 & 0 \\ 0 & L_{5,2} & L_{4,2} & 0 & 0 & 0 \\ 0 & 0 & L_{4,3} & 0 & 0 & 0 \\ 0 & 0 & 0 & L_{3,4} & L_{2,4} & L_{1,4} \\ 0 & 0 & 0 & 0 & L_{2,5} & L_{1,5} \\ 0 & 0 & 0 & 0 & 0 & L_{1,6} \end{vmatrix} = -\Delta \tag{11.7.14}$$

Expanding Δ_1 about the entries in the first column, we find

$$\Delta_1 = L_{6,1}L_{5,2}L_{4,3}L_{3,4}L_{2,5}L_{1,6} \tag{11.7.15}$$

To obtain the elements in the determinant

$$\Delta_1{}^* = \begin{vmatrix} P_{6,1} & P_{5,1} & P_{4,1} & P_{3,1} & P_{2,1} & P_{1,1} \\ P_{6,2} & P_{5,2} & P_{4,2} & P_{3,2} & P_{2,2} & P_{1,2} \\ P_{6,3} & P_{5,3} & P_{4,3} & P_{3,3} & P_{2,3} & P_{1,3} \\ P_{6,4} & P_{5,4} & P_{4,4} & P_{3,4} & P_{2,4} & P_{1,4} \\ P_{6,5} & P_{5,5} & P_{4,5} & P_{3,5} & P_{2,5} & P_{1,5} \\ P_{6,6} & P_{5,6} & P_{4,6} & P_{3,6} & P_{2,6} & P_{1,6} \end{vmatrix} = -\Delta^* \tag{11.7.16}$$

we compute the elements that appear in each row, using (11.7.11). To calculate the elements in the first row of $\Delta_1{}^*$, for example, we write

$$P_{6,1} = -\frac{L'_{6,1}}{\Delta} = -\frac{L_{5,2}L_{4,3}L_{3,4}L_{2,5}L_{1,6}}{\Delta} = \frac{1}{L_{6,1}} = \frac{2}{na} = -P_{1,6}$$

Clearly

$$L'_{5,1} = L'_{4,1} = L'_{3,1} = L'_{2,1} = L'_{1,1} = 0$$

hence we have

$$P_{5\,1} = P_{4,1} = P_{3,1} = P_{2,1} = P_{1,1} = 0$$

Proceeding in this way, we find that

$$\Delta^* = \begin{vmatrix} 0 & 0 & 0 & 0 & 0 & \dfrac{2}{na} \\ 0 & 0 & 0 & 0 & \dfrac{-\cot\phi}{na^2} & \dfrac{-\tan(\phi/2)\cos\phi}{na^2} \\ 0 & 0 & 0 & \dfrac{-1}{na^2\sin i\cos\phi} & \dfrac{-\tan(i/2)}{na^2\cos\phi} & \dfrac{-\tan(i/2)}{na^2\cos\phi} \\ 0 & 0 & \dfrac{1}{na^2\sin i\cos\phi} & 0 & 0 & 0 \\ 0 & \dfrac{\cot\phi}{na^2} & \dfrac{\tan(i/2)}{na^2\cos\phi} & 0 & 0 & 0 \\ \dfrac{-2}{na} & \dfrac{\tan(\phi/2)\cos\phi}{na^2} & \dfrac{\tan(i/2)}{na^2\cos\phi} & 0 & 0 & 0 \end{vmatrix} \tag{11.7.17}$$

The Lagrange planetary equations (11.6.13) can be deduced readily by inspection of the Δ^* determinant.

Exercises

1. Solve the problem of the motion of a point particle of unit mass in a plane, under the influence of a uniform gravitational force, in three different ways as with the simple harmonic oscillator problem discussed in the text. In the treatment by perturbation technique, take the kinetic energy to be the unperturbed Hamiltonian.
2. With reference to solving Exercise 1 by perturbation technique, discuss the possibility of taking the potential energy as the unperturbed Hamiltonian and treating the kinetic energy as a perturbing Hamiltonian.
3. Solve the H-problem discussed in Example 11.2.1 in the text by introducing the transformation to canonical variables given in Exercise 8 of Chapter 10.
4. Evaluate the integrals in (11.4.17) in the text for $\alpha_1 < 0$. Hence verify Equations (11.4.18).
5. In the first equation of (11.4.18), set $r = a(1 - e\cos E)$, where a and e are constants and E is a new variable, and hence deduce Kepler's equation. The constants a and e are defined by the equations

$$r_1 = a(1 - e), \qquad r_2 = a(1 + e)$$

6. Establish Equation (11.7.3) in the text, where i and k can take on the values 1, 2,..., $2r$.
7. Establish Equation (11.7.9) in the text.
8. Complete the verification begun in the text that Δ^* is given by Equation (11.7.17).
9. In Jacobi's method of obtaining a complete integral W of the nonlinear partial differential equation of first order

$$F(q_j\,;p_j) = 0 \qquad \left(p_j = \frac{\partial W}{\partial q_j};\ j = 1, 2, 3\right) \tag{1}$$

an attempt is made to find two distinct relations

$$F_1(q_j\,;p_j) = a_1 \tag{2}$$

$$F_2(q_j\,;p_j) = a_2 \tag{3}$$

where a_1 and a_2 are arbitrary constants, such that (1)–(3) can be solved for the p_j, $j = 1, 2, 3$, in terms of $(q_j;\, a_1, a_2)$ and such that the differential expression

$$dW = \sum_{j=1}^{3} p_j\, dq_j \tag{4}$$

is integrable. The complete integral W of (1) is found by integrating (4) (see Miller [13]).

Show that if (4) is to be an exact differential of W, it is necessary that each of the three Poisson brackets $\{F, F_1\}$, $\{F, F_2\}$, and $\{F_1, F_2\}$ must vanish identically.

10. The expanded forms of the equations

$$\{F, F_1\} = 0 \tag{1}$$

$$\{F, F_2\} = 0 \tag{2}$$

referred to in Exercise 9 are homogeneous linear partial differential equations of first order from which we can attempt to determine F_1 and F_2. It is well known in the theory of partial differential equations that if $F_1(q_j; p_j) = a_1$ and $F_2(q_j; p_j) = a_2$ are two independent integrals of the Jacobi subsidiary equations

$$\frac{dp_1}{\partial F/\partial q_1} = \frac{dp_2}{\partial F/\partial q_2} = \frac{dp_3}{\partial F/\partial q_3} = -\frac{dq_1}{\partial F/\partial p_1} = -\frac{dq_2}{\partial F/\partial p_2} = -\frac{dq_3}{\partial F/\partial p_3}$$

of (1) and (2), then F_1 and F_2 are solutions of (1) and (2) respectively.

Use these facts, together with the information contained in Exercise 9 to find a complete integral of the equation

$$q_1 p_1 p_2 - 3q_3^2 p_1^2 + p_3 - 4 = 0^{\dagger}$$

11. The complete integrals described in Exercises 9 and 10 are not necessarily unique. Review the information contained in Exercise 9 of Chapter 9 and in Exercises 9 and 10 above. Then show that the differential equation

$$4q_1 W p_2^2 + p_1 = 0$$

has the distinct complete integrals

$$W^2 = \alpha q_2 - \alpha^2 q_1^2 + \beta$$

$$W^2(4q_1^2 + a) = (q_2 + b)^2$$

where α, β, a and b are arbitrary constants.

† Exercises 10 and 11 are reprinted from F. H. Miller, "Partial Differential Equations," pp. 125 and 135. Copyright 1941 by Wiley, New York. Used with permission of Wiley.

Chapter 12

The Gravitational Potential

12.1. SPHERICAL HARMONICS APPLIED IN A DIRICHLET PROBLEM

Equations of transformation from the rectangular coordinates (x, y, z) shown in Figure 12.1.1 to the associated spherical coordinates (r, θ, ψ) are

$$x = r \sin\theta \cos\psi$$

$$y = r \sin\theta \sin\psi$$

$$z = r \cos\theta$$

on the domain $0 \leqslant \theta \leqslant \pi$, $0 \leqslant \psi \leqslant 2\pi$, $r \geqslant 0$. In terms of these spherical coordinates, Laplace's equation for the scalar function $V(r, \theta, \psi)$ has the form

$$\nabla^2 V = \frac{\partial^2 V}{\partial r^2} + \frac{2}{r}\frac{\partial V}{\partial r} + \frac{1}{r^2}\frac{\partial^2 V}{\partial \theta^2} + \frac{\cot\theta}{r^2}\frac{\partial V}{\partial \theta} + \frac{1}{r^2 \sin^2\theta}\frac{\partial^2 V}{\partial \psi^2} = 0 \qquad (12.1.1)$$

on the domain $0 < \theta < \pi$, $0 \leqslant \psi \leqslant 2\pi$, $r > 0$.

Any solution V of (12.1.1) is called a *spherical harmonic*. Any solution V_n of (12.1.1) that is homogeneous in x, y, and z of degree n is called a *solid spherical harmonic of degree n*. Thus V_n satisfies the two equations

$$\nabla^2 V_n = 0$$
$$V_n(\lambda x, \lambda y, \lambda z) = \lambda^n V_n(x, y, z)$$

where λ is an arbitrary parameter and n is an integer.

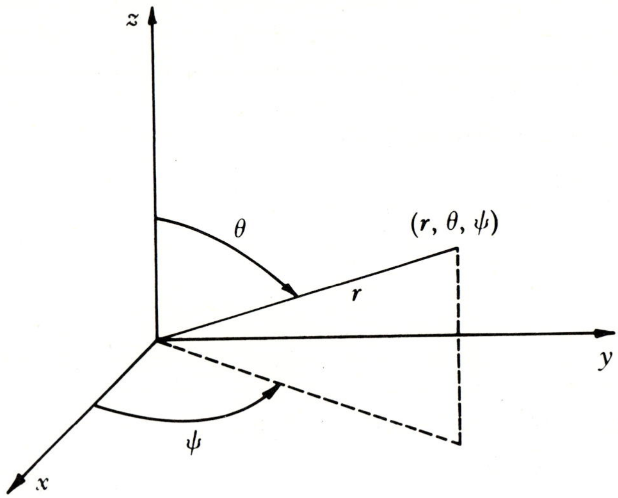

FIGURE 12.1.1

Example 12.1.1. Show that $V_{-1} = r^{-1}$ is a solid spherical harmonic of degree $n = 1$.

Since

$$r^{-1} = f(x, y, z) = (x^2 + y^2 + z^2)^{-1/2}$$

we see that

$$f(\lambda x, \lambda y, \lambda z) = (\lambda^2 x^2 + \lambda^2 y^2 + \lambda^2 z^2)^{-1/2} = \lambda^{-1} f(x, y, z)$$

and V_{-1} is homogeneous of degree -1. From vector analysis we have the formula

$$\nabla^2 f(r) = \frac{2}{r} f'(r) + f''(r)$$

for the Laplacian ∇^2 operating on a function f of the single variable r. If $f(r) = r^{-1}$,

$$\nabla^2 f(r) = \frac{2}{r}\left(\frac{-1}{r^2}\right) + \frac{2}{r^2} = 0$$

Hence r^{-1} satisfies Laplace's equation.

It can also be shown that if V_n is a solid spherical harmonic of degree n, then $r^{-(2n+1)}V_n$ is a solid spherical harmonic of degree $-(n+1)$. To illustrate, since r^{-1} is a solid spherical harmonic of degree -1, then $r^{-(-2+1)}r^{-1} = 1$ is a solid spherical harmonic of degree 0.

If we consider a function $r^n Y(\theta, \psi)$, where Y is a function of θ and ψ alone, it is clear that the function is homogeneous in x, y, z of degree n, for x, y, z appear in θ and ψ only as ratios x/y, y/z, etc. Thus if a function of this form satisfies Laplace's equation, it is a solid spherical harmonic of degree n, Let us designate such a function by V_n. We seek to determine $Y(\theta, \psi)$ so that $\nabla^2 V_n = 0$.

If the function

$$V_n = r^n Y \tag{12.1.2}$$

is substituted into (12.1.1), we obtain the equation

$$\frac{\partial^2 Y}{\partial\theta^2} + \frac{\cos\theta}{\sin\theta}\frac{\partial Y}{\partial\theta} + \csc^2\theta\,\frac{\partial^2 Y}{\partial\psi^2} + n(n+1)Y = 0. \tag{12.1.3}$$

The solutions Y of (12.1.3) are called *surface spherical harmonics of degree n*. They can be determined by a further separation of variables. To this end, let us assume solutions Y of the form

$$Y(\theta, \psi) = \Theta(\theta)\,\Psi(\psi) \tag{12.1.4}$$

and substitute (12.1.4) into (12.1.3) to obtain

$$\Psi\frac{d^2\Theta}{d\theta^2} + \Psi\frac{\cos\theta}{\sin\theta}\frac{d\Theta}{d\theta} + \Theta\csc^2\theta\,\frac{d^2\Psi}{d\psi^2} + n(n+1)\,\Theta\Psi = 0$$

Now multiply each side of this last equation by $\sin^2\theta(\Theta\Psi)^{-1}$. We find that

$$\frac{\sin^2\theta}{\Theta}\frac{d^2\Theta}{d\theta^2} + \frac{\sin\theta\cos\theta}{\Theta}\frac{d\Theta}{d\theta} + \frac{1}{\Psi}\frac{d^2\Psi}{d\psi^2} + n(n+1)\sin^2\theta = 0$$

or, separating variables,

$$\frac{\sin^2\theta}{\Theta}\frac{d^2\Theta}{d\theta^2} + \frac{\sin\theta\cos\theta}{\Theta}\frac{d\Theta}{d\theta} + n(n+1)\sin^2\theta = \frac{-1}{\Psi}\frac{d^2\Psi}{d\psi^2} = m^2$$

where the separation constant m^2, m an integer, has been chosen to lead to solutions for Ψ which have period 2π. We can write the separated equations in the forms

$$\Psi'' + m^2\Psi = 0 \tag{12.1.5}$$

$$\sin^2\theta\,\Theta'' + \sin\theta\cos\theta\,\Theta' + [n(n+1)\sin^2\theta - m^2]\Theta = 0 \tag{12.1.6}$$

The general solution of (12.1.5) is, of course,

$$\Psi = C_1 \cos(m\psi) + C_2 \sin(m\psi) \tag{12.1.7}$$

where C_1 and C_2 are arbitrary constants.

Equation (12.1.6) is referred to as Legendre's associated equation. It is usually studied in the form obtained by setting $s = \cos\theta$. With this substitution, it is readily seen that

$$\Theta' = -(1 - s^2)^{1/2} \frac{d\Theta}{ds}$$

$$\Theta'' = (1 - s^2) \frac{d^2\Theta}{ds^2} - s \frac{d\Theta}{ds}$$

$$\sin\theta \cos\theta\Theta' = -s(1 - s^2) \frac{d\Theta}{ds}$$

so that (12.1.6) can be written in the form

$$(1 - s^2) \frac{d^2\Theta}{ds^2} - 2s \frac{d\Theta}{ds} + \left[n(n + 1) - \frac{m^2}{1 - s^2}\right] \Theta = 0 \tag{12.1.8}$$

where m and n are integers. Equation (12.1.8) is known as the algebraic form of the Legendre associated equation. Let $\Theta_n{}^m$ represent a solution of (12.1.8) corresponding to specific choices of n and m. Then any function $Y_n{}^m$ of the form

$$Y_n{}^m = [a_{n,m} \sin(m\psi) + b_{n,m} \cos(m\psi)]\, \Theta_n{}^m$$

where $a_{n,m}$ and $b_{n,m}$ are arbitrary constants, satisfies (12.1.3) and hence is a surface spherical harmonic of degree n. Furthermore, the function

$$V_n = r^n Y_n{}^m$$

is a solid spherical harmonic of degree n and the function

$$r^{-(2n+1)} V_n = r^{-(n+1)} Y_n{}^m$$

is a solid spherical harmonic of degree $-(n + 1)$.

In potential theory an extremely important class of problems arises in which axial symmetry is involved, about the z-axis say. In such cases, V is a function of r and θ alone and we choose $m = 0$. Hence (12.1.8) reduces to

$$(1 - s^2) \frac{d^2\Theta}{ds^2} - 2s \frac{d\Theta}{ds} + n(n + 1)\Theta = 0 \tag{12.1.9}$$

Equation (12.1.9) is known as Legendre's equation. Before continuing our discussion of spherical harmonics, let us consider some of the properties of certain solutions of (12.1.9).

It can be shown that the polynomials $P_n(s)$ defined by

$$P_n(s) = \frac{1}{2^n} \sum_{j=0}^{N} \frac{(-1)^j (2n-2j)!\, s^{n-2j}}{j!\,(n-j)!\,(n-2j)!} \qquad (n = 0, 1, 2, \ldots) \qquad (12.1.10)$$

where $N = \frac{1}{2}n$ if n is even and $N = \frac{1}{2}(n-1)$ if n is odd, are particular solutions of Legendre's equation. They are called Legendre polynomials. The first several polynomials

$$P_0(s) = 1, \qquad P_1(s) = s$$

$$P_2(s) = \frac{3}{2}\left(s^2 - \frac{1}{3}\right)$$

$$P_3(s) = \frac{5}{2}\left(s^3 - \frac{3}{5}s\right)$$

$$P_4(s) = \frac{35}{8}\left(s^5 - \frac{6}{7}s^2 + \frac{3}{35}\right)$$

are readily obtained from the definition of $P_n(s)$.

Several very important properties of Legendre polynomials will be stated in the form of theorems; the proofs will be left to the exercises at the end of the chapter.

Theorem 12.1.1. (Rodriques's formula)

$$P_n(s) = \frac{1}{2^n n!} \frac{d^n (s^2 - 1)^n}{ds^n} \qquad (12.1.11)$$

Theorem 12.1.2.

$$(1 - 2sz + z^2)^{-1/2} = \sum_{n=0}^{\infty} P_n(s) z^n, \qquad |z| < 1 \qquad (12.1.12)$$

Theorem 12.1.3.

$$\int_{-1}^{1} P_m(s)\, P_n(s)\, ds = \begin{cases} 0 & (m \neq n) \\ 2(2n+1)^{-1} & (m = n) \end{cases}$$

$$\int_{0}^{\pi} P_m(\cos\theta)\, P_n(\cos\theta) \sin\theta\, d\theta = \begin{cases} 0 & (m \neq n) \\ 2(2n+1)^{-1} & (m = n) \end{cases} \qquad (12.1.13)$$

Theorem 12.1.4. $P_n(\cos\theta)$ is a surface spherical harmonic of degree n.

Theorem 12.1.5. The n zeros of the polynomial $P_n(s)$ are all real, distinct, and lie between -1 and $+1$.

Theorem 12.1.6. If $P_n^{(k)}$ represents the kth derivative of P_n with respect to s,

$$| P_n^{(k)}(s)| \leqslant n^{2k}, \qquad | s | < 1 \quad (k = 0, 1, 2,...) \tag{12.1.14}$$

The problem of obtaining solutions to Legendre's associated equation (12.1.8) still remains. Let us designate solutions of it, corresponding to nonnegative integer values of m and n, by the notation $P_n{}^m(s)$. Let us assume solutions in the form

$$P_n{}^m(s) = (1 - s^2)^{m/2}\, u(s) \tag{12.1.15}$$

where $u(s)$ is a function to be determined. If the first and second derivatives

$$\frac{d}{ds} P_n{}^m(s) = (1 - s^2)^{m/2}\, u' - ms(1 - s^2)^{(m/2)-1}\, u$$

$$\frac{d^2}{ds^2} P_n{}^m(s) = (1 - s^2)^{m/2}\, u'' - 2ms(1 - s^2)^{(m/2)-1}\, u'$$
$$-m(1 - s^2)^{(m/2)-1}[1 - (m - 2)\, s^2(1 - s^2)^{-1}]\, u$$

are substituted into (12.1.8), we obtain the differential equation

$$(1 - s^2)\, u'' - 2s(m + 1)\, u' + [n(n + 1) - m(m + 1)]\, u = 0 \tag{12.1.16}$$

for the unknown function u. On the other hand, if we employ the formula

$$\frac{d^m(vw)}{dx^m} = \sum_{j=0}^{m} C(m, j)\, v^{(m-j)} w^{(j)} \qquad \left[v^{(m-j)} = \frac{d^{m-j}v}{dx^{m-j}},\ w^{(j)} = \frac{d^j w}{dx^j}\right]$$

for differentiating the product function $v(x)\, w(x)$ m times, where $C(m, j)$, $j = 1, 2,..., m$ are the binomial coefficients, we find, by setting

$$w = 1 - s^2$$

that

$$\frac{d^m}{ds^m} [(1 - s^2)\, P_n''] = (1 - s^2)\, P_n^{(m+2)} - 2smP_n^{(m+1)} - m(m - 1)\, P_n^{(m)}$$

$$\frac{d^m}{ds^m} (-2sP_n') = -2sP_n^{(m+1)} - 2mP_n^{(m)}$$

Thus, differentiating (12.1.8) m times, we have

$$(1 - s^2)\frac{d^2}{ds^2}\left(\frac{d^m}{ds^m}P_n\right) - 2s(m+1)\frac{d}{ds}\left(\frac{d^m}{ds^m}P_n\right) + [n(n+1) - m(m+1)]\frac{d^m}{ds^m}P_n = 0 \qquad (12.1.17)$$

Comparing (12.1.16) and (12.1.17), we note that if we choose

$$u = \frac{d^m P_n}{ds^m}$$

then Equation (12.1.16) will be satisfied. Hence the function

$$P_n{}^m(s) = (1 - s^2)^{m/2}\, u(s) = (1 - s^2)^{m/2}\frac{d^m}{ds^m}P_n(s) \qquad (12.1.18)$$

is a solution of (12.1.8) for specified nonnegative integers n and m. Each $P_n{}^m(s)$ is called a *Legendre associated function of degree m and order n.* Clearly, we have $P_n{}^m(s) = 0$, $m > n$.

For given values of m and n,

$$Y_n{}^m(s, \psi) = [a_{n,m}\sin(m\psi) + b_{n,m}\cos(m\psi)]\, P_n{}^m(s) \qquad (12.1.19)$$

is a surface spherical harmonic of degree n and order m. When $m = 0$, $Y_n{}^0(s, \psi) = b_{n0}P_n(s)$ and we speak of the $P_n(s)$ as *zonal harmonics.* When $0 < m < n$, the $Y_n{}^m(s, \psi)$ are called *tesseral harmonics.* When $m = n$, the $Y_n{}^m(s, \psi)$ are called *sectoral harmonics.*

Several important properties of surface spherical harmonics and Legendre associated functions are stated as theorems, with the proofs left to the exercises at the end of this chapter.

Theorem 12.1.7. If $Y_n{}^m(s, \psi)$, $Y_{n'}^{m'}(s, \psi)$ are two surface spherical harmonics of degrees n and n' and orders m and m', show that

$$\int_0^{2\pi}\int_{-1}^{+1} Y_n{}^m(s, \psi)\, Y_{n'}^{m'}(s, \psi)\, ds\, d\psi = 0$$

if either $m = m'$ or $n = n'$, but not both.

Theorem 12.1.8.

$$\int_{-1}^{+1} P_n{}^m(s)\, P_k{}^m(s)\, ds = \begin{cases} 0 & (n \neq k) \\ \dfrac{2}{2n+1}\dfrac{(n+m)!}{(n-m)!} & (n = k) \end{cases}$$

Theorem 12.1.9.

$$\int_0^{2\pi} \cos(m\psi)\, d\psi = 0, \qquad \int_0^{2\pi} \sin(m\psi) \cos(n\psi)\, d\psi = 0$$

$$\int_0^{2\pi} \cos(m\psi) \cos(n\psi)\, d\psi = 0, \qquad \int_0^{2\pi} \sin(m\psi) \sin(n\psi)\, d\psi = 0 \qquad (m \neq n)$$

where m and n are any intergers.

As we have seen, the solid spherical harmonics

$$r^n Y_n{}^m(s, \psi), \quad r^{-(n+1)} Y_n{}^m(s, \psi)$$

of degrees n and $-(n + 1)$, respectively, are particular solutions of Laplace's equation. Since Laplace's equation is a linear partial differential equation, the sum of any finite number of solutions is also a solution. Hence functions of the form

$$V = r^n \sum_{m=0}^{n} [a_{n,m} \sin(m\psi) + b_{n,m} \cos(m\psi)]\, P_n{}^m(s)$$

$$V = r^{-(n+1)} \sum_{m=0}^{n} [A_{n,m} \sin(m\psi) + B_{n,m} \cos(m\psi)]\, P_n{}^m(s) \tag{12.1.20}$$

where $A_{n,m}$ and $B_{n,m}$ are arbitrary constants, are also particular solutions of Laplace's equation.

We can extend the results on linear combinations of functions to infinite sets $\{U_0, U_1, U_2, \ldots, U_n, \ldots\}$. Thus we cite, but do not prove, the following theorem [14]:

Theorem 12.1.10. If each function of an indefinite set $\{U_0, U_1, \ldots, U_n, \ldots\}$ satisfies a linear homogeneous differential equation $LU_n = 0$, then the function

$$U = \sum_{n=0}^{\infty} C_n U_n$$

also satisfies $LU = 0$ provided that

(1) the constants C_n and the functions U_n are such that the infinite series converges to U throughout some domain of the independent variables,

(2) the infinite series is differentiable for all derivatives involved in L.

Theorem 12.1.10 is, of course, general and is not restricted to any particular linear differential equation. Neither is it restricted as to the choice of coordinate system. Now, in many physical problems—chiefly electrostatics, magnetostatics, gravitation, heat flow, and hydrodynamics—one is interested in solutions to Laplace's equation. The particular curvilinear coordinate system u_i, $i = 1, 2, 3$, in which one expresses the Laplacian will usually be determined by the geometric considerations of the particular problem to be studied. In these studies, one is often interested in finding a potential function V which is continuous everywhere in a domain D, either exterior to or interior to a smooth closed surface S on which certain boundary conditions are to be met. The function V is to satisfy Laplace's equation. The symbol V is used here to designate a potential function because such usage is common in the literature. It is not to be confused with our former use of V for the total potential energy of a system of n particles.

The boundary conditions may be stated in the form

$$p \frac{\partial V}{\partial n} + hV = g \tag{12.1.21}$$

where p, h, and g are prescribed constants or functions of the coordinates u_i, $i = 1, 2, 3$, and $\partial V/\partial n$ represents the change in V in the direction of the normal to the surface S. It is assumed that p and h are not simultaneously zero. When $p = 0$ at every point of S, the boundary conditions are said to be Dirichlet; when $h = 0$ at every point of S, the boundary conditions are said to be Neumann.

If V satisfies certain conditions of regularity, $\nabla^2 V = 0$, and conditions (12.1.21) are satisfied, the solution for the interior problem will either be unique or unique to within an additive constant. Hence the fields derivable from V will be uniquely determined. In this regard the following theorem is stated. Its proof is left as an exercise.

Theorem 12.1.11. Let V satisfy these conditions of regularity in a domain bounded by a closed surface S:

(1) V and its first partial derivatives are continuous on the surface S,

(2) the second partial derivatives of V are continuous in D.

Then if V satisfies $\nabla^2 V = 0$ and Equation (12.1.21), it is unique unless $h = 0$ at every point of S. If $h = 0$ at every point of S, the solution is unique to within an additive constant.

This theorem also applies to the exterior problem with the additional requirement that rV and $r^2(\partial^2 V/\partial u_i^2)$, $i = 1, 2, 3$, are bounded for all r

greater than some fixed number. If, for example, $V = \mu/r$ when expressed in the coordinates (r, θ, ψ), then rV,

$$r^2 \frac{\partial^2 V}{\partial r^2} = \frac{2\mu}{r}, \qquad r^2 \frac{\partial^2 V}{\partial \theta^2} = r^2 \frac{\partial^2 V}{\partial \psi^2} = 0$$

are all bounded for all r greater then some number. If $p = 0$ everywhere on S, one of the demands in (1) can be relaxed. We need require only that V be continuous on S and that its derivative be continuous in D as implied in (2).

In the following material, we will be concerned primarily with Laplace's equation in spherical coordinates. In this connection, we note that functions V of the form

$$V = \sum_{n=0}^{\infty} r^n \sum_{m=0}^{n} [a_{n,m} \sin(m\psi) + b_{n,m} \cos(m\psi)]\, P_n{}^m(s) \tag{12.1.22a}$$

$$V = \sum_{n=0}^{\infty} r^{-(n+1)} \sum_{m=0}^{n} [A_{n,m} \sin(m\psi) + B_{n,m} \cos(m\psi)]\, P_n{}^m(s) \tag{12.1.22b}$$

can be shown to meet the demands of Theorem 12.1.10 if the constants are chosen in a manner to be prescribed. Hence V will satisfy Laplace's equation. The proof of this statement is left to the exercises. Since $\nabla^2 V(r, \theta, \psi)$ is not defined along the polar axis and at $r = 0$, we write that, for $a > 0$, (12.1.22a) satisfies

$$\nabla^2 V(r, \theta, \psi) = 0 \qquad (0 < r \leqslant a, 0 < \theta < \pi, 0 \leqslant \psi \leqslant 2\pi)$$

over the indicated domain, whereas (12.1.22b) satisfies

$$\nabla^2 V(r, \theta, \psi) = 0 \qquad (r \geqslant a, 0 < \theta < \pi, 0 \leqslant \psi \leqslant 2\pi)$$

over the indicated domain. We now illustrate the previous theory by considering a Dirichlet boundary value problem.

Example 12.1.2. We seek a spherical harmonic function $V(r, \theta, \psi)$ in the domain $r > a$, $0 \leqslant \theta \leqslant \pi$, $0 \leqslant \psi \leqslant 2\pi$, such that $V(r, \theta, \psi)$ assumes prescribed values $V_0(\theta, \psi)$ on the sphere $r = a$. We assume that V_0 is continuous. Specifically, we require that

1. V satisfy $\nabla^2 V(r, \theta, \psi) = 0$ in the domain $r > a$, $0 < \theta < \pi$, $0 \leqslant \psi \leqslant 2\pi$,
2. $\lim_{r\to a} V(r, \theta, \psi) = V_0(\theta, \psi)$; $\lim_{r\to\infty} V(r, \theta, \psi) = 0$,
3. V is to be continuous on the domain $r \geqslant a, 0 \leqslant \theta \leqslant \pi, 0 \leqslant \psi \leqslant 2\pi$;

partial derivatives of second order are to be continuous on the domain $r > a$, $0 \leqslant \theta \leqslant \pi$, $0 \leqslant \psi \leqslant 2\pi$.

4. rV, $r^2(\partial^2 V/\partial r^2)$, $r^2(\partial^2 V/\partial \theta^2)$, $r^2(\partial^2 V/\partial \psi^2)$ are bounded.

Since $\lim_{r\to\infty} V(r, \theta, \psi) = 0$, we seek a solution V of the form

$$V = \sum_{n=0}^{\infty} \left(\frac{a}{r}\right)^{n+1} \sum_{m=0}^{n} P_n{}^m(s)[A_{n,m} \sin(m\psi) + B_{n,m} \cos(m\psi)] \quad (12.1.23)$$

where a is introduced into the coefficients for convenience in the subsequent calculation. In what follows, it will be convenient to write (12.1.23) in the form

$$V = \sum_{n=0}^{\infty} \left(\frac{a}{r}\right)^{n+1} B_{n,0}P_n(s) + \sum_{n=0}^{\infty} \left(\frac{a}{r}\right)^{n+1} \sum_{m=1}^{n} B_{n,m}P_n{}^m(s) \cos m\psi$$

$$+ \sum_{n=0}^{\infty} \left(\frac{a}{r}\right)^{n+1} \sum_{m=1}^{n} A_{n,m}P_n{}^m(s) \sin m\psi \quad (12.1.24)$$

Clearly the condition $\lim_{r\to\infty} V(r, \theta, \psi) = 0$ is satisfied by (12.1.23) as is the boundedness demand (4) as well. We must still satisfy, on the boundary $r = a$, the condition

$$V_0(\theta, \psi) = \sum_{n=0}^{\infty} B_{n,0}P_n(s) + \sum_{n=0}^{\infty} \sum_{m=1}^{n} B_{n,m}P_n{}^m(s) \cos m\psi$$

$$+ \sum_{n=0}^{\infty} \sum_{m=1}^{n} A_{n,m}P_n{}^m(s) \sin m\psi \quad (12.1.25)$$

The notions of Theorem 12.1.7 can be extended and it can be shown that the $\{Y_n{}^m\}$ form a complete† orthonormal set. Thus a continuous function of two variables can be developed in a series of these surface spherical harmonics that will converge to the function on the domain of definition, provided the constant coefficients appearing in the series are chosen according to a certain formal procedure. Since $V_0(\theta, \psi)$ is a continuous function of θ and ψ on the sphere $r = a$, it can be so developed. The formal procedure for determining the coefficients $A_{n,m}$ and $B_{n,m}$ appears in the following development.

† If there exists no function $f(x)$, except the identically zero function, with the property that

$$\int_a^b f(x)\, \Phi_n(x)\, dx = 0$$

for all members of an orthogonal system $\{\Phi_n(x)\}$, the set $\{\Phi_n(x)\}$ is said to be complete. Similar definitions hold for functions of more than one variable.

We first integrate both sides of (12.1.25) over ψ, applying the orthogonality properties of the trigonometric functions cited in Theorem 12.1.9, to obtain

$$\int_0^{2\pi} V_0(\theta, \psi)\, d\psi = 2\pi \sum_{n=0}^{\infty} B_{n,0} P_n(s) \tag{12.1.26}$$

Now multiply by $P_k(s)$ and use the orthogonality properties cited in Theorem 12.1.8. Integrating, we find that

$$\begin{aligned} 2\pi \sum_{n=0}^{\infty} B_{n,0} \int_{-1}^{1} P_n(s)\, P_k(s)\, ds &= 2\pi B_{k,0} \left(\frac{2}{2k+1}\right) \\ &= \int_{-1}^{1} \int_0^{2\pi} V_0(\theta, \psi)\, P_k(s)\, d\psi\, ds \end{aligned}$$

Hence

$$B_{n,0} = \frac{2n+1}{4\pi} \int_{-1}^{1} \int_0^{2\pi} V_0(\theta, \psi)\, P_n(s)\, d\psi\, ds \tag{12.1.27}$$

To determine $B_{n,m}$, $m > 0$, we multiply both sides of (12.1.25) by $\cos(k\psi)$ and integrate on ψ to obtain

$$\begin{aligned} \int_0^{2\pi} V_0(\theta, \psi) \cos(k\psi)\, d\psi &= \sum_{n=0}^{\infty} \sum_{m=1}^{n} B_{n,m} P_n^{\,m}(s) \int_0^{2\pi} \cos(m\psi) \cos(k\psi)\, d\psi \\ &= \sum_{n=0}^{\infty} B_{n,k} P_n^{\,k}(s) \int_0^{2\pi} \cos^2(k\psi)\, d\psi = \pi \sum_{n=0}^{\infty} B_{n,k} P_n^{\,k}(s) \end{aligned}$$

We next multiply each side by $P_l^{\,k}(s)$ and integrate over s to obtain

$$\begin{aligned} \int_{-1}^{1} \int_0^{2\pi} V_0(\theta, \psi)\, P_l^{\,k}(s) \cos(k\psi)\, d\psi &= \pi \sum_{n=0}^{\infty} B_{n,k} \int_{-1}^{1} P_n^{\,k}(s) P_l^{\,k}(s)\, ds \\ &= \pi B_{l,k} \left(\frac{2}{2l+1}\right) \frac{(l+k)!}{(l-k)!} \end{aligned}$$

Changing l to n and k to m, we write

$$B_{n,m} = \frac{(2n+1)(n-m)!}{2\pi(n+m)!} \int_{-1}^{1} \int_0^{2\pi} V_0(\theta, \psi)\, P_n^{\,m}(s) \cos(m\psi)\, d\psi\, ds \tag{12.1.28}$$

Similarly, we can determine the $A_{n,m}$. We find that

$$A_{n,m} = \frac{2n+1}{2\pi} \frac{(n-m)!}{(n+m)!} \int_{-1}^{1} \int_0^{2\pi} V_0(\theta, \psi)\, P_n^{\,m}(s) \sin(m\psi)\, d\psi\, ds \tag{12.1.29}$$

If we choose the constants from formulas (12.1.27)–(12.1.29), it can be shown that the series (12.1.24) is uniformly convergent on the domain $r \geqslant a$, $0 \leqslant \theta \leqslant \pi$, $0 \leqslant \psi \leqslant 2\pi$, and hence V as given by this series is continuous. Series (12.1.25) represents V evaluated at $r = a$; that is, $V_0(\theta, \psi)$. Since V is continuous everywhere in the domain $r \geqslant a$, $0 \leqslant \theta \leqslant 2\pi$, $0 \leqslant \psi \leqslant 2\pi$, as $r \to a^+$, $V(r, \theta, \psi) \to V_0(\theta, \psi)$.

We have thus seen how requirement (2) of our boundary value problem is satisfied by the series (12.1.24) with constants determined as indicated. We have also seen that the series V is continuous on the domain $r \geqslant a$, $0 \leqslant \theta \leqslant \pi$, $0 \leqslant \psi \leqslant 2\pi$. In the exercises at the end of this chapter, it is to be shown that V satisfies Laplace's equation. In establishing this result, it has to be shown that the series whose terms are the functions

$$U_n = r^{-(n+1)} \sum_{m=0}^{n} Y_n^{\,m}$$

and the several series whose terms are the partial derivatives of first and second order of the U_n are uniformly convergent. Hence the second-order partial derivatives of V are continuous functions on the required domain. It is now easy to see that V and its derivatives have the requisite continuity properties required in part (3) of the example. Hence V is the unique solution to our boundary value problem.

For easy reference, a table of $P_n^{\,m}(\cos\theta)$ through the fourth order is given (see Table 12.1.1).

12.2. GRAVITATIONAL POTENTIAL OUTSIDE OF AN IRREGULAR BODY OF MASS M

We have seen from Gauss's law (1.1.3) that

$$\iint_S \mathbf{g} \cdot d\mathbf{S} = -4\pi k^2 M \qquad \left(M = \iiint_\tau \sigma\, d\tau\right) \tag{12.2.1}$$

where $\mathbf{g}$ is the entire gravity field at a point P. Here the point P is considered to lie on a closed surface S which encloses the total mass M, the mass being distributed throughout the volume τ. The mass density is represented by the symbol σ. Since the Newtonian gravity field $\mathbf{g}$ is conservative, $\mathbf{g}$ is derivable from a scalar potential V. Thus we write

$$\mathbf{g} = \nabla V \tag{12.2.2}$$

TABLE 12.1.1

EXPLICIT FORM OF THE FIRST FIVE LEGENDRE ASSOCIATED FUNCTIONS

n \ m	0	1	2	3	4
0	1				
1	$\cos\theta$	$\sin\theta$			
2	$\frac{3}{2}\cos^2\theta - \frac{1}{2}$	$3\sin\theta\cos\theta$	$3\sin^2\theta$		
3	$\frac{5}{2}(\cos^3\theta - \frac{3}{5}\cos\theta)$	$\frac{15}{2}\sin\theta(\cos^2\theta - \frac{1}{5})$	$15\sin^2\theta\cos\theta$	$15\sin^3\theta$	
4	$\frac{35}{8}(\cos^4\theta - \frac{6}{7}\cos^2\theta + \frac{3}{35})$	$\frac{35}{2}\sin\theta(\cos^3\theta - \frac{3}{7}\cos\theta)$	$\frac{105}{2}\sin^2\theta(\cos^2\theta - \frac{1}{7})$	$105\sin^3\theta\cos\theta$	$105\sin^4\theta$

If the divergence theorem is applied to (12.2.1), we have

$$\iint_S \mathbf{g} \cdot d\mathbf{S} = \iiint_\tau \nabla \cdot \mathbf{g}\, d\tau = -4\pi k^2 \iiint_\tau \sigma\, d\tau$$

It follows that

$$\iiint_\tau (\nabla \cdot \mathbf{g} + 4\pi k^2 \sigma)\, d\tau = 0$$

and

$$\nabla \cdot \mathbf{g} = -4\pi k^2 \sigma$$

Now applying (12.2.2), we have

$$\nabla^2 V = -4\pi k^2 \sigma \tag{12.2.3}$$

Equation (12.2.3) is known as Poisson's equation. In particular, where the density function $\sigma = 0$, we have Laplace's equation

$$\nabla^2 V = 0 \tag{12.2.4}$$

At points exterior to the body, the mass density σ is zero; thus, at external points, the gravitational potential satisfies $\nabla^2 V = 0$. The gravitational potential due to the irregular body must vanish as we recede to an infinitely great distance from it. Thus, at external points, we can represent the potential V in the form (12.1.23), where we will assume that the constants $A_{n,m}$ and $B_{n,m}$ are bounded, as is always the case in problems of physical interest. We can write

$$\begin{aligned}
V = \frac{aB_{0,0}}{r} \Big\{ 1 &+ \left(\frac{a}{r}\right) [C_{1,0}P_1(s) + C_{1,1}P_1{}^1(s) \cos \psi + S_{1,1}P_1{}^1(s) \sin \psi] \\
&+ \left(\frac{a}{r}\right)^2 [C_{2,0}P_2(s) + C_{2,1}P_2{}^1(s) \cos \psi + C_{2,2}P_2{}^2(s) \cos 2\psi \\
&+ S_{2,1}P_2{}^1(s) \sin \psi + S_{2,2}P_2{}^2(s) \sin 2\psi] \\
&+ \left(\frac{a}{r}\right)^3 [C_{3,0}P_3(s) + C_{3,1}P_3{}^1(s) \cos \psi + C_{3,2}P_3{}^2(s) \cos 2\psi \\
&+ C_{3,3}P_3{}^3(s) \cos 3\psi + S_{3,1}P_3{}^1(s) \sin \psi + S_{3,2}P_3{}^2(s) \sin 2\psi \\
&+ S_{3,3}P_3{}^3(s) \sin 3\psi] \\
&+ \left(\frac{a}{r}\right)^4 [C_{4,0}P_4(s) + C_{4,1}P_4{}^1(s) \cos \psi + C_{4,2}P_4{}^2(s) \cos 2\psi \\
&+ C_{4,3}P_4{}^3(s) \cos 3\psi + C_{4,4}P_4{}^4(s) \cos 4\psi + S_{4,1}P_4{}^1(s) \sin \psi \\
&+ S_{4,2}P_4{}^2(s) \sin 2\psi + S_{4,3}P_4{}^3(s) \sin 3\psi + S_{4,4}P_4{}^4(s) \sin 4\psi] \\
&+ \sum_{n=5}^{\infty} \left(\frac{a}{r}\right)^n \sum_{m=0}^{n} P_n{}^m(s)[C_{n,m} \cos(m\psi) + S_{n,m} \sin(m\psi)] \Big\}
\end{aligned} \tag{12.2.5}$$

where the new notation $C_{n,m}$, $S_{n,m}$ has been introduced for the constants, for we have factored out the $B_{0,0}$ term. These new constants are related to the former constants in an obvious way. As in (12.1.23),the constant a is introduced into the coefficients for convenience. When we apply these results later to describe the Earth's gravitational field, the letter a will represent the Earth's equatorial radius.

We have seen that V as given by (12.2.5) satisfies $\nabla^2 V = 0$. If its value, or $\partial V/\partial n$, is prescribed on the body surface, and the coefficients are evaluated in such a way as to satisfy this boundary condition, then V will be the unique solution to the boundary value problem. The external gravity field can then be obtained from V by differentiation.

12.3. THE EARTH AND ITS GRAVITY FIELD

Let us now turn to the problem of determining the external gravitational potential of the Earth. Since the potential must fall off as $r \to \infty$, we can choose a representation of V in the form (12.2.5). Since $\lim_{r\to\infty} V$ is prescribed to be zero on the infinite sphere, in principle, we need only prescribe the bounding surface of the Earth and the V-distribution on this surface (the Dirichlet problem) and determine the coefficients in the series (12.2.5) in such a way that V satisfies these boundary conditions. Actually, the matter is not so simple because we do not have precise knowledge about the shape of the Earth.

In early antiquity the Earth was recognized as being essentially spherical. The Earth's shadow cast during lunar eclipses was circular in outline, and sailors knew that, regardless of the direction of approach, the tip of the mast of an oncoming vessel appeared first on the horizon. The first reliable measurement of the Earth's circumference was made by Eratosthenes (276–194 B.C.). Gradually the Earth's shape came to be recognized as approximately the shape of a rotation ellipsoid, obtained by rotating an ellipse on its minor axis. The Earth's surface is far from being even, for there are elevations and depressions. In reality, the configuration of the Earth's surface, with its unevenness above and below sea level, is extraordinarily complex. The task of measuring this surface and the gravity field of the Earth is difficult, and its fulfillment depends on a highly specialized science called geodesy.

If gravity measurements are carried out at different points on the Earth's surface—for example, in low-level lands, in mountains, and at sea—a mélange of data is obtained. On a mountaintop, for instance, the gravity g must be smaller than in the level lands at the same latitude because mountain stations are farther from the Earth's center of gravity than are the lowland stations. At sea, the gravity is smaller than on the continents, for the density of seawater is less than that of the Earth's crust. The measured values of g are, roughly speaking, approximately 980 cm/sec^2. From the point of view of application, however, the differences between the measured values and that quoted are not negligible.

In order to be able to interpret these data in terms of an overall picture of the Earth's gravity field, some sort of mathematical model or reference figure of the Earth is required. If we ignore the influence of the atmosphere and the tidal effects caused by the Sun and the Moon, we have to study an Earth, the sea level of which, appears to be in equilibrium in the total field of its Newtonian potential V and its centrifugal potential Ω.†

In formulating a mathematical model, we are thus led to assume that the free surface of the oceans is in equilibrium. If this is so, then there must not be any component of the total field parallel to the water surface. Hence the sea surface is an equipotential surface on which the total potential

$$\Phi = V + \Omega \tag{12.3.1}$$

is constant. This equipotential surface, constituted by sea level, may be imagined to stretch over the continents. Its position and course can be computed if the direction of gravity—that is, of the plumb line—is known everywhere. Determining the course of such a surface of reference through a given point and its position relative to the equipotential surface of the sea is called *leveling*. The continuation of sea level across continents as a mathematical surface of reference is termed the *geoid*.

Let $Ox'y'z'$ represent an inertial equatorial rectangular reference frame with origin at geocenter. Let $Oxyz$ represent an equatorial rectangular reference frame which rotates with the Earth and coincided with $Ox'y'z'$ at time $t = 0$. A particle of unit mass on the Earth's surface must partake of the Earth's rotation. Let (x, y, z) refer to a point on the Earth's surface and (x', y', z') refer to the corresponding point on the Earth's surface assumed not be rotate.
The equations

$$\begin{pmatrix} x \\ y \\ z \end{pmatrix} = \begin{pmatrix} \cos \omega t & -\sin \omega t & 0 \\ \sin \omega t & \cos \omega t & 0 \\ 0 & 0 & 1 \end{pmatrix} \begin{pmatrix} x' \\ y' \\ z' \end{pmatrix}$$

relate the primed and unprimed coordinates at some instant t, where ω

† Small time-dependent changes in the geopotential result from mass displacements associated with tides and also with atmospheric variations (see, for example, Kozai [14a]).

is the angular speed of the Earth's rotation. Taking two time derivatives of (x, y, z), we obtain the equations

$$\ddot{x} = -\omega^2(x' \cos \omega t - y' \sin \omega t) = -\omega^2 x$$
$$\ddot{y} = -\omega^2(x' \sin \omega t + y' \cos \omega t) = -\omega^2 y$$
$$\ddot{z} = 0$$

Thus the magnitude of the centrifugal acceleration is

$$(\ddot{x}^2 + \ddot{y}^2)^{1/2} = \omega^2(x^2 + y^2)^{1/2} = \omega^2 r \sin\theta$$

This acceleration, which is directed outward along a perpendicular to the z-axis, is derivable from the centrifugal potential

$$\Omega = \frac{\omega^2(x^2 + y^2)}{2} = \frac{\omega^2 r^2 \sin^2\theta}{2} \tag{12.3.2}$$

the magnitude of whose gradient is $\omega^2 r \sin\theta$. We can now replace (12.3.1) with the equation

$$\Phi = V + \frac{\omega^2 r^2 \sin^2\theta}{2} \tag{12.3.3}$$

It is clear that Equation (12.3.2) will apply equally well for the rotation potential Ω of a homogeneous oblate spheroidal mass distribution, rotating about its minor axis, say the z-axis. Thus we write

$$\Omega = \frac{\omega^2 r^2 \sin^2\theta}{2} \tag{12.3.4}$$

where now, of course, (r, θ) are the coordinates of a point on the spheroidal surface.

The physically logical reference figure for the Earth is that of a rotating fluid in equilibrium, but this model is not too tractable mathematically. The theoretical investigation of rotating fluid models is still being pursued. A more mathematically tractable reference figure, generally used in practice, is that of an ellipsoid of revolution of major axis $2a$ and minor axis $2b$, with the properties:

1. The volume of the ellipsoid is equal to the volume of the geoid. The ellipsoid flattening $f = (a - b)/a$ is taken to be 1/297.0.

2. Containing the same mass E as the Earth, the ellipsoid shall rotate about its minor axis with the same angular velocity $\boldsymbol{\omega}$ as the Earth and its center of gravity shall coincide with the Earth's. It shall have equatorial symmetry.

3. The surface of the ellipsoid shall be an equipotential surface to the sum of its Newtonian potential V and its centrifugal potential Ω. Designate by Φ this total potential, so that $\Phi = V + \Omega$. We require that on the surface of the ellipsoid $\Phi = V + (\omega^2 r^2 \sin^2 \theta)/2 =$ constant.

The ellipsoid of revolution described above will have a gravitational potential V of the form

$$V = \frac{aB_{0,0}}{r}\left[1 + \sum_{n=2,4,6,\ldots}^{\infty} \left(\frac{a}{r}\right)^n C_{n,0} P_n(\cos\theta)\right] \tag{12.3.5}$$

if z is chosen to coincide with the axis of rotation. Equation (12.3.5) follows from the fact that (1) V is independent of ψ (hence we take $m = 0$) and (2) the definition of $P_n(s)$ shows that for n odd, $P_n(s) = -P_n(-s)$, which violates the assumption of equatorial symmetry. Then (12.3.5) follows directly from (12.2.5).

By imposing the boundary condition $\Phi =$ constant, we can relate the constants $C_{n,0}$ to the flattening f of the ellipsoid of reference, to the rotation rate ω, to the semimajor axis a, and to the equatorial gravity g_e. The gravity field g over the surface of the ellipsoid is given approximately by $\partial\Phi/\partial r$. To a high order of accuracy, the radial spherical coordinate r_e of the surface of the ellipsoid may be obtained from the boundary condition $\Phi = V + \Omega$ by demanding that r_e reduce the equation to an identity. The various parameters just cited are related, and the essential results which are found after carrying out some rather lengthy calculations appear in the following equations (see Kaula [15]):†

$$f = \frac{a-b}{a}, \qquad m = \frac{\omega^2 a}{g_e}$$

$$aB_{0,0} = a^2 g_e\left[1 - f + \frac{3}{2}m - \frac{15}{4}mf + O(f^3)\right]$$

$$g = g_e\left[1 + \left(\frac{5}{2}m - f - \frac{17}{4}mf\right)\sin^2\phi_g + \left(\frac{1}{8}f^2 - \frac{5}{8}mf\right)\sin^2 2\phi_g + O(f^3)\right]$$

$$r_e = a\left[1 - \left(f + \frac{3}{2}f^2\right)\sin^2\phi_g + \frac{3}{2}f^2\sin^4\phi_g + O(f^3)\right] \tag{12.3.6}$$

$$C_{2,0} = -\frac{2}{3}f\left(1 - \frac{1}{2}f\right) + \frac{1}{3}m\left(1 - \frac{3}{2}m - \frac{2}{7}f\right) + O(f^3)$$

$$C_{4,0} = \frac{4}{35}f(7f - 5m) + O(f^3)$$

$$V = \frac{GE}{r}\left[1 + C_{2,0}\left(\frac{a}{r}\right)^2 P_2(\cos\theta) + C_{4,0}\left(\frac{a}{r}\right)^4 P_4(\cos\theta) + O(f^3)\right]$$

† The notation $O(f^3)$ is to be read "terms of order f^3."

In (12.3.6), ϕ_g represents the geodetic latitude, not the geocentric latitude $\phi' = \pi/2 - \theta$. A useful approximate formula for relating the two quantities is given by

$$\tan \phi_g = \left(1 + \frac{26.734232}{r}\right) \tan\left(\frac{\pi}{2} - \theta\right) \qquad (r \text{ in statute miles})$$

It will be seen later that

$$aB_{0,0} = GM$$

where G is the gravitational constant and M is the total mass of the gravitating body.

Only if the Earth were truly a rotating fluid in equilibrium would its figure be essentially an ellipsoid of revolution. There are, however, deviations from equilibrium and hence irregularities in the shape of the geoid. There are undulations of the geoid above a best-fitting ellipsoid. These undulations, called the geoid heights, show little if any correlation with the surface topographical features. These geoid heights may be deduced from a knowledge of the gravity anomalies Δg, which are the differences between observed gravity values g_0, properly reduced to sea level, and the theoretical gravity calculated from the formula in (12.3.6). If the gravity formula fits well, the only phenomena causing gravity anomalies are the disturbing mass anomalies Δm, visible or invisible, topographic or subcrustal. Thus the shape of the geoid can be computed from the free-air gravity anomalies over a complete level surface. Generally speaking, the geoid is approximately an ellipsoid that bulges at the equator and is slightly flattened at the poles. The order of magnitude of the geoid heights is relatively small, probably not exceeding 50 meters.

If the shape of the geoid and the values of gravity over it were completely known, it would be possible to infer the field at all external points without making any special assumption about the state of the interior. It is this field that we would consider to be the overall gravity field of the Earth and it would serve as a basis for interpolation to points at which gravity measurements had not yet been made. The knowledge of the actual gravity field at the point could only be determined by actual measurement.

The shape of the geoid can also be computed from the Earth's external gravitational potential. The reason that gravity anomalies appear is that the potential represented by the formula (12.3.5) is not a complete representation. We can think of the general potential given by (12.2.5) as one that permits us to accomodate departures of the Earth's gravitational potential from the potential given by the reference ellipsoid. Thus contributions from coefficients other than the $C_{n,0}$ can be expected to appear.

It is a purely geometrical transformation to take the potential function $V(r, \theta, \psi)$ given by (12.2.5) and express it as a function of the orbital elements $a, e, i, \omega, \Omega, M$ (see Kaula [15]). Then $V(a, e, i, \omega, \Omega, M)$ is in a form suitable for use in the Lagrange planetary equations. Analyses of secular and long-period perturbations of satellite orbits based on the use of observed data and the Lagrange planetary equations have yielded accurate determinations of the coefficients of the zonal harmonics up to the fourteenth order. Determinations of the coefficients of the tesseral and sectoral harmonics (through $C_{15,14}$ and $S_{13,14}$) based on the analysis of short-period perturbations of satellite orbits have also been obtained. Typical results are given in Table 12.3.1 (see [16]). The

TABLE 12.3.1

COEFFICIENTS OF ZONAL, TESSERAL, AND SECTORAL HARMONICS

n	$C_{n,0} \times 10^6$	n	m	$\bar{C}_{n,m} \times 10^6$	$\bar{S}_{n,m} \times 10^6$
2	−1082.645	2	2	2.379	−1.351
3	2.546	3	1	1.936	0.266
4	1.649	3	2	0.734	−0.538
5	0.210	3	3	0.561	1.620
6	−0.646	4	1	−0.572	−0.469
7	0.333	4	2	0.330	0.661
8	0.270	4	3	0.851	−0.190
9	0.053	4	4	−0.053	0.230

harmonic coefficients $C_{2,1}$ and $S_{2,1}$ from observations of latitude variations, are known to be vanishing small (see Kaula [17]). The coefficients $\bar{C}_{n,m}$ and $\bar{S}_{n,m}$ appearing in Table 12.3.1 are related to the coefficients $C_{n,m}$ and $S_{n,m}$ appearing in (12.2.5), by the formulas

$$C_{n,m} = T\bar{C}_{n,m}, \qquad S_{n,m} = T\bar{S}_{n,m}, \qquad T = \left[\frac{(2n+1)(n-m)!\,(2-\delta_{0,m})}{(n+m)!}\right]^{1/2}$$

where the Kronecker delta $\delta_{0,m}$ is to be interpreted in the usual sense.

As we shall see later, the coefficients $C_{1,0}$, $C_{1,1}$ and $S_{1,1}$ in (12.2.5), will be zero if the origin of coordinates is chosen to be at the center of mass. We will not make use of this fact for the moment, nor will we take advantage of the empirical facts that $C_{2,1}$ and $S_{2,1}$ vanish. We can write the external gravitational potential of the Earth through the fourth harmonic as

$$V = V_0 + V_1 + V_2 + V_3 + V_4 \tag{12.3.7}$$

where, if we substitute for the $P_n{}^m$ from Table 12.2.1, we can, with a slight rearrangement of order of the terms, write

$$V_0 = \frac{aB_{0,0}}{r} \tag{12.3.8}$$

$$V_1 = \left(\frac{a}{r}\right)^2 B_{0,0}(C_{1,0} \cos\theta + C_{1,1} \sin\theta \cos\psi + S_{1,1} \sin\theta \sin\psi) \tag{12.3.9}$$

$$\begin{aligned} V_2 = \left(\frac{a}{r}\right)^3 B_{0,0}\Big[&C_{2,0}\left(1 - \frac{3}{2}\sin^2\theta\right) + 3C_{2,2}\sin^2\theta\cos 2\psi + \frac{3}{2}C_{2,1}\sin 2\theta\cos\psi \\ &+ \frac{3}{2}S_{2,1}\sin 2\theta \sin\psi + 3S_{2,2}\sin^2\theta\sin 2\psi\Big] \end{aligned} \tag{12.3.10}$$

$$\begin{aligned} V_3 = \left(\frac{a}{r}\right)^4 B_{0,0}\Big[&C_{3,0}\left(\cos^3\theta - \frac{3}{2}\sin^2\theta\cos\theta\right) + 15S_{3,3}\sin^3\theta\sin 3\psi \\ &+ 15S_{3,2}\sin^2\theta\cos\theta\sin 2\psi - 2S_{3,1}\left(\frac{3}{4}\sin^3\theta - 3\sin\theta\cos^2\theta\right)\sin\psi \\ &+ 15C_{3,3}\sin^3\theta\cos 3\psi + 15C_{3,2}\sin^2\theta\cos\theta\cos 2\psi \\ &- 2C_{3,1}\left(\frac{3}{4}\sin^3\theta - 3\sin\theta\cos^2\theta\right)\cos\psi\Big] \end{aligned} \tag{12.3.11}$$

$$\begin{aligned} V_4 = \left(\frac{a}{r}\right)^5 B_{0,0}\Big\{&C_{4,0}\left[\frac{35}{8}\left(\cos^4\theta - \frac{6}{7}\cos^2\theta + \frac{3}{35}\right)\right] \\ &+ C_{4,1}\left[\frac{35}{2}\sin\theta\left(\cos^3\theta - \frac{3}{7}\cos\theta\right)\right]\cos\psi \\ &+ C_{4,2}\left[\frac{105}{2}\sin^2\theta\left(\cos^2\theta - \frac{1}{7}\right)\right]\cos 2\psi + C_{4,3}(105\sin^3\theta\cos\theta)\cos 3\psi \\ &+ C_{4,4}(105\sin^4\theta)\cos 4\psi + S_{4,1}\left[\frac{35}{2}\sin\theta\left(\cos^3\theta - \frac{3}{7}\cos\theta\right)\right]\sin\psi \\ &+ S_{4,2}\left[\frac{105}{2}\sin^2\theta\left(\cos^2\theta - \frac{1}{7}\right)\right]\sin 2\psi + S_{4,3}(105\sin^3\theta\cos\theta)\sin 3\psi \\ &+ S_{4,4}(105\sin^4\theta)\sin 4\psi\Big\} \end{aligned} \tag{12.3.12}$$

The problem of the motion of an artificial satellite of the Earth was not given much serious attention until 1957. At that time, little was known about the magnitudes of the coefficients of the tesseral and sectoral harmonics in the Earth's gravity potential. In fact, a number of independent investigators were well along with their theoretical studies of the motion of a particle about an oblate Earth, in which they were representing the Earth's potential V by the zeroth and second zonal harmonics, before it became clear from analysis of the observations on certain satellites that the contribution of the third zonal harmonic to the motion was not entirely negligible. When this "pear-shaped" aspect of the Earth's field arising from the contribution of the third harmonic became known, many of the investigators modified their treatments to permit the inclusion of not only the third zonal harmonic but also the fourth and fifth zonal harmonics. It was pretty well known at this time (1957–1960) that the contributions of the third, fourth, and fifth zonal harmonics were of an order higher than the contribution due to the second zonal harmonic, but the values of the coefficients $C_{3,0}$, $C_{4,0}$, $C_{5,0}$ were not very well established. No reliable information was available for the tesseral or sectoral coefficients except that the observations on orbiting satellites indicated that these coefficients must be small, certainly no more than first order with respect to $C_{2,0}$.

Up to the present, relatively few attempts have been made to include tesseral and sectoral harmonics in the development of an oblate Earth theory for satellites. In fact, little attention has been paid to the inclusion of the fifth zonal harmonic in most of the theories. Generally speaking, the gravity fields that have been assumed in even the more elaborate studies to date are described essentially by a potential V of the form

$$V = \frac{aB_{0,0}}{r}\left[1 + \left(\frac{a}{r}\right)^2 \frac{C_{2,0}}{2}(3\cos^2\theta - 1) + \left(\frac{a}{r}\right)^3 \frac{C_{3,0}}{2}(5\cos^3\theta - 3\cos\theta) + \left(\frac{a}{r}\right)^4 \frac{C_{4,0}}{8}(35\cos^4\theta - 30\cos^2\theta + 3)\right] \tag{12.3.13}$$

where only the contributions of the zeroth, second, third, and fourth zonal harmonics are taken into account. If we neglect the third zonal harmonic, assume $aB_{0,0} = GE$, and introduce the notation

$$J = -\frac{3}{2}C_{2,0}, \qquad K = \frac{15}{2}C_{4,0} \tag{12.3.14}$$

we obtain the potential

$$V = \frac{GE}{r}\left[1 - \frac{J}{3}\left(\frac{a}{r}\right)^2 (3\cos^2\theta - 1) + \frac{K}{30}\left(\frac{a}{r}\right)^4 (35\cos^4\theta - 30\cos^2\theta + 3)\right] \tag{12.3.15}$$

The gravity field $\mathbf{g}$ is obtained from the equation

$$\mathbf{g} = \nabla V \tag{12.3.16}$$

In the literature, the negative of the potential in (12.3.15) is sometimes adopted. When this is done, the gravity field is obtained from the equation

$$\mathbf{g} = -\nabla V \tag{12.3.17}$$

12.4. AN ALTERNATIVE DESCRIPTION OF THE GRAVITATIONAL POTENTIAL: PART 1

Thus far we have viewed the gravitational potential of an irregular body of mass M as a function V that satisfies Laplace's equation and that is uniquely determined when certain Dirichlet conditions are prescribed. We can deduce an equivalent expression for the potential from more fundamental considerations. To do so, we consider a rigid body of mass M and irregular shape. We will assume a continuous distribution of matter throughout a volume τ but not necessarily a continuous macroscopic density (see Figure 12.4.1).

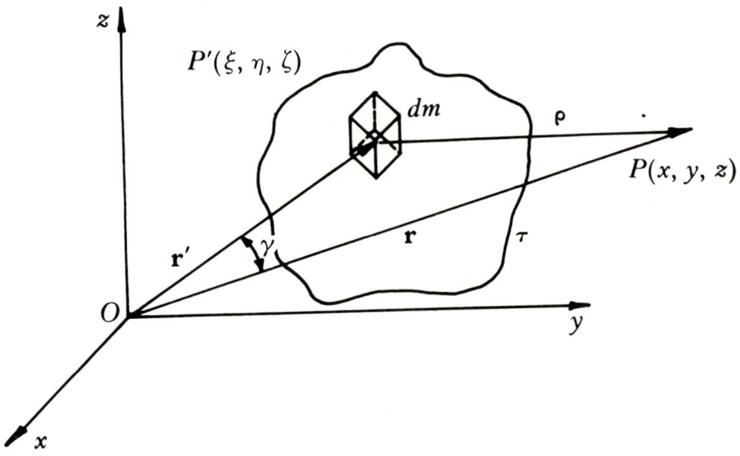

FIGURE 12.4.1

Let $P(x, y, z)$ and $P'(\xi, \eta, \zeta)$ represent points in the rectangular reference frame $Oxyz$. The acceleration field $d\mathbf{F}$ at point P due to an increment of mass dm located at the point P' is given by

$$d\mathbf{F} = \frac{-G\,dm}{\rho^3}\,\boldsymbol{\rho} = \frac{-G\,dm}{\rho^3}(\mathbf{r} - \mathbf{r}') \qquad (\rho = |\boldsymbol{\rho}|) \tag{12.4.1}$$

where $G = k^2$ is the universal gravitational constant and the vector $\boldsymbol{\rho} \equiv \mathbf{r} - \mathbf{r}'$ is drawn from P to P'. The total field $\mathbf{F}$ at P due to the entire mass M is

$$\mathbf{F} = -G \iiint_M \frac{\mathbf{r} - \mathbf{r}'}{\rho^3}\,dm = X\mathbf{i} + Y\mathbf{j} + Z\mathbf{k} \tag{12.4.2}$$

where $\iiint_M$ is to be read "the integral over M of $\cdots$," and

$$X = -G \iiint_M \frac{(x-\xi)}{\rho^3}\, dm$$

$$Y = -G \iiint_M \frac{(y-\eta)}{\rho^3}\, dm \qquad (12.4.3)$$

$$Z = -G \iiint_M \frac{(z-\zeta)}{\rho^3}\, dm$$

Equations (12.4.3) hold even if the point P is inside the mass M. In other words, we can permit the value $\rho = 0$ without having the integrands become indeterminate. To see this, write

$$dm = \sigma\, d\tau$$

where σ is the mass density and $d\tau$ is the volume element. Take a spherical coordinate system with origin at P as shown in Figure 12.4.2,

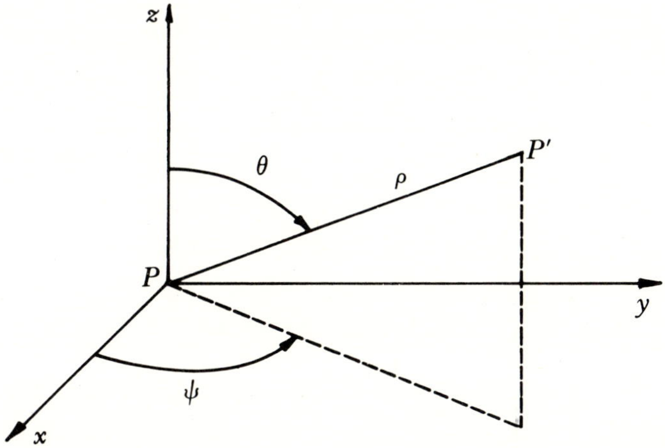

FIGURE 12.4.2

where the rectangular axes are parallel to those in Figure 12.4.1. The element of mass dm is given by

$$dm = \sigma\, d\tau = \sigma\rho^2 \sin\theta\, d\rho\, d\theta\, d\psi$$

Hence, using the equations of transformation

$$x - \xi = \rho \sin\theta \cos\psi$$

$$y - \eta = \rho \sin\theta \sin\psi \qquad (12.4.4)$$

$$z - \zeta = \rho \cos\theta$$

we see that

$$\begin{aligned} X &= -G \iiint_M \sigma \sin^2 \theta \cos \psi \, d\rho \, d\theta \, d\psi \\ Y &= -G \iiint_M \sigma \sin^2 \theta \sin \psi \, d\rho \, d\theta \, d\psi \\ Z &= -G \iiint_M \sigma \cos \theta \sin \theta \, d\rho \, d\theta \, d\psi \end{aligned} \tag{12.4.5}$$

and clearly there is no indeterminacy in the three integrands of (12.4.5) when $\rho = 0$. Thus the integrals X, Y, and Z may represent the gravitational acceleration both inside and outside the body.

We now turn to the potential function V. Since it will be convenient to have V positive, we define the potential at $P(x, y, z)$, due to the total mass M, to be

$$V = G \iiint_M \frac{dm}{\rho} \tag{12.4.6}$$

In spherical coordinates, with origin at P, we can write

$$V = G \iiint_M \sigma\rho \sin \theta \, d\rho \, d\theta \, d\psi \tag{12.4.7}$$

Equation (12.4.6) represents the potential for both exterior and interior points. The gravity field $\mathbf{F}$, given in (12.4.2), is derivable from (12.4.6) by taking the gradient of V as it appears there. We have

$$\begin{aligned} \nabla V &= \frac{\partial V}{\partial x} \mathbf{i} + \frac{\partial V}{\partial y} \mathbf{j} + \frac{\partial V}{\partial z} \mathbf{k} \\ &= G \left[\mathbf{i} \iiint_M \frac{\partial}{\partial x} \left(\frac{1}{\rho} \right) dm + \mathbf{j} \iiint_M \frac{\partial}{\partial y} \left(\frac{1}{\rho} \right) dm + \mathbf{k} \iiint_M \frac{\partial}{\partial z} \left(\frac{1}{\rho} \right) dm \right] \end{aligned}$$

Since

$$\rho^2 = (x - \xi)^2 + (y - \eta)^2 + (z - \zeta)^2$$

we have

$$\frac{\partial \rho}{\partial x} = \frac{x - \xi}{\rho}, \qquad \frac{\partial \rho}{\partial y} = \frac{y - \eta}{\rho}, \qquad \frac{\partial \rho}{\partial z} = \frac{z - \zeta}{\rho}$$

and so

$$\begin{aligned} \nabla V &= -G \left[\mathbf{i} \iiint_M \frac{x - \xi}{\rho^3} dm + \mathbf{j} \iiint_M \frac{y - \eta}{\rho^3} dm + \mathbf{k} \iiint_M \frac{z - \zeta}{\rho^3} dm \right] \\ &= X\mathbf{i} + Y\mathbf{j} + Z\mathbf{k} = \mathbf{F} \end{aligned} \tag{12.4.8}$$

In Theorem 12.1.2, we have the result that

$$(1 - 2sz + z^2)^{-1/2} = \sum_{n=0}^{\infty} P_n(s)\, z^n \qquad (|\, z \,| < 1) \tag{12.4.9}$$

As applied below, Callandreau [18] has shown that the series in z converges if $|\, z \,| = 1$. In this extended sense, we will write the result

$$(1 - 2sz + z^2)^{-1/2} = \sum_{n=0}^{\infty} P_n(s)\, z^n \qquad (|\, z \,| \leqslant 1) \tag{12.4.10}$$

Consider now the potential at some point $P(x, y, z)$ interior to the mass M. Construct a sphere through P with center at the origin O. The potential at $P(x, y, z)$ is composed of contributions from elements of mass interior to or on the sphere and elements of mass exterior to this sphere. For the contribution of elements interior to or on the sphere, we note that, where $r' = |\, \mathbf{r}' \,|$, $r'/r \leqslant 1$, whereas, for the contributions from elements exterior to the sphere, $r'/r > 1$. In particular, for all points whose constructed spheres contain the total mass M, we have $r'/r < 1$.

We will now develop the potential for the case that $r' \leqslant r$. The corresponding development for $r' > r$ would be entirely analogous. With reference to Figure 12.4.1, the distance ρ may be written

$$\rho = (r'^2 + r^2 - 2r'r \cos \gamma)^{1/2}$$

or

$$\rho^{-1} = r^{-1} \left[1 - \frac{2r'}{r} \cos \gamma + \left(\frac{r'}{r} \right)^2 \right]^{-1/2}$$

where γ is the angle between $\mathbf{r}'$ and $\mathbf{r}$. We now use (12.4.10) to write

$$\rho^{-1} = r^{-1} \sum_{n=0}^{\infty} \left(\frac{r'}{r} \right)^n P_n(\cos \gamma) \tag{12.4.11}$$

Then (12.4.6) can be written

$$\begin{aligned} V &= G \iiint_M \left[\sum_{n=0}^{\infty} r^{-1} \left(\frac{r'}{r} \right)^n P_n(\cos \gamma) \right] dm \\ &= G \sum_{n=0}^{\infty} \frac{1}{r^{n+1}} \iiint_M (r')^n P_n(\cos \gamma)\, dm \end{aligned} \tag{12.4.12}$$

If we designate the respective terms in the sum by V_i, $i = 0, 1, 2,\ldots m,\ldots$, the zeroth and first harmonics are

$$V_0 = \frac{G}{r} \iiint_M P_0(\cos\gamma)\, dm = \frac{G}{r} \iiint_M dm = \frac{GM}{r}$$

$$V_1 = \frac{G}{r^2} \iiint_M r' P_1(\cos\gamma)\, dm = \frac{G}{r^2} \iiint_M r' \cos\gamma\, dm$$

With reference to Figure 12.4.1, we can represent unit vectors in the directions OP' and OP by $\mathbf{r}_1'$ and $\mathbf{r}_1$ respectively, and write

$$\begin{aligned} \mathbf{r}_1' &= l\mathbf{i} + m\mathbf{j} + n\mathbf{k} = \frac{\xi}{r'}\mathbf{i} + \frac{\eta}{r'}\mathbf{j} + \frac{\zeta}{r'}\mathbf{k} \\ \mathbf{r}_1 &= \lambda\mathbf{i} + \mu\mathbf{j} + \nu\mathbf{k} = \frac{x}{r}\mathbf{i} + \frac{y}{r}\mathbf{j} + \frac{z}{r}\mathbf{k} \end{aligned} \tag{12.4.13}$$

where (l, m, n) and (λ, μ, ν) are triads of direction cosines associated with these unit vectors. Since

$$\cos\gamma = \mathbf{r}_1' \cdot \mathbf{r}_1 = l\lambda + m\mu + n\nu = \frac{1}{r'}(\lambda\xi + \mu\eta + \nu\xi)$$

we have

$$\begin{aligned} V_1 &= \frac{G}{r^2} \iiint_M r' \cos\gamma\, dm \\ &= \frac{G}{r^2}\left(\lambda \iiint_M \xi\, dm + \mu \iiint_M \eta\, dm + \nu \iiint_M \zeta\, dm\right) \\ &= \frac{GM}{r^3}(x\bar{\xi} + y\bar{\eta} + z\bar{\zeta}) \end{aligned} \tag{12.4.14}$$

where $(\bar{\xi}, \bar{\eta}, \bar{\zeta})$ are the coordinates of the center of mass. Thus

$$V_1 = 0 \tag{12.4.15}$$

if the origin is chosen at the center of mass of the body.

The second harmonic is

$$\begin{aligned} V_2 &= \frac{G}{r^3} \iiint_M (r')^2 P_2(\cos\gamma)\, dm = \frac{G}{2r^3} \iiint_M (r')^2(3\cos^2\gamma - 1)\, dm \\ &= \frac{G}{2r^3} \iiint_M [3(\lambda\xi + \mu\eta + \nu\zeta)^2 - (\xi^2 + \eta^2 + \zeta^2)]\, dm \\ &= \frac{G}{2r^3}\Big[(3\lambda^2 - 1) \iiint_M \xi^2\, dm + (3\mu^2 - 1) \iiint_M \eta^2\, dm \\ &\quad + (3\nu^2 - 1) \iiint_M \zeta^2\, dm + 6\lambda\mu \iiint_M \xi\eta\, dm + 6\mu\nu \iiint_M \eta\zeta\, dm \\ &\quad + 6\lambda\nu \iiint_M \xi\zeta\, dm\Big] \end{aligned}$$

Now

$$3\lambda^2 - 1 = 3\lambda^2 - (\lambda^2 + \mu^2 + \nu^2) = 2\lambda^2 - \mu^2 - \nu^2 = \frac{2x^2 - y^2 - z^2}{r^2}$$

$$3\mu^2 - 1 = 3\mu^2 - (\lambda^2 + \mu^2 + \nu^2) = -\lambda^2 + 2\mu^2 - \nu^2 = \frac{-x^2 + 2y^2 - z^2}{r^2}$$

$$3\nu^2 - 1 = 3\nu^2 - (\lambda^2 + \mu^2 + \nu^2) = -\lambda^2 - \mu^2 + 2\nu^2 = \frac{-x^2 - y^2 + 2z^2}{r^2}$$

and

$$6\lambda\mu = \frac{6xy}{r^2}, \qquad 6\mu\nu = \frac{6yz}{r^2}, \qquad 6\lambda\nu = \frac{6xz}{r^2}$$

so that V_2 may be written in the form

$$\begin{aligned} V_2 = \frac{G}{2r^5}\Big[&(2x^2 - y^2 - z^2) \iiint_M \xi^2\, dm + (-x^2 + 2y^2 - z^2) \iiint_M \eta^2\, dm \\ &+ (-x^2 - y^2 + 2z^2) \iiint_M \zeta^2\, dm + 6xy \iiint_M \xi\eta\, dm \\ &+ 6yz \iiint_M \eta\zeta\, dm + 6xz \iiint_M \xi\zeta\, dm\Big] \end{aligned} \tag{12.4.16}$$

The integrals appearing in the expression for V_2 are of the form

$$I_{ijk} = \iiint_M \xi^i \eta^j \zeta^k\, dm$$

where i, j and k are nonnegative integers. Integrals of this type are called *inertial integrals* and, in this notation, the potential of an irregular body is given by

$$V = V_0 + V_1 + V_2 + G \sum_{n=3}^{\infty} \frac{1}{r^{n+1}} \iiint_M (r')^n P_n(\cos\gamma)\, dm$$

where

$$V_0 = \frac{G}{r} I_{000} = \frac{GM}{r}$$

$$V_1 = \frac{G}{r^3}(I_{100}x + I_{010}y + I_{001}z)$$

$$\begin{aligned} V_2 = \frac{G}{2r^5}\Big[&I_{200}(2x^2 - y^2 - z^2) + I_{020}(-x^2 + 2y^2 - z^2) + I_{002}(-x^2 - y^2 + 2z^2) \\ &+ 6I_{110}xy + 6I_{011}yz + 6I_{101}xz\Big] \end{aligned} \tag{12.4.17}$$

In particular, if the origin is chosen at the center of mass of the irregular body, we have

$$V = V_0 + V_2 + G \sum_{n=3}^{\infty} \frac{1}{r^{n+1}} \iiint_M (r')^n P_n(\cos \gamma)\, dm \qquad (12.4.18)$$

In a similar fashion, the higher harmonics may be developed in terms of the inertial integrals (see Exercise 15 at end of chapter).

12.5. PRINCIPAL AXES

Any real quadratic form may be represented in matrix form. For example,

$$x^2 + 2xy + 4xz + 3y^2 + yz + 7z^2$$

may be represented by the matrix product $\mathbf{r}A\mathbf{r}^T$, where

$$\mathbf{r} = (x, y, z), \qquad \mathbf{r}^T = \begin{pmatrix} x \\ y \\ z \end{pmatrix}$$

$$A = \begin{pmatrix} 1 & 1 & 1 \\ 1 & 3 & \frac{1}{2} \\ 2 & \frac{1}{2} & 7 \end{pmatrix}$$

since

$$\begin{aligned} \mathbf{r}A\mathbf{r}^T &= (x, y, z) \begin{pmatrix} 1 & 1 & 2 \\ 1 & 3 & \frac{1}{2} \\ 2 & \frac{1}{2} & 7 \end{pmatrix} \begin{pmatrix} x \\ y \\ z \end{pmatrix} \\ &= (x + y + 2z \quad x + 3y + \tfrac{1}{2}z \quad 2x + \tfrac{1}{2}y + 7z) \begin{pmatrix} x \\ y \\ z \end{pmatrix} \\ &= x^2 + 2xy + 4xz + 3y^2 + yz + 7z^2 \end{aligned}$$

In the following discussion, we will have occasion to invoke a well-known theorem from higher algebra. Since the groundwork needed to establish the theorem is lengthy, we simply state it without proof (see Birkhoff and Maclane [19]).

Theorem 12.5.1. Any real quadratic form $\mathbf{r}A\mathbf{r}^T$ may be reduced by orthogonal transformations of variables to a diagonal form

$$\lambda_1 z_1{}^2 + \lambda_2 z_2{}^2 + \cdots + \lambda_n z_n{}^2$$

in which the coefficients λ_i are the roots of the characteristic equation $| A - \lambda I | = (\lambda_1 - \lambda)(\lambda_2 - \lambda) \cdots (\lambda_n - \lambda) = 0$, of A, and each λ_i, $i = 1,\ldots, n$, is real.

In the statement of the theorem, I is the identity matrix; $\mathbf{r}$, $\mathbf{r}^T$, and A are the matrices associated with the real quadratic form, in the sense of the example.

Consider now the irregularly shaped body of mass M represented in Figure 12.5.1. The angle α is the angle between a unit vector $\mathbf{u}$ in the

FIGURE 12.5.1

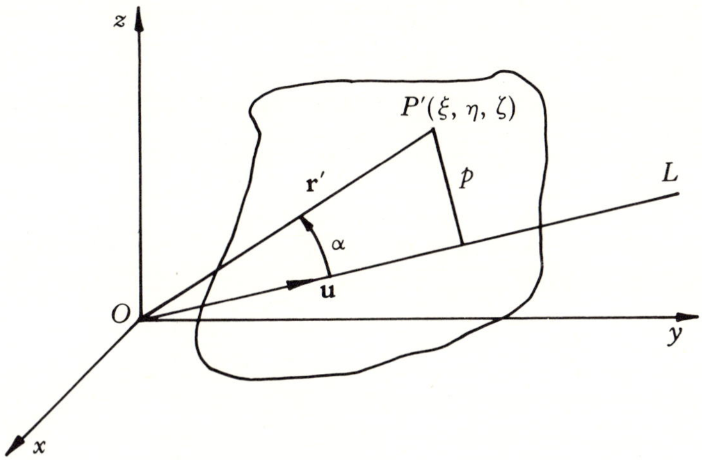

direction of an arbitrary line L through the origin O and a vector $\mathbf{r}'$ from O to an increment of mass dm at P'; unit vectors $\mathbf{i}$, $\mathbf{j}$, and $\mathbf{k}$ are in the usual sense and p is the perpendicular distance from dm to the line L. The moment of inertia I_L of the body with respect to the axis L is defined to be

$$I_L = \iiint_M p^2 \, dm \tag{12.5.1}$$

Now $\mathbf{u} \times \mathbf{r}' = r' \sin \alpha \mathbf{n}_1 = p\mathbf{n}_1$, where $\mathbf{n}_1$ is a unit vector perpendicular to the plane of $\mathbf{u}$ and $\mathbf{r}'$. Thus we see that

$$p^2 = (\mathbf{u} \times \mathbf{r}')^2$$

where

$$\mathbf{u} \times \mathbf{r}' = \begin{vmatrix} \mathbf{i} & \mathbf{j} & \mathbf{k} \\ l^* & m^* & n^* \\ \xi & \eta & \zeta \end{vmatrix} = (m^*\zeta - n^*\eta)\mathbf{i} + (n^*\xi - l^*\zeta)\mathbf{j} + (l^*\eta - m^*\xi)\mathbf{k}$$

and l^*, m^*, and n^* are direction cosines of $\mathbf{u}$. If follows that

$$(\mathbf{u} \times \mathbf{r}')^2 = m^{*2}(\xi^2 + \zeta^2) + n^{*2}(\xi^2 + \eta^2) + l^{*2}(\eta^2 + \zeta^2) \\ - 2m^*n^*\eta\zeta - 2l^*n^*\xi\zeta - 2l^*m^*\xi\eta$$

so that (12.5.1) can be written in the form

$$I_L = l^{*2}I_x + m^{*2}I_y + n^{*2}I_z - 2l^*m^*P_{xy} - 2l^*n^*P_{xz} - 2m^*n^*P_{yz} \tag{12.5.2}$$

where the integrals

$$I_x = \iiint_M (\eta^2 + \zeta^2)\, dm$$

$$I_y = \iiint_M (\xi^2 + \zeta^2)\, dm$$

$$I_z = \iiint_M (\xi^2 + \eta^2)\, dm$$

are called the moments of inertia about the respective coordinate axes and the integrals

$$P_{xy} = \iiint_M \xi\eta\, dm$$

$$P_{xz} = \iiint_M \xi\zeta\, dm$$

$$P_{yz} = \iiint_M \eta\zeta\, dm$$

are called products of inertia.

Let $P(x, y, z)$ designate any point on the line L and let $\mathbf{r}_L$ be a vector directed along L to the point (x, y, z). Then

$$l^* = \frac{x}{r_L}, \qquad m^* = \frac{y}{r_L}, \qquad n^* = \frac{z}{r_L} \qquad (r_L = |\, \mathbf{r}_L \,|)$$

If we substitute these expressions into (12.5.2), we obtain the equation

$$I_x x^2 + I_y y^2 + I_z z^2 - 2P_{xy}xy - 2P_{xz}xz - 2P_{yz}yz = I_L r_L^2 \tag{12.5.3}$$

Now if L is permitted to range over all directions around the point O, then I_L will, in general, change in value. But since r_L is not as yet specified, we can, for any given line L, choose r_L such that $I_L r_L^2$ remains equal to some constant, say K. Since I_x, I_y, and I_z are all positive, we see that we may, in principle, construct an ellipsoid about O. In fact, we can construct a family of ellipsoids about the point for different choices of the parameter K. This family of ellipsoids is called the family of ellipsoids of inertia for the point O. Since $r_L^2 = x^2 + y^2 + z^2$, the tip of $\mathbf{r}_L$, for a given K, sweeps out a given ellipsoid surface.

Since the point O was quite arbitrary, we have thus established that at every point O there exists an ellipsoid of inertia for the body M. Furthermore, in view of Theorem 12.5.1, we see that we can diagonalize the particular real quadratic form given by (12.5.3). Thus by rotating the axes we can arrive at a set of axes, called principal axes, for which the products of inertia will vanish. The procedure for carrying out such a rotation of axes is illustrated in the following example.

Example 12.5.1. A uniform lamina has the shape of a right triangle with perpendicular sides a and b. Find the principal axes at the point of intersection of a and b (see Figure 12.5.2).

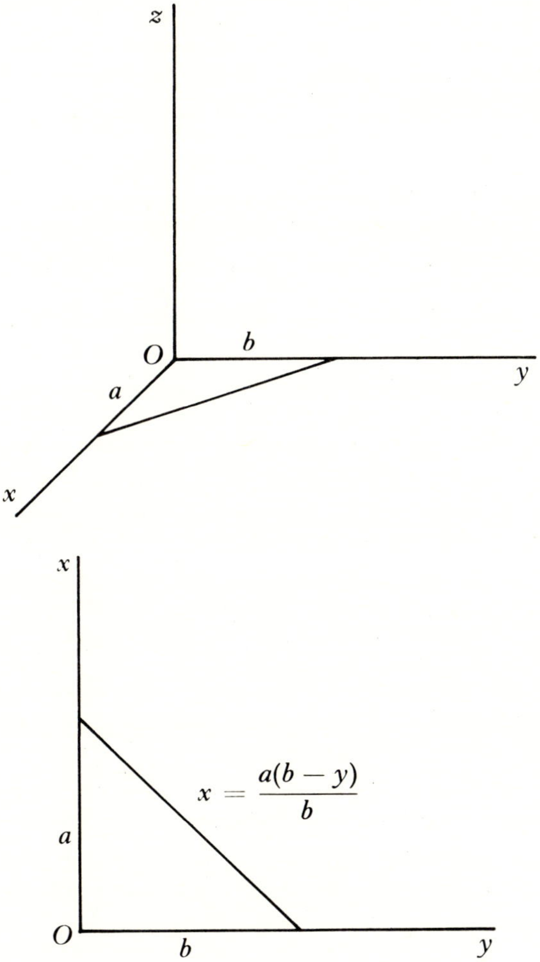

FIGURE 12.5.2

We designate the mass of the lamina by M, its surface area by S, and its mass density by ρ, assumed constant. The mass element is given by

$$dm = \rho\, d\xi\, d\eta$$

We seek to determine the real quadratic form corresponding to (12.5.3). When the values of the definite integrals

$$I_x = \iiint_M (\eta^2 + \zeta^2)\, dm = \rho \iint_S \eta^2\, d\xi\, d\eta = \frac{\rho}{3}\int_0^a \frac{b^3}{a^3}(a - \xi)^3\, d\xi = \frac{Mb^2}{6}$$

$$I_y = \iiint_M (\xi^2 + \zeta^2)\, dm = \frac{Ma^2}{6}$$

$$I_z = \iiint_M (\xi^2 + \eta^2)\, dm = M\frac{(a^2 + b^2)}{6}$$

$$P_{xz} = \iiint_M \xi\zeta\, dm = 0$$

$$P_{yz} = \iiint_M \eta\zeta\, dm = 0$$

$$P_{xy} = \iiint_M \xi\eta\, dm = \rho \iint_S \xi\eta\, d\xi\, d\eta = \frac{\rho b^2}{2a^2}\int_0^a (\xi^3 - 2a\xi^2 + a^2\xi)\, d\xi = \frac{Mab}{12}$$

are substituted into (12.5.3), we obtain the equation

$$\frac{Mb^2}{6}x^2 + \frac{Ma^2}{6}y^2 + M\frac{(a^2 + b^2)}{6}z^2 - \frac{Mab}{6}xy = I_L r_L^2 \qquad (12.5.4)$$

which reduces to the real quadratic form

$$b^2x^2 + a^2y^2 + (a^2 + b^2)\,z^2 - abxy = \frac{6}{M} \qquad (12.5.5)$$

when we choose $K = r_L^2 I_L = 1$. The ellipsoid (12.5.5) obtained by the choice $K = 1$ is called Poinsot's *ellipsoid of inertia* of the body at point O.

The matrix A associated with (12.5.5) is clearly

$$A = \begin{pmatrix} b^2 & -\dfrac{ab}{2} & 0 \\ -\dfrac{ab}{2} & a^2 & 0 \\ 0 & 0 & a^2 + b^2 \end{pmatrix}$$

whence the characteristic equation is

$$|A - \lambda I| = \begin{vmatrix} b^2 - \lambda & -\dfrac{ab}{2} & 0 \\ -\dfrac{ab}{2} & a^2 - \lambda & 0 \\ 0 & 0 & (a^2 + b^2) - \lambda \end{vmatrix}$$

$$= (a^2 + b^2 - \lambda)\left[(b^2 - \lambda)(a^2 - \lambda) - \frac{a^2b^2}{4}\right] = 0 \qquad (12.5.6)$$

Clearly $\lambda = a^2 + b^2$ is a root of (12.5.6). The remaining two roots are to be obtained from the quadratic factor. They are given by

$$\lambda = \frac{(a^2 + b^2) \pm [(a^2 + b^2)^2 - 3a^2b^2]^{1/2}}{2}$$

The three values of λ now permit us to determine a triad of basis vectors (characteristic vectors) that orient a new coordinate system in which the matrix representing the real quadratic form is a diagonal matrix. To see this, it will be useful to establish a theorem that relates to characteristic vectors. Before introducing a statement of the theorem, we observe that the quadratic form

$$h(x, y, z) = ax^2 + by^2 + cz^2 + 2dxy + 2exz + 2fyz - g$$

where $a, b, c, \ldots, g$ are real constants, represents a surface when we set $h(x, y, z)$ equal to a constant. A vector $\mathbf{u}$ normal to the surface at a point (x, y, z) on the surface is obtained by evaluating $\nabla h(x, y, z)$ at the point; that is,

$$\mathbf{u} = \left[\frac{\partial h}{\partial x}\mathbf{i} + \frac{\partial h}{\partial y}\mathbf{j} + \frac{\partial h}{\partial z}\mathbf{k}\right]_{(x,y,z)}$$

$$= 2(ax + dy + ez)\mathbf{i} + 2(dx + by + fz)\mathbf{j} + 2(ex + fy + cz)\mathbf{k}$$

The matrix A representing the real quadratic form $h(x, y, z)$ is

$$A = \begin{pmatrix} a & d & e \\ d & b & f \\ e & f & c \end{pmatrix}$$

We now state and prove the following theorem.

Theorem 12.5.2. If $\mathbf{r}^T$ and $\mathbf{u}^T$ are column matrices that represent the position vector $\mathbf{r}$ to the surface $h(x, y, z) = 0$ and the vector $\mathbf{u}$, respectively, then $2A\mathbf{r}^T = \mathbf{u}^T$.

Proof. The proof follows immediately from the equation

$$2A\mathbf{r}^T = 2\begin{pmatrix} a & d & e \\ d & b & f \\ e & f & c \end{pmatrix}\begin{pmatrix} x \\ y \\ z \end{pmatrix} = 2\begin{pmatrix} ax + dy + ez \\ dx + by + fz \\ ex + fy + cz \end{pmatrix} = \mathbf{u}^T$$

Thus the vector $A\mathbf{r}^T$ is a vector in the direction of the normal to the surface. For an ellipsoidal surface, the principal axes are determined by the condition that $A\mathbf{r}^T$ and the position vector $\mathbf{r}^T$ be parallel. This condition leads to an equation for determining a characteristic vector in the direction of a principal axis, namely,

$$A\mathbf{r}^T = \lambda\mathbf{r}^T$$

where λ is a scalar constant. The scalar equations of condition

$$(a - \lambda)\,x + dy + ez = 0$$

$$dx + (b - \lambda)\,y + fz = 0$$

$$ex + fy + (c - \lambda)\,z = 0$$

have nontrivial solutions for coordinates (x, y, z) if and only if the determinant of the coefficients vanishes; that is, $|A - \lambda I| = 0$. This condition is the one from which the λ's in Theorem 12.5.1 are determined.

If λ_1 and λ_2 are two distinct real roots of $|A - \lambda I| = 0$, then the corresponding characteristic vectors are orthogonal. To show this, we consider two characteristic vectors $\mathbf{r}_1{}^T$ and $\mathbf{r}_2{}^T$ corresponding to the distinct values of λ_1 and λ_2 respectively. We may write the bilinear form

$$\mathbf{r}_2(A\mathbf{r}_1{}^T) = \mathbf{r}_2\lambda_1\mathbf{r}_1{}^T = \lambda_1\mathbf{r}_2\mathbf{r}_1{}^T$$

in the alternative form

$$\mathbf{r}_2(A\mathbf{r}_1{}^T) = (\mathbf{r}_2A)\,\mathbf{r}_1{}^T = \lambda_2\mathbf{r}_2\mathbf{r}_1{}^T$$

Since $\lambda_1 \neq \lambda_2$, we conclude that $\mathbf{r}_2\mathbf{r}_1{}^T = 0$ and hence the vectors $\mathbf{r}_2{}^T$ and $\mathbf{r}_1{}^T$ are orthogonal.

If all three values of λ are the same, the ellipsoids of inertia are concentric spheres centered at the point O, and any axis through O is a principal axis. If any two values of λ are the same, the ellipsoids of inertia are ellipsoids of revolution, and any axis in the circular cross section through O is a principal axis.

To find the characteristic vectors for the example we are considering, we first write

$$A\mathbf{r}^T = \lambda\mathbf{r}^T$$

or

$$\begin{pmatrix} b^2 & -\dfrac{ab}{2} & 0 \\ -\dfrac{ab}{2} & a^2 & 0 \\ 0 & 0 & a^2+b^2 \end{pmatrix} \begin{pmatrix} x \\ y \\ z \end{pmatrix} = \lambda \begin{pmatrix} x \\ y \\ z \end{pmatrix}$$

The equations of condition

$$x(b^2 - \lambda) = \frac{ab}{2} y \tag{1}$$

$$y(a^2 - \lambda) = \frac{ab}{2} x \tag{2}$$

$$\lambda z = (a^2 + b^2)\, z \tag{3}$$

can be used to determine three characteristic vectors by substituting, successively, the three values of λ obtained earlier.

From the third equation of condition, it is found that z is arbitrary for $\lambda_1 = a_2 + b_2$. We must also satisfy simultaneously the equations

$$-xa^2 = \frac{ab}{2} y, \qquad -yb^2 = \frac{ab}{2} x$$

Clearly these two conditions can be satisfied only if $x = y = 0$. Thus characteristic vectors corresponding to λ_1 are $(0, 0, z)$. Similarly, it is found that characteristic vectors corresponding to

$$\lambda_2 = \frac{(a^2 + b^2)^2}{2} + \frac{[(a^2 + b^2)^2 - 3a^2b^2]^{1/2}}{2}$$

and

$$\lambda_3 = \frac{(a^2 + b^2)^2}{2} - \frac{[(a^2 + b^2)^2 - 3a^2b^2]^{1/2}}{2}$$

are $(\alpha y, y, 0)$ and $(\beta y, y, 0)$ respectively, where

$$\alpha \equiv \frac{a^2 - b^2}{ab} + \left[\frac{(a^2 - b^2)^2}{a^2b^2} + 1\right]^{1/2}$$

and

$$\beta \equiv \frac{a^2 - b^2}{ab} - \left[\frac{(a^2 - b^2)^2}{a^2b^2} + 1\right]^{1/2}$$

Since

$$\left[\frac{(a^2 - b^2)^2}{a^2b^2} + 1\right]^{1/2} > \frac{a^2 - b^2}{ab}$$

for any a and b, it follows that α is positive and β is negative.

The principal axes x', y', z' are shown in Figure 12.5.3. The x'- and y'-axes lie in the xy-plane. In particular, if $a = b$, we have $\alpha = 1$

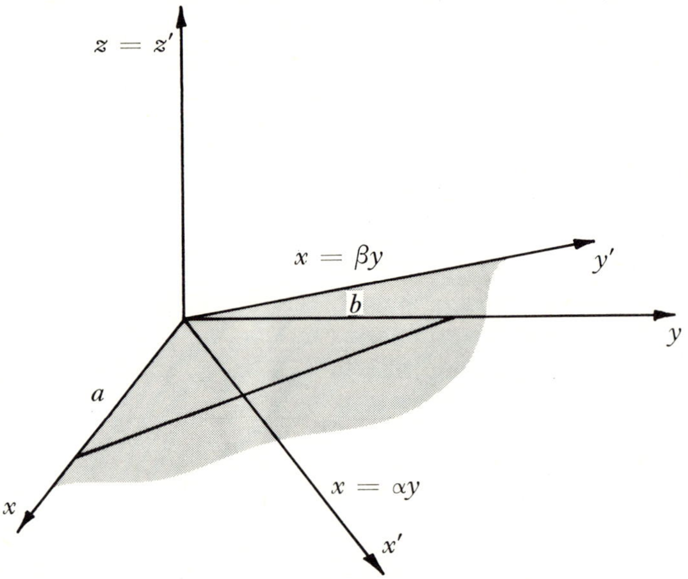

FIGURE 12.5.3

and $\beta = -1$, so that the principal axes are determined by the characteristic vectors $(y, y, 0)$, $(-y, y, 0)$, and $(0, 0, z)$. The diagonalized quadratic form is

$$\lambda_2(x')^2 + \lambda_3(y')^2 + \lambda_1(z')^2 = \frac{6}{M} \tag{12.5.7}$$

and the moments of inertia referenced to the principal axes are

$$I_{x'} = \frac{M}{6}\lambda_2, \qquad I_{y'} = \frac{M}{6}\lambda_3, \qquad I_{z'} = \frac{M}{6}\lambda_1 \tag{12.5.8}$$

12.6. AN ALTERNATIVE DESCRIPTION OF THE GRAVITATIONAL POTENTIAL: PART 2

In view of the discussion in Section 12.5, we see that at every point O there exist principal axes for an irregular body M. Thus let us assume that the rectangular reference system shown in Figure 12.4.1 is referred to principal axes for the body which, for convenience, we continue to designate by x, y, and z. With respect to these principal axes, the products of inertia vanish. If we designate the principal moments of inertia by the symbols A, B, and C, we have, from the definition,

$$A = \iiint_M (\eta^2 + \zeta^2)\, dm, \qquad B = \iiint_M (\xi^2 + \zeta^2)\, dm$$
$$C = \iiint_M (\xi^2 + \eta^2)\, dm \tag{12.6.1}$$

From (12.6.1), we can obtain expressions for I_{200}, I_{020}, and I_{002} (referred to principal axes) in terms of A, B, and C. If we solve the three equations

$$A = I_{020} + I_{002}, \qquad B = I_{200} + I_{002}, \qquad C = I_{200} + I_{020}$$

for I_{200}, I_{020}, and I_{002}, we find that

$$I_{200} = \frac{B + C - A}{2}, \qquad I_{002} = \frac{B - C + A}{2}$$
$$I_{020} = \frac{A - B + C}{2} \tag{12.6.2}$$

Collecting the coefficients of x^2, y^2, and z^2 in (12.4.17), we obtain (noting that we are dealing with principal axes)

$$V_2 = \frac{G}{2r^5} [(2I_{200} - I_{002} - I_{020})\, x^2 + (2I_{020} - I_{200} - I_{002})\, y^2 + (2I_{002} - I_{200} - I_{020})\, z^2]$$

The coefficients, expressed in terms of A, B, and C, are given by

$$2I_{200} - I_{002} - I_{020} = B + C - 2A$$
$$2I_{020} - I_{200} - I_{002} = A + C - 2B$$
$$2I_{002} - I_{200} - I_{020} = A + B - 2C$$

Hence V_2 may be written in the form

$$V_2 = \frac{G}{2r^3} \left[A + B + C - 3\, \frac{(Ax^2 + By^2 + Cz^2)}{r^2} \right] \tag{12.6.3}$$

But

$$\frac{x}{r} = \lambda, \qquad \frac{y}{r} = \mu, \qquad \frac{z}{r} = \nu$$

are the direction cosines of the line segment OP (see Figure 12.4.1); hence we have

$$V_2 = \frac{G}{2r^3}[A + B + C - 3(A\lambda^2 + B\mu^2 + C\nu^2)] \tag{12.6.4}$$

Reference to (12.5.2) shows that $A\lambda^2 + B\mu^2 + C\nu^2$ represents the moment of inertia I of the body M with respect to the line through OP. Thus we write

$$V_2 = \frac{G}{2r^3}(A + B + C - 3I) \tag{12.6.5}$$

and (referred to principal axes with origin at the center of mass)

$$V = \frac{GM}{r} + \frac{G}{2r^3}(A + B + C - 3I) + G\sum_{n=3}^{\infty}\frac{1}{r^{n+1}}\iiint_M (r')^n P_n(\cos\gamma)\, dm \tag{12.6.6}$$

Consider the spherical coordinate systems and the transformations shown in Figure 12.6.1. Designate the origin O as the center of mass of the body of mass M. The primed coordinates refer to the attracting point $P'(\xi, \eta, \zeta)$.

$$\xi = r'\sin\theta'\cos\psi'$$
$$\eta = r'\sin\theta'\sin\psi'$$
$$\zeta = r'\cos\theta'$$

$$x = r\sin\theta\cos\psi$$
$$y = r\sin\theta\sin\psi$$
$$z = r\cos\theta$$

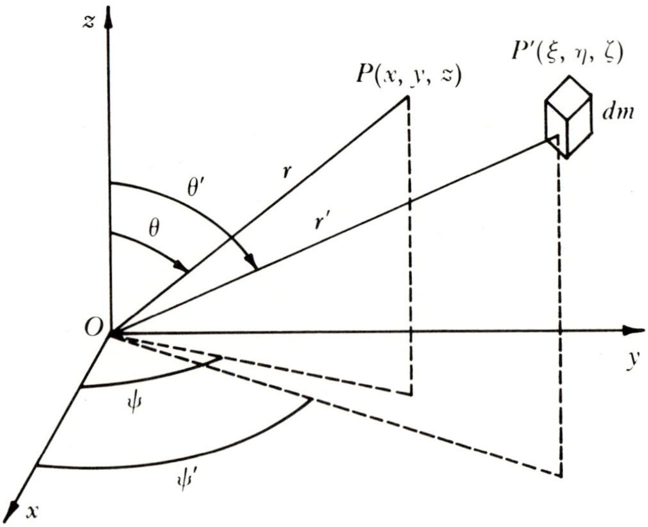

FIGURE 12.6.1

Assume again that x, y, and z are principal axes for the body at point O so that V_2 is given by (12.6.3). If the equations of transformation are substituted into (12.6.3), we find, after some algebraic simplification,

$$V_2 = \frac{G}{2r^3}\left[\left(C - \frac{A+B}{2}\right)(1 - 3\cos^2\theta) - \frac{3}{2}(A - B)\sin^2\theta\cos 2\psi\right] \tag{12.6.7}$$

In particular, if the body is rotationally symmetric about the z-axis, $A = B$, and V_2 reduces to

$$V_2 = \frac{G}{2r^3}(C - A)(1 - 3\cos^2\theta) \tag{12.6.8}$$

Briefly then, for an irregular body, with origin at the center of mass of the body and referred to principal axes for the body at the center of mass, we have

$$V = \frac{GM}{r} + \frac{G}{2r^3}\left[\left(C - \frac{A+B}{2}\right)(1 - 3\cos^2\theta) - \frac{3}{2}(A - B)\sin^2\theta\cos 2\psi\right]$$
$$+ G\sum_{n=3}^{\infty}\frac{1}{r^{n+1}}\iiint_M (r')^n P_n(\cos\gamma)\,dm \tag{12.6.9}$$

If, in addition, the irregular body has rotational symmetry about z, then

$$V = \frac{GM}{r} + \frac{G}{2r^3}(C - A)(1 - 3\cos^2\theta) + G\sum_{n=3}^{\infty}\frac{1}{r^{n+1}}\iiint_M (r')^n P_n(\cos\gamma)\,dm \tag{12.6.10}$$

Note carefully that no assumption about the mass density σ is involved in (12.6.9). In (12.6.10) the only assumption about σ is that it does not depend on ψ.

12.7. AN ALTERNATIVE DESCRIPTION OF THE GRAVITATIONAL POTENTIAL: PART 3

The expressions for V_0, V_1, V_2, and V_3 given in (12.4.17) and Exercise 15 at the end of the chapter were developed without any appeal to the concept of principal axes. Still without appeal to the notion of principal axes, we may use the transformation equations given in Figure 12.6.1 to express V_1, V_2, and V_3 in terms of the spherical coordinates (r, θ, ψ). The relations

$$x^2 = \tfrac{1}{2}r^2\sin^2\theta(1 + \cos 2\psi), \qquad y^2 = \tfrac{1}{2}r^2\sin^2\theta(1 - \cos 2\psi), \qquad z^2 = r^2\cos^2\theta$$

$$xy = \tfrac{1}{2}r^2\sin^2\theta\sin 2\psi, \qquad xz = r^2\sin\theta\cos\theta\cos\psi, \qquad yz = r^2\sin\theta\cos\theta\sin\psi$$

may be used to do this. After some simplification, we find that

$$2x^2 - y^2 - z^2 = r^2(\tfrac{3}{2}\sin^2\theta - 1 + \tfrac{3}{2}\sin^2\theta\cos 2\psi)$$
$$-x^2 + 2y^2 - z^2 = r^2(\tfrac{3}{2}\sin^2\theta - 1 - \tfrac{3}{2}\sin^2\theta\cos 2\psi)$$
$$-x^2 - y^2 + 2z^2 = r^2(-3\sin^2\theta + 2)$$

Thus after regrouping terms, we obtain

$$V_1 = \frac{G}{r^2}(I_{100}\sin\theta\cos\psi + I_{010}\sin\theta\sin\psi + I_{001}\cos\theta) \tag{12.7.1}$$

$$V_2 = \frac{G}{2r^3}\Big[(I_{200} + I_{020} - 2I_{002})\left(\frac{3}{2}\sin^2\theta - 1\right) + \frac{3}{2}(I_{200} - I_{020})\sin^2\theta\cos 2\psi + 3I_{101}\sin 2\theta\cos\psi + 3I_{011}\sin 2\theta\sin\psi + 3I_{110}\sin^2\theta\sin 2\psi\Big] \tag{12.7.2}$$

In particular, if the axes are principal, then $I_{101} = I_{011} = I_{110} = 0$, $I_{200} - I_{020} = B - A$, and V_2 reduces to (12.6.7).

In a similar way, one may obtain an expression for V_3 in spherical coordinates. It is found that

$$\begin{aligned} V_3 = \frac{G}{2r^4}\Big[&(2I_{003} - 3I_{201} - 3I_{021})\left(\cos^3\theta - \frac{3}{2}\sin^2\theta\cos\theta\right) \\ &+ \frac{5}{4}(3I_{210} - I_{030})\sin^3\theta\sin 3\psi + 15I_{111}\cos\theta\sin^2\theta\sin 2\psi \\ &+ (I_{030} + I_{210} - 4I_{012})\left(\frac{3}{4}\sin^3\theta - 3\sin\theta\cos^2\theta\right)\sin\psi \\ &+ (I_{300} + I_{120} - 4I_{102})\left(\frac{3}{4}\sin^3\theta - 3\sin\theta\cos^2\theta\right)\cos\psi \\ &+ \frac{15}{2}(I_{201} - I_{021})\sin^2\theta\cos\theta\cos 2\psi + \frac{5}{4}(I_{300} - 3I_{120})\sin^3\theta\cos 3\psi\Big] \end{aligned} \tag{12.7.3}$$

We may now compare Equations (12.4.17), (12.7.1), (12.7.2), and (12.7.3) with (12.3.8), (12.3.9), (12.3.10), and (12.3.11). We note that $B_{0,0} = GM/a$, justifying our earlier assumption. In addition, we have

$$B_{0,0}C_{1,1} = \frac{G}{a^2}I_{100}, \qquad B_{0,0}S_{1,1} = \frac{G}{a^2}I_{010}, \qquad B_{0,0}C_{1,0} = \frac{G}{a^2}I_{001}$$

$$B_{0,0}C_{2,0} = \frac{G}{2a^3}(2I_{002} - I_{200} - I_{020}), \qquad B_{0,0}C_{2,2} = \frac{G}{4a^3}(I_{200} - I_{020})$$

$$B_{0,0}C_{2,1} = \frac{G}{a^3}I_{101}, \qquad B_{0,0}S_{2,1} = \frac{G}{a^3}I_{011}, \qquad B_{0,0}S_{2,2} = \frac{G}{2a^3}I_{110}$$

$$B_{0,0}C_{3,0} = \frac{G}{2a^4}(2I_{003} - 3I_{201} - 3I_{021}), \qquad B_{0,0}S_{3,3} = \frac{G}{24a^4}(3I_{210} - I_{030})$$

$$B_{0,0}S_{3,2} = \frac{G}{2a^4} I_{111}, \qquad B_{0,0}S_{3,1} = \frac{G}{4a^4}(4I_{012} - I_{030} - I_{210})$$

$$B_{0,0}C_{3,1} = \frac{G}{4a^4}(4I_{102} - I_{300} - I_{120}), \qquad B_{0,0}C_{3,2} = \frac{G}{4a^4}(I_{201} - I_{021})$$

$$B_{0,0}C_{3,3} = \frac{G}{24a^4}(I_{300} - 3I_{201}) \tag{12.7.4}$$

We see that the constants $B_{0,0}C_{n,m}$ and $B_{0,0}S_{n,m}$ appearing in the representation of V in solid spherical harmonics are linear combinations of the inertial integrals. If the axes of reference are principal axes, then $C_{2,1} = S_{2,1} = S_{2,2} = 0$.

We may consider the inertial integrals when the body possesses symmetry about some axis, which we designate by z. For this case, the mass density σ is a function of r' and θ' alone. Thus we write, where the primed notation again refers to the attracting point,

$$\sigma = \sigma(r', \theta')$$
$$dm = \sigma r'^2 \sin\theta' \, dr' \, d\theta' \, d\psi'$$

Since

$$\xi^i \eta^j \zeta^k \, dm = (r')^{i+j+k+2} \sigma(r', \theta') \sin^{i+j+1}\theta' \cos^i\psi' \sin^j\psi' \cos^k\theta' \, dr' \, d\theta' \, d\psi'$$

the inertial integrals I_{ijk} become

$$I_{ijk} = \int_{r'} \int_{\theta'} \sigma(r', \theta')(r')^{i+j+k+2} \sin^{i+j+1}\theta' \cos^k\theta' \int_0^{2\pi} \cos^i\psi' \sin^j\psi' \, d\psi' \, dr' \, d\theta'$$

If we define

$$\mathscr{L}_{ij} = \int_0^{2\pi} \cos^i\psi' \sin^j\psi' \, d\psi' \tag{12.7.5}$$

we have

$$\mathscr{L}_{00} = \int_0^{2\pi} d\psi' = 2\pi \qquad \text{if} \quad i = j = 0$$

$$\mathscr{L}_{01} = \int_0^{2\pi} \sin\psi' \, d\psi' = 0 \qquad \text{if} \quad i = 0, \quad j = 1$$

$$\mathscr{L}_{10} = \int_0^{2\pi} \cos\psi' \, d\psi' = 0 \qquad \text{if} \quad i = 1, \quad j = 0$$

Suppose i is odd, say $i = s + 1$, where s is even, so that $\cos^i\psi' = \cos^s\psi' \cos\psi'$, where $\cos^s\psi'$ may be expressed in even powers of $\sin\psi'$.

Then $\mathscr{L}_{ij}$ may be decomposed into a sum of integrals, the integrand of each one of which may be expressed in the form $\sin^q \psi' \cos \psi'$, where q is a positive integer. The antiderivatives $(q+1)^{-1} \sin^{q+1}\psi'$ then vanish at either limit. A similar type of reasoning applies if j is odd. Thus, for either i or j odd, $\mathscr{L}_{ij} = 0$. Hence we have

$$I_{ijk} = 0 \qquad (i \quad \text{or} \quad j \text{ odd}) \tag{12.7.6}$$

Exercises

1. Verify that c, $ax + by + cz$ and $x^2 - y^2 + yz$ are solid spherical harmonics of degrees 0, 1, and 2, respectively, where a, b, and c are constants.
2. Show that if V_n is a solid spherical harmonic of degree n, then $r^{-(2n+1)}V_n$ is a solid spherical harmonic of degree $-(n+1)$.
3. Verify that the polynomials $P_n(s)$ of Equation (12.1.10) are particular solutions of Legendre's equation.
4. Prove Theorem 12.1.1.
5. Prove Theorem 12.1.2.
6. Prove Theorem 12.1.3.
7. Show that $P_n(\cos\theta)$ is a surface spherical harmonic of degree n.
8. Show that the n zeros of the polynomial $P_n(s)$ are all real, distinct, and lie between -1 and $+1$.
9. Show that

$$|P_n^{(k)}(s)| \leqslant n^{2k} \qquad k = 0, 1, 2, \ldots,$$

where

$$P_n^{(k)}(s) \equiv \frac{d^k P_n(s)}{ds^k} \qquad |s| \leqslant 1$$

10. Prove Theorem 12.1.7.
11. Prove Theorem 12.1.8.
12. Prove Theorem 12.1.9.
13. Prove Theorem 12.1.11.
14. Show that V as given in Equation (12.1.22b), with constants determined by (12.1.27)–(12.1.29), satisfies Laplace's equation by showing that the conditions of Theorem 12.1.10 are satisfied.

15. Show that

$$V_3 = \frac{G}{2r^7}[x(2x^2 - 3y^2 - 3z^2)\,I_{300} + y(-3x^2 + 2y^2 - 3z^2)\,I_{030}$$
$$+ z(-3x^2 - 3y^2 + 2z^2)\,I_{003} + y(12x^2 - 3y^2 - 3z^2)\,I_{210}$$
$$+ z(12x^2 - 3y^2 - 3z^2)\,I_{201} + x(-3x^2 + 12y^2 - 3z^2)\,I_{120}$$
$$+ z(-3x^2 + 12y^2 - 3z^2)\,I_{021} + x(-3x^2 - 3y^2 + 12z^2)\,I_{102}$$
$$+ y(-3x^2 - 3y^2 + 12z^2)\,I_{012} + 30xyz\,I_{111}]$$

16. Three point masses, rigidly connected together are situated as follows:

$4m$ at $(a, -a, a)$

$3m$ at $(-a, a, a)$

$2m$ at (a, a, a)

Find a set of principal axes at the point $(0, 0, a)$ and the moments of inertia about these axes.†

17. Find the principal axes of a quadrant of an ellipse of semiaxes a and b at the center of the ellipse.

† From R. A. Becker, "Introduction to Theoretical Mechanics," p. 314. Copyright 1954 by McGraw-Hill, New York. Used with permission of McGraw-Hill Book Company.

Chapter 13

Gravity and Drag Effects on the Orbit of an Artificial Earth Satellite

13.1. INTRODUCTION

Deviations of the motion of the center of mass of an artificial Earth satellite from a two-body elliptic orbit are caused by the Earth's oblateness and the drag (that is, resistance) of its atmosphere. The study of these deviations is usually considered the main problem of satellite orbital theory because for most Earth satellites other perturbing influences, such as radiation pressure and magnetic fields, are of minor importance compared to the perturbing effects of flattening and drag.

Since the advent of artificial Earth satellites, more than forty authors have published general perturbation theories for predicting artificial satellite motions. The various theories, which illustrate almost every facet of the known perturbation techniques, differ in such respects as choice of parameters, orders carried in the potential function and the subsequent development, completeness, accuracy, and computer adaptability. The approaches used and the results obtained have been compiled and compared for a large number of these theories (see Arsenault *et al.* [20]). Because our principal purpose in this chapter is to show the reader how the concepts discussed in earlier material may be applied in a straightforward way to the study of the motion of artificial

Earth satellites, the discussion, in general, is confined to certain first-order results for gravity and drag perturbations, considered separately. The approximate first-order gravitational theory given here is due to Dr. Leland E. Cunningham, Department of Astronomy, University of California, Berkeley, California. The essential features of Cunningham's treatment also appear in a work by Sterne [21], which includes Cunningham's final equations for the perturbed elements a, e, i, Ω, ω and the mean anomalistic motion. In the drag treatment, we follow Sterne [21]. In conclusion, brief outlines of two analytical treatments of the main problem are given. One is due to Brouwer and Hori and the other to Lane. In these theories, second-order effects are considered and the combined effects of flattening and drag are included in the equations of motion (see Brouwer and Hori [22] and Lane [23]).

13.2. A FIRST-ORDER GRAVITATIONAL THEORY FOR ARTIFICIAL EARTH SATELLITES

Consider an artificial satellite in motion about the Earth, influenced only by the Earth's gravitational potential V. We take the center of mass of the Earth to be the origin of an equatorial system in which the x_q-axis is directed toward the first point of Aries. It is convenient to use spherical coordinates (r, α, δ) defined as follows: r is the radial distance from the center of mass of the Earth, α is the right ascension of the artificial satellite, and δ is the geocentric declination of the artificial satellite.

In our treatment of the motion, we will neglect the terms in Equation (12.3.13) that involve coefficients of higher order than $C_{2,0}$ and write

$$V = -\left(\frac{\mu}{r} + R\right) \tag{13.2.1}$$

where

$$R = \frac{Ga_{2,0}}{2r^3}(3\sin^2\delta - 1) \tag{13.2.2}$$

$$a_{2,0} = Ea_e^{\,2}C_{2,0} \tag{13.2.3}$$

Here E and a_e refer to the Earth's mass and equatorial radius respectively. The quantity $\mu = GE$ is roughly equal to $4(10)^{20}\ \text{cm}^3\,\text{sec}^{-2}$. The negative sign preceding the right-hand side of (13.2.1) indicates that the gravity field exterior to the Earth is given by $-\nabla V$.†

† The most complete gravitational theory available at time of writing is due to Aksnes [23a]. He used the Hori perturbation method (see Section 11.6) to obtain a development through second order in $C_{2,0}$ which holds for all allowable values of the eccentricity and inclination angle.

The term $-\mu/r$ in (13.2.1) produces elliptic motion and R is the contribution due to the second harmonic in the gravitational potential of the Earth. A theory of the motion of a satellite about the Earth, in which the perturbing function R is as defined in (13.2.2), is referred to as a first-order gravitational theory (see Exercise 3, Chapter 8).

We will examine the perturbations of the elements ω, Ω, i, a, e, M by means of the Lagrange planetary equations (8.3.11). If the perturbations of the elements have been determined as functions of time, the coordinates (r, α, δ), at each instant in the perturbed orbit, can be determined as functions of time by substituting the instantaneous values of the perturbed orbital elements into the equations

$$M = E - e \sin E \tag{13.2.4a}$$

$$r = a(1 - e \cos E) \tag{13.2.4b}$$

$$\tan \frac{w}{2} = \left(\frac{1+e}{1-e}\right)^{1/2} \tan \frac{E}{2} \tag{13.2.4c}$$

$$\sin \delta = \sin i \sin(w + \omega) \tag{13.2.4d}$$

$$\tan(\alpha - \Omega) = \cos i \tan(w + \omega) \tag{13.2.4e}$$

derived earlier in Chapters 3 and 6 and renumbered here for convenience. To determine $(\dot{r}, \dot{\alpha}, \dot{\delta})$ at each instant in the perturbed orbit, we first form the partial derivatives of (13.2.4) with respect to time, considering the elements to have their unperturbed values. The resultant expressions when evaluated at the instantaneous values of the perturbed orbital elements yield the values of $(\dot{r}, \dot{\alpha}, \dot{\delta})$ in the perturbed orbit at the instant.

Explicit expressions for the partial derivatives of R with respect to orbital elements a, e, i, ω, Ω, M are needed for use in (8.3.11). To determine these expressions, we proceed as follows. We note first that the dependences of r, w, δ, and R upon the orbital elements can be expressed in the form

$$r = r(a, e, M)$$

$$w = w(e, M)$$

$$\delta = \delta(i, \omega, e, M)$$

$$R = R(r, \delta) = R(a, e, i, \omega, M)$$

Thus R does not depend on Ω explicitly and we have $\partial R/\partial \Omega = 0$. The calculation of the remaining partial derivatives of R is simplified by using the results given in Exercise 12, Chapter 7. Except for $\partial R/\partial M$, the procedures are straightforward. After carrying them out, we find that

$$\left(\frac{\partial R}{\partial a}\right)_M = -\frac{3}{a}R$$

$$\frac{\partial R}{\partial e} = -\frac{3Ga_{2,0}}{2(1-e^2)\,r^3}[\cos w(1+e\cos w)(1-3\sin^2 i\sin^2 u) - \sin^2 i\sin 2u\sin w(2+e\cos w)]$$

$$\frac{\partial R}{\partial i} = \frac{3Ga_{2,0}}{r^3}\sin i\cos i\sin^2 u$$

$$\frac{\partial R}{\partial \omega} = \frac{3Ga_{2,0}}{2r^3}\sin^2 i\sin 2u$$

$$\frac{\partial R}{\partial \Omega} = 0$$

$$\frac{\partial R}{\partial M} = \frac{1}{n}\frac{dR}{dt} \tag{13.2.5}$$

The last equation in (13.2.5) can be obtained by noting that, in terms of elements in an instantaneous osculating ellipse, we can write

$$\frac{\partial r}{\partial M} = \frac{ae\sin w}{(1-e^2)^{1/2}} = \frac{\dot r}{n}$$

$$\frac{\partial \delta}{\partial M} = \sin i\cos u\sec\delta\,\frac{\partial w}{\partial M} = \frac{\dot\delta}{n}$$

Hence we have

$$\frac{\partial R}{\partial M} = \frac{\partial R}{\partial r}\frac{\partial r}{\partial M} + \frac{\partial R}{\partial \delta}\frac{\partial \delta}{\partial M} = \frac{1}{n}\left(\frac{\partial R}{\partial r}\dot r + \frac{\partial R}{\partial \delta}\dot\delta\right) = \frac{1}{n}\frac{dR}{dt}$$

Consider that Equations (13.2.5) have been substituted into the variational equations (8.3.11). In developing a first-order theory, we will (except for a slight modification in handling the secular perturbation in M) expand the right-hand members of the variational equations in a Taylor series about initial values a_0, e_0, i_0, ω_0 of the corresponding orbital elements and retain only the first term in the expansion. Thus, in integrating the variational equations, we will treat the elements a, e, i, ω as constant and equal to their initial values a_0, e_0, i_0, and ω_0 during the integration, and so the time will appear in the integrands only through the disturbed true anomaly. Now, we saw in Exercise 9,

Chapter 7, that $\dot{w}$ for the disturbed motion differs from $\dot{w}$ for the unperturbed motion only by terms that involve the partial derivatives of the perturbing function as a factor. Thus, in developing a first-order theory, we can replace time as the variable of integration by the undisturbed true anomaly w, through the differential condition [see (2.6.7)]

$$dt = \frac{r^2}{h}\,dw \tag{13.2.6}$$

We then treat the true anomaly as it appears in the integrands as undisturbed in carrying out the integrations. Where w appears in the integrated results, it will be understood to relate to the undisturbed motion and it can be calculated as a function of t from the two-body equations when initial values of the orbital elements are prescribed.

If use is made of (13.2.5) and (13.2.6) and the relation

$$r = p^*/(1 + e\cos w)$$

the variational equations for $(i)^{\cdot}$ and $\dot{\Omega}$ take the form

$$\frac{di}{dw} = \frac{3Ga_{2,0}}{2\mu p^{*2}} \sin i \cos i\,(1 + e\cos w)\sin 2(\omega + w) \tag{13.2.7}$$

$$\frac{d\Omega}{dw} = \frac{3Ga_{2,0}}{\mu p^{*2}} \cos i\,(1 + e\cos w)\sin^2(\omega + w) \tag{13.2.8}$$

The trigonometric identities

$$\begin{aligned} 2\sin A\cos B &= \sin(A+B) + \sin(A-B) \\ 2\cos A\sin B &= \sin(A+B) - \sin(A-B) \\ 2\cos A\cos B &= \cos(A+B) + \cos(A-B) \\ 2\sin A\sin B &= \cos(A-B) - \cos(A+B) \end{aligned} \tag{13.2.9}$$

for any two numbers A and B can be used to express (13.2.7) and (13.2.8) in forms more convenient to integrate. Using (13.2.9), we find that

$$\begin{aligned} \frac{di}{dw} = \frac{3Ga_{2,0}}{8\mu p^{*2}} \sin 2i[2\sin(2\omega + 2w) + e\sin(2\omega + 3w) \\ + e\sin(2\omega + w)] \end{aligned}$$

$$\begin{aligned} \frac{d\Omega}{dw} = \frac{3Ga_{2,0}}{4\mu p^{*2}} \cos i[2 + 2e\cos w - e\cos(2\omega + w) - 2\cos(2\omega + 2w) \\ - e\cos(2\omega + 3w)] \end{aligned}$$

Integrating these last two equations under the assumption that a, e, i, and ω are constant, we obtain

$$i = K_i - \frac{Q}{8}\sin 2i_0\left[\cos(2\omega_0 + 2w) + \frac{e_0}{3}\cos(2\omega_0 + 3w) + e_0\cos(2\omega_0 + w)\right] \tag{13.2.10}$$

$$\Omega = K_\Omega + \frac{Q}{4}\cos i_0\left[2w + 2e_0\sin w - e_0\sin(2\omega_0 + w) - \sin(2\omega_0 + 2w) - \frac{e_0}{3}\sin(2\omega_0 + 3w)\right] \tag{13.2.11}$$

where

$$Q = \frac{3Ga_{2,0}}{\mu p_0^{*2}}, \qquad p_0^* = a_0(1 - e_0^2)$$

and K_i and K_Ω are constants of integration which are determined by the initial conditions. For typical satellites, $Q \approx 2C_{2,0} \approx -0.002$.

The procedures used to evaluate the perturbations of i and Ω can also be used to evaluate the first-order perturbations of a, e, and ω. The details are left as an exercise. We find that

$$\begin{aligned} a = K_a - \frac{a_0 Q}{1 - e_0^2}\Bigg[& e_0\left(1 + \frac{e_0^2}{4}\right)\left(1 - \frac{3}{2}\sin^2 i_0\right)\cos w \\ &+ \frac{e_0^2}{2}\left(1 - \frac{3}{2}\sin^2 i_0\right)\cos 2w \\ &+ \frac{e_0^3}{12}\left(1 - \frac{3}{2}\sin^2 i_0\right)\cos 3w + \frac{e_0^3}{16}\sin^2 i_0\cos(2\omega_0 - w) \\ &+ \frac{3}{4}e_0\left(1 + \frac{e_0^2}{4}\right)\sin^2 i_0\cos(2\omega_0 + w) \\ &+ \frac{1}{2}\left(1 + \frac{3}{2}e_0^2\right)\sin^2 i_0\cos(2\omega_0 + 2w) \\ &+ \frac{9}{12}e_0\left(1 + \frac{e_0^2}{4}\right)\sin^2 i_0\cos(2\omega_0 + 3w) \\ &+ \frac{3}{8}e_0^2\sin^2 i_0\cos(2\omega_0 + 4w) \\ &+ \frac{e_0^3}{16}\sin^2 i_0\cos(2\omega_0 + 5w)\Bigg] \end{aligned} \tag{13.2.12}$$

$$e = K_e - \frac{Q}{2}\Big[\Big(1 + \frac{e_0^2}{4}\Big)\Big(1 - \frac{3}{2}\sin^2 i_0\Big)\cos w + \frac{e_0}{2}\Big(1 - \frac{3}{2}\sin^2 i_0\Big)\cos 2w$$

$$+ \frac{e_0^2}{12}\Big(1 - \frac{3}{2}\sin^2 i_0\Big)\cos 3w + \frac{e_0^2}{16}\sin^2 i_0 \cos(2\omega_0 - w)$$

$$+ \frac{1}{4}\Big(1 + \frac{11}{4}e_0^2\Big)\sin^2 i_0 \cos(2\omega_0 + w) + \frac{5}{4}e_0 \sin^2 i_0 \cos(2\omega_0 + 2w)$$

$$+ \frac{1}{12}\Big(7 + \frac{17}{4}e_0^2\Big)\sin^2 i_0 \cos(2\omega_0 + 3w) + \frac{3}{8}e_0 \sin^2 i_0 \cos(2\omega_0 + 4w)$$

$$+ \frac{e_0^2}{16}\sin^2 i_0 \cos(2\omega_0 + 5w)\Big] \tag{13.2.13}$$

$$\omega = K_\omega - \frac{Q}{2}\Big\{\Big(2 - \frac{5}{2}\sin^2 i_0\Big)w + \frac{1}{e_0}\Big[1 - \frac{3}{2}\sin^2 i_0 + \frac{e_0^2}{4}\Big(7 - \frac{17}{2}\sin^2 i_0\Big)$$

$$+ \frac{e_0^2}{8}\sin^2 i_0 \cos 2\omega_0\Big]\sin w + \frac{1}{2}\Big(1 - \frac{3}{2}\sin^2 i_0\Big)\sin 2w$$

$$+ \frac{e_0}{12}\Big(1 - \frac{3}{2}\sin^2 i_0\Big)\sin 3w$$

$$- \frac{1}{e_0}\Big[\frac{e_0^2}{2} + \frac{1}{4}\Big(1 - \frac{7}{2}e_0^2\Big)\sin^2 i_0\Big]\sin(2\omega_0 + w)$$

$$- \frac{1}{2}\Big(1 - \frac{5}{2}\sin^2 i_0\Big)\sin(2\omega_0 + 2w)$$

$$- \frac{1}{3e_0}\Big[\frac{e_0^2}{2} - \Big(\frac{7}{4} + \frac{19}{16}e_0^2\Big)\sin^2 i_0\Big]\sin(2\omega_0 + 3w)$$

$$+ \frac{3}{8}\sin^2 i_0 \sin(2\omega_0 + 4w) + \frac{e_0}{16}\sin^2 i_0 \sin(2\omega_0 + 5w)\Big\} \tag{13.2.14}$$

where K_a, K_e, and K_ω are constants of integration which are determined by the initial conditions.

We now turn to the Lagrange planetary equation for M, which can be written in the differential form

$$dM = n\,dt - \frac{n}{\mu h}\mathscr{C}\,dw \tag{13.2.15}$$

where

$$\mathscr{C} = \frac{p^*}{e}r^2\frac{\partial R}{\partial e} + 2a^2r^2\Big(\frac{\partial R}{\partial a}\Big)_M \tag{13.2.16}$$

as before, is understood to be expressed in terms of w. To carry out the integration of (13.2.15), we will first express n as a sum of its unperturbed part, plus a perturbation of first order. To do this, we note first that the variational equation

$$\dot{a} = \frac{2na^2}{\mu}\frac{\partial R}{\partial M} = \frac{2a^2}{\mu}\frac{dR}{dt} \tag{13.2.17}$$

can readily be integrated to yield the exact equation

$$\frac{1}{a} + \frac{2R}{\mu} = C_E \tag{13.2.18}$$

where, if α represents the total energy per unit mass of the satellite, the constant C_E is related to α through the equation $\mu C_E = -2\alpha$. [It is also possible to show that the component of the orbital angular momentum per unit mass along the z_q-axis is constant for the motion (see Exercise 3 at end of chapter).] Using (13.2.18), we can now write

$$n = \mu^{1/2}a^{-3/2} = \mu^{1/2}C_E^{3/2}\left[1 - \frac{F}{C_E r^3}(1 - 3\sin^2\delta)\right]^{3/2} \qquad \left(F = -\frac{Ga_{2,0}}{\mu}\right) \tag{13.2.19}$$

If the bracketed quantity in (13.2.19) is expanded binomially and terms of order higher than the first are dropped, we find that

$$n = \mu^{1/2}C_E^{3/2} - \frac{3\mu^{1/2}F(1 - 3\sin^2\delta)}{2a^{1/2}r^3} \tag{13.2.20}$$

We have, in (13.2.20), a suitable partition of n in the form

$$n = n_r - \frac{3\mu^{1/2}F(1 - 3\sin^2\delta)}{2a^{1/2}r^3} \tag{13.2.21}$$

where the constant n_r is given by

$$n_r = \mu^{1/2}C_E^{3/2} = n_0\left[1 + \frac{F}{a_0^2}\left(\frac{a_0}{r_0}\right)^3(1 - 3\sin^2\delta_0)\right]^{3/2} \tag{13.2.22}$$

and the zero subscript indicates evaluation at the initial conditions. Returning to (13.2.15), we see that

$$dM = n_r\,dt - \frac{3\mu^{1/2}F(1 - 3\sin^2\delta)}{2a^{1/2}r^3}\,dt - \frac{n}{\mu h}\mathscr{C}\,dw \tag{13.2.23}$$

In the second term on the right-hand side of (13.2.23), replace dt by $(r^2/h)\,dw$ and then express the coefficient of dw in terms of w. Next, grouping the second term with the third in performing the subsequent integration over w, we obtain

$$M = K_M + n_r t + \frac{Q(1 - e_0^2)^{1/2}}{2e_0} \Big\{ \Big[\Big(1 - \frac{3}{2} \sin^2 i_0\Big) (1 - \frac{1}{4} e_0^2)$$
$$+ \frac{e_0^2}{8} \sin^2 i_0 \cos 2\omega_0 \Big] \sin w + \frac{e_0}{2} \Big(1 - \frac{3}{2} \sin^2 i_0\Big) \sin 2w$$
$$+ \frac{e_0^2}{12} \Big(1 - \frac{3}{2} \sin^2 i_0\Big) \sin 3w - \frac{1}{4} \Big(1 + \frac{3}{2} e_0^2\Big) \sin^2 i_0 \sin(2\omega_0 + w)$$
$$+ \frac{1}{12} \Big(7 - \frac{1}{4} e_0^2\Big) \sin^2 i_0 \sin(2\omega_0 + 3w) + \frac{3}{8} e_0 \sin^2 i_0 \sin(2\omega_0 + 4w)$$
$$+ \frac{1}{16} e_0^2 \sin^2 i_0 \sin(2\omega_0 + 5w) \Big\} \tag{13.2.24}$$

where K_M is a constant of integration determined by the initial conditions.

The first-order gravitational perturbations are given by (13.2.10)–(13.2.14) and (13.2.24). It is clear from these equations that, to first order, there are no secular perturbations of the elements a, e, and i. They do have periodic perturbations, however. The amplitudes of the periodic components vary, of course, with the individual terms but, typically, we are considering amplitudes not in excess of one part in 10^3 in a, 0.001 in e, and one minute of arc in i. Similar orders of magnitude hold for the amplitudes of the periodic terms in the remaining elements.

The most significant perturbations are the secular effects that appear in Ω and ω. Designating by S_ω and S_Ω the coefficients of the secular motions of the node and the perigeal line respectively, we write

$$S_\omega = -\frac{1}{2} Q \Big(2 - \frac{5}{2} \sin^2 i_0\Big) \tag{13.2.25}$$

$$S_\Omega = \frac{1}{2} Q \cos i_0 \tag{13.2.26}$$

The angle i_c at which the perigeal line is stationary is called the critical angle. Its value ($\approx 63°.26'$) is determined by the condition that $2 - \frac{5}{2} \sin^2 i_0 = 0$. If $i_c < i \leqslant 90°$ the perigeal line regresses; if $0 \leqslant i < i_c$ the perigeal line advances. Since Q is negative, the secular motion of node is a regression of the nodal line in the equatorial plane. Close-Earth satellites make about 15 revolutions per day; hence typical secular motions in ω and Ω are roughly about 5 deg per day. Since n_r, as given by (13.2.22), differs from n_0, there is also a secular perturbation in the mean anomaly which must be taken into account in determining the average anomalistic, nodal, and sidereal periods of the motion.

13.3. A FIRST-ORDER DRAG THEORY FOR ARTIFICIAL EARTH SATELLITES

As an artificial satellite moves through the Earth's atmosphere, its orbital motion is perturbed by the resisting force of the atmosphere. This resisting force, which is one of the most important perturbing influences in the altitude regime from 150 to 600 km, is called the drag force. The main effects of the force are to bring about secular decreases in the elements a and e that cause the orbit of the satellite to contract; eventually the satellite plunges toward Earth (see Exercise 15, Chapter 7).

The drag force $\mathbf{F}_D$, per unit mass of satellite, can be well represented by the equation

$$\mathbf{F}_D = -\frac{1}{2} C_D \frac{A}{m} \rho V \mathbf{V} \tag{13.3.1}$$

where m is the mass of the satellite, $\mathbf{V}$ is the velocity of the atmosphere relative to the satellite (called the ambient velocity), and $V = |\mathbf{V}|$. In (13.3.1), C_D is a dimensionless drag coefficient, empirically determined to have a value close to 2, while A represents the effective cross-sectional area of the satellite, which is to be found by averaging all possible projected areas of the satellite onto a plane perpendicular to $\mathbf{V}$. For the case of a body whose surface intersects a straight line no more than twice, A is equal to one-fourth of the surface area of the satellite. The symbol ρ, which represents the local atmospheric density, depends on the position and the time in a very complicated manner.

To study the effect of drag on the motion of a satellite, we will introduce (13.3.1) as a perturbation into the Lagrange variational equations for a, e, i, ω, Ω, M given in (7.8.2). Since it is convenient here to work with the eccentric anomaly E rather than w, we rewrite the relevant variational equations in terms of E as follows:

$$\dot{a} = \frac{2a^2}{(\mu p^*)^{1/2}} fT \tag{13.3.2a}$$

$$\dot{e} = \left(\frac{p^*}{\mu}\right)^{1/2} \frac{1}{f} \left[\frac{2(1-e^2)\cos E}{1 - e\cos E} T + (1-e^2)^{1/2} \sin E\, N\right] \tag{13.3.2b}$$

$$(i)^{\cdot} = \frac{r \cos u}{(\mu p^*)^{1/2}} F''_\zeta \tag{13.3.2c}$$

$$\dot{\Omega} = \frac{r \sin u}{(\mu p^*)^{1/2} \sin i} F''_\zeta \tag{13.3.2d}$$

$$\dot{\omega} = (-\cos i)\,\dot{\Omega} + \frac{1}{ef}\left(\frac{p^*}{\mu}\right)^{1/2}\left[\frac{2(1-e^2)^{1/2}\sin E}{1-e\cos E}\,T - (e+\cos E)\,N\right] \tag{13.3.2e}$$

$$\dot{M} = n - \frac{1}{ef}\left(\frac{p^*}{\mu}\right)^{1/2}\left[\frac{2\sin E(1-e^3\cos E)}{1-e\cos E}\,T - (1-e_2)^{1/2}(\cos E - e)\,N\right] \tag{13.3.2f}$$

$$f = [(1-e^2)(1+e\cos E)]^{1/2}\,(1-e\cos E)^{-1/2} \tag{13.3.2g}$$

We must now express the components N, T, and F_ζ'' of $\mathbf{F}_D$ in a form suitable for use in (13.3.2). To begin, we can write (see Figure 7.5.1)

$$\mathbf{F}_D = -\frac{1}{2}C_D\frac{A}{m}\rho V\mathbf{V} = R\mathbf{r}_1 + S\mathbf{s}_1 + F_\zeta''\mathbf{k}'' \tag{13.3.3}$$

where we seek to express ρV in terms of the orbital elements, and $\mathbf{V}$ as a linear combination of the basis vectors $(\mathbf{r}_1, \mathbf{s}_1, \mathbf{k}'')$. When this has been done, the components R, S, and F_ζ'' are determined by equating coefficients in (13.3.3). The components N and T can then be obtained readily from (7.5.15).†

The ambient velocity $\mathbf{V}$ is given by

$$\mathbf{V} = \mathbf{v} - \mathbf{V}_A \tag{13.3.4}$$

where the instantaneous velocity $\mathbf{v}$ of the satellite, relative to the inertial system $Ox_q y_q z_q$ at geocenter, is given by $\mathbf{v} = \dot{r}\mathbf{r}_1 + r\dot{w}\mathbf{s}_1$ and $\mathbf{V}_A$ is the velocity of a point in the atmosphere relative to $Ox_q y_q z_q$. The atmosphere is assumed to rotate with the Earth with the angular speed l about the z_q-axis. Thus $\mathbf{V}_A$ is parallel to the equatorial plane and can be expressed in the form

$$\mathbf{V}_A = lr\cos\delta[-\sin(\alpha-\Omega)\,\mathbf{i}' + \cos(\alpha-\Omega)\,\mathbf{j}^*] \tag{13.3.5}$$

where $\mathbf{i}'$ is a unit vector directed along the line of nodes toward the ascending node and $\mathbf{j}^* = \mathbf{k}\times\mathbf{i}'$. It is clear from (2.5.3) and (2.5.4) that the $(\mathbf{i}', \mathbf{j}^*, \mathbf{k})$ basis is related to the $(\mathbf{r}_1, \mathbf{s}_1, \mathbf{k}'')$ basis through the transformation

$$\begin{pmatrix}\mathbf{i}'\\ \mathbf{j}^*\\ \mathbf{k}\end{pmatrix} = \begin{pmatrix}\cos u & -\sin u & 0\\ \cos i\sin u & \cos i\cos u & -\sin i\\ \sin i\sin u & \sin i\cos u & \cos i\end{pmatrix}\begin{pmatrix}\mathbf{r}_1\\ \mathbf{s}_1\\ \mathbf{k}''\end{pmatrix} \tag{13.3.6}$$

† Note that, in this section, the symbols V and R are used differently than they are in Section 13.2. A certain amount of duplication of notation is unavoidable here, and elsewhere. Where such duplication occurs, the meaning to ascribe to a given symbol will be clear from the context.

If (13.3.6) and the identities

$$\cos\delta\cos(\alpha - \Omega) = \cos u$$
$$\cos\delta\sin(\alpha - \Omega) = \sin u\cos i$$

are used, we can write

$$\mathbf{V}_A = lr\cos i\mathbf{s}_1 - lr\cos u\sin i\mathbf{k}'' \tag{13.3.7}$$

Hence

$$\mathbf{V} = \dot{r}\mathbf{r}_1 + (r\dot{w} - lr\cos i)\,\mathbf{s}_1 + lr\cos u\sin i\mathbf{k}'' \tag{13.3.8}$$

and

$$V^2 = v^2 - 2r^2\dot{w}l\cos i + l^2r^2(\cos^2 i + \cos^2 u\sin^2 i) \tag{13.3.9}$$

After expressing v^2, r^2, and $r^2\dot{w}$ in terms of the orbital elements, we find that

$$V = \left(\frac{\mu}{a}\right)^{1/2}\left(\frac{1 + e\cos E}{1 - e\cos E}\right)^{1/2}(1 - \gamma)^{1/2} \tag{13.3.10}$$

where

$$\gamma = 2(\tau - s) \tag{13.3.11}$$

$$\tau = \frac{d(1 - e\cos E)}{1 + e\cos E} \tag{13.3.12}$$

$$d = \frac{l}{n}(1 - e^2)^{1/2}\cos i \tag{13.3.13}$$

and

$$s = \frac{l^2(1 - e\cos E)^2}{2n^2(1 + e\cos E)^2}(\cos^2 i + \cos^2 u\sin^2 i)(1 - e^2\cos^2 E) \tag{13.3.14}$$

For most satellites, τ and s have numerical values that do not exceed 0.07 and 0.002 respectively. Thus the numerical value of γ is positive and is usually less than 0.14. Neglecting terms of order s, we can write

$$V = \left(\frac{\mu}{a}\right)^{1/2}\left(\frac{1 + e\cos E}{1 - e\cos E}\right)^{1/2}(1 - \tau) \tag{13.3.15}$$

If, in (13.3.8), r, $\dot{r}$, and $\dot{w}$ are expressed in terms of the orbital elements and $\dot{E}$, then $\mathbf{F}_D$ can be expressed in the form

$$\mathbf{F}_D = -b\rho V\left\{ae\sin E\dot{E}\mathbf{r}_1 + \left[a(1 - e^2)^{1/2}\dot{E} - \frac{la}{n}\cos i(1 - e\cos E)^2\dot{E}\right]\mathbf{s}_1 + \frac{la}{n}\cos u\sin i(1 - e\cos E)^2\dot{E}\mathbf{k}_1\right\} \tag{13.3.16}$$

where $b = C_D A/2m$. The components R, S, and F''_ζ are given by the equations

$$R = -bae\rho V \sin E\, \dot{E} \tag{13.3.17}$$

$$S = -ba\rho V(1 - e^2)^{1/2}\, C\, \dot{E} \tag{13.3.18}$$

$$F''_\zeta = -b\rho V \frac{la}{n} \sin i \cos u(1 - e \cos E)^2\, \dot{E} \tag{13.3.19}$$

where

$$C = 1 - \frac{d(1 - e \cos E)^2}{(1 - e^2)} \tag{13.3.20}$$

If (7.5.15) is used, we find that

$$N = -bae\rho V \frac{d}{f} \sin E(1 - e \cos E)\, \dot{E} \tag{13.3.21}$$

$$T = -ba \frac{\rho V}{f} (1 - e^2)^{1/2} \frac{[e^2 \sin^2 E + (1 - e^2)\, C]}{1 - e \cos E}\, \dot{E} \tag{13.3.22}$$

If the factor $e^2 \sin^2 E + (1 - e^2)\, C$ in (13.3.22) is examined, it can be shown that, neglecting terms of order e^4 and higher,

$$e^2 \sin^2 E + (1 - e^2)\, C = (1 - \tau)(1 - e^2 \cos^2 E) \tag{13.3.23}$$

For most artificial satellites, e is less than 0.2. Thus terms of order e^4 may be neglected because they are smaller than terms of order s, which have already been dropped. With this understanding, we take

$$T = -ba(1 - e^2)^{1/2} \rho \frac{V}{f} (1 - \tau)(1 + e \cos E)\, \dot{E} \tag{13.3.24}$$

If Equations (13.3.19), (13.3.21), and (13.3.24) are introduced into (13.3.2) and V is expressed in terms of the orbital elements through (13.3.15), the following equations are obtained:

$$\frac{da}{dE} = -2b\rho a^2 \frac{(1 + e \cos E)^{3/2}}{(1 - e \cos E)^{1/2}} (1 - \tau)^2 \tag{13.3.25a}$$

$$\frac{de}{dE} = -2b\rho a(1 - e^2) \left(\frac{1 + e \cos E}{1 - e \cos E}\right)^{1/2} (1 - \tau)^2 \cos E$$
$$-b\rho aed \frac{(1 - e \cos E)^{3/2}}{(1 + e \cos E)^{1/2}} (1 - \tau) \sin^2 E \tag{13.3.25b}$$

$$\frac{di}{dE} = -\frac{b}{2} l\rho \frac{a}{n} \sin i(1 - e^2)^{-1/2} (1 + e \cos E)^{1/2} \times (1 - e \cos E)^{5/2} (1 + \cos 2u)(1 - \tau) \quad (13.3.25c)$$

$$\frac{d\Omega}{dE} = -\frac{b}{2} l\rho \frac{a}{n} (1 - e^2)^{-1/2} \sin 2u(1 + e \cos E)^{1/2} (1 - e \cos E)^{5/2} (1 - \tau) \quad (13.3.25d)$$

$$\frac{d\omega}{dE} = -(\cos i)\, \dot{\Omega} - 2b\rho \frac{a}{e} (1 - e^2)^{1/2} \sin E \left(\frac{1 + e \cos E}{1 - e \cos E}\right)^{1/2} (1 - \tau) \times \left[1 - \frac{d}{2(1 - e^2)} (1 - e \cos E)(2 - e^2 - e \cos E)\right] \quad (13.3.25e)$$

$$\frac{dM}{dE} = n + 2b\rho \frac{a}{e} (1 - e^3 \cos E) \left(\frac{1 + e \cos E}{1 - e \cos E}\right)^{1/2} \sin E(1 - \tau) \times \left[1 - \frac{d(1 - e \cos E)}{2(1 - e^3 \cos E)} (2 - e^2 - e \cos E)\right] \quad (13.3.25f)$$

All of the Equations in (13.3.25) agree exactly with those given by Sterne except the equation for de/dE, which agrees exactly with Sterne if the terms of order $e^3 d$, which it contains, are dropped.

The secular and periodic perturbations in the orbital elements, which take place during one revolution because of drag, can be evaluated by numerical integration of (13.3.25). In the definite integrals that arise, the four orbital elements a, e, i, and ω appear. The elements a, e, and i can be assumed to retain their initial values over an interval of 2π in E. Allowance can be made for the secular motion of ω due to the Earth's flattening during the same interval. Experimentally determined values of ρ, at intervals in E of 15 deg or so, can be used in forming the integrands. Any simple numerical integration scheme can be used to evaluate these definite integrals. By performing occasional quadratures over widely separated revolutions, one can obtain the secular and periodic perturbations in the orbital elements as functions of E. For intermediate times, interpolation can be used. In this way, the first-order drag perturbations for the orbital elements a, e, i, Ω, ω, M can be obtained over most of the lifetime of the satellite.

13.4. ANALYTICAL REPRESENTATIONS OF AIR DENSITY IN THE VARIATIONAL EQUATIONS

In order to integrate (13.3.25) analytically, some representation of the space and time dependence of ρ must be assumed. It is not a simple matter, however, to describe a realistic mathematical model of the atmosphere because of the complex nature of the density distribution and our limited knowledge of its space-time behavior. A number of models of the atmosphere have been introduced that help to clarify the dependence of density on position and time in the altitude regime between 150 and 750 km, herein referred to as the upper atmosphere. These models are constantly being improved upon in an attempt to reflect the current state of knowledge of the observed variations of the upper atmosphere. These observed variations are usually divided into the four types outlined below. In making the following comments, we assume that the four types of variations are independent and that the total variation of the upper atmosphere is a linear superposition of the four individual variations.†

The first type of variation is a diurnal (that is, daily) variation of the density at a point in the atmosphere, for a given level of intensity of solar flux. This variation, which increases with altitude in the upper atmosphere, is only significant above 200 km. As a point above the Earth moves from darkness into daylight, the density at the point may, depending on its latitude and altitude, increase by a factor of about ten.

The second type of variation is a long-period (11-year) variation of density, at a given point in the atmosphere. This variation appears to be associated with the periodic (11-year) variation of the intensity of the sunspot activity and hence with the 11-year variation of the intensity of solar flux impinging on the atmosphere. This latter variation is referred to as the 11-year solar cycle. The density everywhere in the Earth's atmosphere (not only in that part of the atmosphere being irradiated by the Sun) fluctuates according to the level of solar activity. The maximum density (roughly ten times the mean density) occurs when sunspot activity is at its peak.

The third type of variation is a nonperiodic, unpredictable variation of density at a point in the atmosphere. This variation appears to be primarily associated with large, but short-lived changes in the Earth's magnetic field (magnetic storms), arising from unpredictable violent eruptions on the surface of the Sun (solar flares). This variation is referred to as the variation with geomagnetic activity. The atmospheric density, at a given altitude at midlatitudes, may increase by a factor of about ten from its geomagnetic quiescent value, but such an increase usually lasts only about 24 hours.

The fourth type of variation is a semiannual variation. Over a period of one year, the density at a point in the upper atmosphere falls to its least

† Only a brief sketch of the properties of the upper atmosphere is given here. For a more detailed discussion of this subject, the reader is referred to Jacchia [24].

value in July and rises to its largest value in October, followed by a less pronounced minimum and maximum in January and April, respectively. Although the range of the semiannual variation is, generally speaking, less than one-twentieth the range of the 11-year solar cycle, it is still pronounced enough to make it readily observable over a period of several years.

Over a short period of time, say 12 hours, during which there are no fluctuations in atmospheric density due to magnetic storms, the diurnal variation at a point in the atmosphere at midlatitudes is the most important of the variational effects. Over a much longer period, say 3 or 4 years, the 11-year solar cycle is the most important effect influencing atmospheric density.

Thus far, in describing a first-order theory for drag perturbations, we have considered only an empirical representation of ρ. Despite the great variability in density with respect to time, certain analytical representations of ρ have been used with considerable success in carrying out integrations of (13.3.25) over both an orbital period and the whole lifetime of a satellite. The mathematical models of air density, which we are led to most naturally, are either the power-law or exponential-law variations of density with altitude, as will be clear from the following considerations.

The general conditions of equilibrium for a gas are contained in the fundamental equation of hydrostatics

$$\mathbf{F} = \frac{1}{\rho} \nabla p \tag{13.4.1}$$

where

$\mathbf{F}$ = the resultant force per unit mass

ρ = mass density of the gas

p = pressure in the gas

If $\mathbf{F} = \nabla V$, where V is a scalar potential, by forming the dot product of each side of (13.4.1) with a differential displacement, we can write

$$dV = \frac{1}{\rho}\, dp \tag{13.4.2}$$

If, in addition, the perfect gas law is assumed, we have

$$p = \rho H^* g \qquad \left(H^* = \frac{RT}{Mg}\right) \tag{13.4.3}$$

where g is the value of gravity at a distance r from geocenter and the "atmospheric scale height" H^* is the ratio of the product of the universal gas constant R and the temperature T to the product of g and the molecular weight M of the gas.

Representing the gravitational potential of the Earth by the first term of (12.3.15), we write

$$V = \frac{GE}{r} \tag{13.4.4}$$

where E is the Earth's mass. Then, using (13.4.3) and (13.4.4), we can write Equation (13.4.2) as

$$-\rho dr = H^* \, d\rho \tag{13.4.5}$$

Our knowledge of the way in which H^* varies in space and time is still far from satisfactory. It does, however, appear to increase slowly with height in the height regime from 150 to 400 km. Thus, as a first approximation, we can assume a linear variation

$$H^* = H_0^* + \beta^*(r - q_0) \tag{13.4.6}$$

with height, where β^* is the scale height gradient and H_0^* is the value of H^* at the reference level q_0. If (13.4.5) is integrated under the assumptions that H^* satisfies (13.4.6) and the perfect gas law holds, it can be shown that the air density satisfies the almost power law

$$\rho = \rho_0 \left(\frac{r}{q_0}\right)^2 \left(\frac{q_0 - s^*}{r - s^*}\right)^{n^*} \tag{13.4.7a}$$

where ρ_0 is the value of air density at reference level q_0,

$$s^* = q_0 - H_0^*(\beta^*)^{-1}$$

is the intercept of the line (13.4.6) with the r-axis, and $n^* = 1 + (\beta^*)^{-1}$. If it is assumed that H^* is constant, it is a simple matter to show that (13.4.7a) passes over to the almost exponential law

$$\lim_{\beta^* \to 0} = \rho_0 \left(\frac{r}{q_0}\right)^2 \exp\left[-\frac{(r - q_0)}{H^*}\right] \tag{13.4.7b}$$

Alternative forms of (13.4.7a) suggest themselves, for the factor r/q_0 is close to unity in the region of interest. For example, the functions

$$\rho = \rho_0 \left(\frac{r}{q_0}\right)\left(\frac{q_0 - s^*}{r - s^*}\right)^{n^*} \tag{13.4.8}$$

$$\rho = \rho_0 \left(\frac{q_0 - s^*}{r - s^*}\right)^{n^*} \tag{13.4.9}$$

give values of air density that, in general, differ from those given by (13.4.7a) by less than 3 percent over the entire band 200–400 km (see Lane *et al.* [25]).

A power law

$$\rho = \rho_0 \left(\frac{q_0 - s}{r - s} \right)^n \tag{13.4.10}$$

similar to (13.4.9) can be derived directly from the definition

$$H d\rho = -\rho \, dr \tag{13.4.11}$$

of the "density scale height" H if a linear variation

$$H = \beta(r - s) \tag{13.4.12}$$

with height is assumed. Here $s = q_0 - H_0\beta^{-1}$, H_0 is the density scale height at q_0, and $\beta = n^{-1}$ is the density scale height gradient. If H is assumed to be constant, (13.4.10) leads, in the limit as $\beta \to 0$, to the exponential representation of air density

$$\rho = \rho_0 \exp\left[-\frac{(r - q_0)}{H} \right] \tag{13.4.13}$$

In the preceding development, it has been tacitly assumed that the Earth's atmosphere is spherically symmetrical. Actually, the Earth's atmosphere tends to be oblate; that is, it is more closely a function of the height above the Earth's surface then it is a function of the distance from geocenter. In the power-law models of air density, the effect of the Earth's oblateness can be allowed for by replacing q_0, except in the r/q_0 factor, by the radial spherical coordinate of the surface of the spheroid that passes through q_0 at some geocentric declination δ_0. In the exponential versions, atmospheric oblateness can be allowed for by taking $z = r - q_0$ to be the difference between the satellite's altitude and the reference altitude, both referenced to the surface of a spheroidal Earth. To illustrate, when atmospheric oblateness is allowed for, Equation (13.4.10) takes the form

$$\rho = C^* \left[\frac{1 + K \sin^2(w + \omega)}{r - s} \right]^n \tag{13.4.14}$$

where

$$K = -\frac{f q_0 \sin^2 i}{q_0 - s(1 - f \sin^2 \delta_0)}$$

$$C^* = \rho_0 \frac{[q_0 - s(1 - f \sin^2 \delta_0)]^n}{(1 - f \sin^2 \delta_0)}$$

$$f = \text{ellipticity of the Earth}$$

The corresponding exponential law is

$$\rho = \rho_0 \exp\left(-\frac{z}{H}\right) \tag{13.4.15}$$

To give some idea of what is involved when a relatively simple atmospheric density representation, such as (13.4.15), is used in (13.3.25), we outline briefly the procedure used by Sterne [21] to integrate the variational equation for a over one orbital period (perigee to perigee) through terms of order e^3.

When the definition of τ is introduced into (13.3.25a), the equation can be written in the form

$$da = -2ba^2\rho(1-d)^2 \frac{[1+(1+d)\,e\cos E/(1-d)]^2}{(1-e^2\cos^2 E)^{1/2}}\,dE \tag{13.4.16}$$

Sterne represents z, in (13.4.15), by the expression

$$\begin{aligned} z = r - a_e[1 - f\sin^2 i \sin^2(w+\omega)] - a(1+e) \\ + a_e(1 - f\sin^2 i \sin^2 \omega) \end{aligned} \tag{13.4.17}$$

where a_e is the equatorial radius of the Earth, and, if b_e is the polar radius of the Earth, the ellipticity $f = (a_e - b_e)/a_e \approx 0.0034$; ρ_0 is taken to be the density at perigee and is designated by ρ_π. For typical satellites, most of the drag occurs near perigee. If the satellite is within 5 deg of perigee, then (13.4.17) represents the difference between the altitude of the satellite above the Earth's surface and the altitude of perigee above the Earth's surface to within about 3 parts in 10^4.

If the true anomaly w is replaced by the eccentric anomaly E, then z can be written in the form

$$\begin{aligned} z = Hc(1-\cos E) + Hq\cos 2\omega\,\frac{(1-e^2)\sin^2 E}{(1-e\cos E)^2} \\ + Hq(1-e^2)^{1/2}\,\frac{\sin 2\omega \sin E(\cos E - e)}{(1-e\cos E)^2} \end{aligned} \tag{13.4.18}$$

where $Hc = ae$ and $Hq = a_e f \sin^2 i$. The quantity q is usually less than 0.4.

The change Δa, in a, during one revolution is obtained by integrating (13.4.16) over one complete cycle in E. This leads to the equation

$$\Delta a = -2ba^2(1-d)^2 \rho_\pi \exp(-c) \int_{-\pi}^{\pi} \frac{[1+(1+d)\,e\cos E/(1-d)]^2}{(1-e^2\cos^2 E)^{1/2}}$$

$$\times \exp\left[-q\cos 2\omega \frac{(1-e^2)\sin^2 E}{(1-e\cos E)^2} - q(1-e^2)^{1/2}\right.$$

$$\left.\times \sin 2\omega \frac{\sin E(\cos E - e)}{(1-e\cos E)^2}\right] \exp(c\cos E)\, dE \tag{13.4.19}$$

where, since the integration is to be carried out over only one orbital period (perigee to perigee), variations in the orbital elements during the integration can be neglected. For most satellites, i is essentially constant for the satellite's entire lifetime. Typically, during one revolution the changes in ω, a, and e are less than 0.01 radian, 0.00002 gr, and 0.000017 respectively.

Apart from the factor $\exp(c\cos E)$, the integrand of (13.4.19) can be expanded in a series in powers of e, the coefficients of which contain powers of the trigonometric functions $\sin E$ and $\cos E$. Thus (13.4.19) can be written as a sum of integrals of the types

$$\int_{-\pi}^{\pi} \sin E \cos^m E \exp(c\cos E)\, dE \tag{13.4.20}$$

and

$$\int_{-\pi}^{\pi} \cos^m E \exp(c\cos E)\, dE \tag{13.4.21}$$

where m is a nonnegative integer. The integrands in (13.4.20) are odd functions of E and the definite integrals from $-\pi$ to π will vanish. Since definite integrals of the form appearing in (13.4.21) are linear combinations of modified Bessel functions of the first kind of integral order, they can be expressed as linear combinations of the modified Bessel functions of the first kind, $I_0(c)$ and $I_1(c)$, of order zero and one respectively, through the use of well-known recurrence relations (see Exercise 12, Chapter 4).

If the details of integrating (13.4.19) are carried out and terms of order higher than e^3, q_1e^2, and q_2e are dropped, it is a straightforward matter to show that

$$\Delta a = -4ba^2(1-d)^2 \pi\rho_\pi \exp(-c)[H_0I_0(c) + H_1I_1(c)] \tag{13.4.22}$$

where

$$H_0 = 1 + e^2\left(j^2 + \frac{1}{2}\right) - \frac{e^3 j}{c} + \frac{q_1 e}{c}\left[2(j+1) - \frac{3e}{c}\left(j^2 + 4j + \frac{7}{2}\right)\right]$$
$$+ \frac{3q_2}{c^2}\left[1 - 8\frac{e}{c}(j+2)\right]$$

$$H_1 = 2ej - \frac{e^2}{c}\left(j^2 + \frac{1}{2}\right) + \left(1 + \frac{2}{c^2}\right) je^3$$
$$+ \frac{q_1}{c}\left[1 - \frac{4e}{c}(j+1) + e^2\left(1 + \frac{6}{c^2}\right)\left(j^2 + 4j + \frac{7}{2}\right)\right]$$
$$- \frac{6q_2}{c^2}\left[\frac{1}{c} - \left(1 + \frac{8}{c^2}\right) e(j+2)\right] \qquad (13.4.23)$$

$$q_1 = (1 - e^2)\left(-q \cos 2\omega + \frac{1}{2} q^2 \sin^2 2\omega\right)$$

$$q_2 = \frac{1}{2}(1 - e^2)^2 q^2 \cos 4\omega$$

$$j = \frac{1 + d}{1 - d}$$

Equation (13.4.23) provides us with an expression for the change in the semimajor axis during one revolution. Many other authors have developed expressions of a similar nature for the variations in semimajor axis and eccentricity during one orbital period. Several authors have extended integrations of this type to cover orbital variations during the entire lifetime of the satellite (see Lane *et al.* [25] and King–Hele [26]). Equations such as (13.4.22) have also been used to obtain valuable information about the density of the Earth's atmosphere. The change in the orbital period of a satellite is an observable quantity, and it is essentially proportional to the change in semimajor axis during one revolution. Thus if the change in period has been determined experimentally, then (13.4.22) can be used to obtain an estimate of the density near perigee. Data obtained in this way have contributed significantly to our understanding of the atmospheric density distribution.

13.5. JOINT TREATMENT OF GRAVITY AND DRAG PERTURBATIONS

In most of the analytical studies of the motion of a close-Earth satellite in an atmosphere, the perturbations due to drag and to the Earth's oblateness are considered separately. The joint perturbing effects are then deduced by some type of superposition of the separate perturbations. Such treatments leave unresolved questions, because there is little mathematical or physical justification for using a superposition approach to a dynamical system in which the equations of motion are nonlinear and the perturbing forces are intricately coupled. There have been two notable exceptions to the foregoing approach. In 1961 Brouwer and Hori used the methods of canonical transformations and variation of parameters to develop a theory that gives a joint treatment of oblateness and drag perturbations in a single solution (see Brouwer and Hori [22]). Later, in 1965, Lane also gave a description of the motion of an artificial Earth satellite under the combined influences of oblateness and drag (see Lane [23]). Lane's development has the general features of the procedures introduced by Brouwer and Hori. However, he replaced the nonrotating spherical exponential model atmosphere used in the Brouwer–Hori theory by a nonrotating spherical power function model of the atmosphere. In both treatments, the Delaunay set of canonical variables are used. These variables are defined by the equations

$$\begin{aligned} L &= (\mu a)^{1/2}, & l &= n(t - T) \\ G &= L(1 - e^2)^{1/2}, & g &= \omega \\ H &= G \cos i, & h &= \Omega \end{aligned} \tag{13.5.1}$$

Whenever convenient, we will also use the notation $(L_j; l_j)$, $j = 1, 2, 3$, to describe the Delaunay variables in the order L, G, H, l, g, h. In their analytical approach to the coupled problem of oblateness and drag, Brouwer and Hori used an earlier solution of Brouwer [27], for the motion of an artificial satellite in the gravitational field of the Earth without drag, as their point of departure.

The drag-free solution of Brouwer is developed in powers of the small parameter

$$k_2 = \frac{a_e^2 C_{2,0}}{2}$$

Short-period terms (that is, terms containing trigonometric functions of the mean anomaly, with a typical period of approximately 2 hours) are retained to first order in k_2. Long-period terms (that is, terms containing trigonometric functions of the argument of pericenter, with a typical period of approximately 2 months) and secular terms are retained to second order in k_2. The same order is maintained when drag effects are included in the theory.

For the drag-free problem, the equations of motion are

$$\dot{L}_j = \frac{\partial F(L, G, H, l, g, -)}{\partial l_j}$$
$$\dot{l}_j = -\frac{\partial F(L, G, H, l, g, -)}{\partial L_j} \qquad (j = 1, 2, 3) \qquad (13.5.2)$$

where the dash is used to indicate the absence of the angle variable h from the Hamiltonian F. The potential function used in forming F is $-V$ [see (13.2.1)], expressed in Delaunay variables.†

Brouwer first introduced a canonical transformation to a new primed set of variables L', G', H', l', g', h' through the equations [see (10.3.18)]

$$L_j = \frac{\partial S(L_j'; l_j)}{\partial l_j}, \qquad l_j' = \frac{\partial S(L_j'; l_j)}{\partial L_j'} \qquad (j = 1, 2, 3) \qquad (13.5.3a)$$

where the alternative notation $(L_j'; l_j')$, $j = 1, 2, 3$, is used to denote the single-primed variables in the order L', G', H', l', g', h'. To obtain the generating function $S(L_j'; l_j)$, Brouwer used an approximation method introduced by von Zeipel [28–30] to study the motions of minor planets perturbed by both Jupiter and the Sun. In applying the von Zeipel method, Brouwer introduced (13.5.3a) into both sides of the demand

$$F(L, G, H, l, g, -) = F^*(L', G', H', -, g', -) \qquad (13.5.3b)$$

that the new Hamiltonian F^* be free of l' and h'. Brouwer then expanded both members of (13.5.3b) in Taylor series through terms of order k_2^2. In this way he was able to construct a generating function S, which, through short-period terms of first order in k_2 and long-period and secular terms of second order in k_2, ensures that the new Hamiltonian $F^*(L', G', H', -, g', -)$ does not contain the variables l' and h'. The equations of motion in the new variables L', G', H', l', g', h' are

$$\dot{L}_j' = \frac{\partial F^*}{\partial l_j'}, \qquad \dot{l}_j' = -\frac{\partial F^*}{\partial L_j'} \qquad (j = 1, 2, 3) \qquad (13.5.4)$$

Repeating the von Zeipel procedure, Brouwer then obtained an approximate generator S^* of a second canonical transformation to the double-primed variables L'', G'', H'', l'', g'', h''. The generating function S^* is chosen in such a way that the final Hamiltonian $F^{**}(L''$,

† Brouwer extended the treatment outlined here to take account of the perturbing effects of higher harmonics in the Earth's gravitational potential [27].

G'', H'', –, –, –) is independent of the angle variables l'', g'', h''. The final canonical equations of motion are

$$\dot{L}_j'' = 0, \qquad \dot{l}_j'' = -\frac{\partial F^{**}}{\partial L_j''} \qquad (j = 1, 2, 3) \tag{13.5.5}$$

where $(L_j''; l_j'')$, $j = 1, 2, 3$, are used for the double-primed variables. Then L'', G'', and H'' are constants of the motion and l'', g'', and h'' are linear functions of the time. The solution to the drag-free problem is then obtained in terms of the initial conditions and the time by inverting the von Zeipel transformation equations.

When the drag-free problem is extended to include the dissipative effects of an atmosphere, the Newtonian equations of motion, in an equatorial rectangular coordinate system $Ox_1x_2x_3$ with origin O at geocenter, have the form

$$\ddot{x}_j = -\frac{\partial V}{\partial x_j} + X_j \qquad (j = 1, 2, 3) \tag{13.5.6}$$

where X_j represents the component of the drag acceleration along x_j. Equation (13.5.6) can readily be written in a "canonical" form if we set $\xi_j = \dot{x}_j$ and $\eta_j = x_j$. It is readily verified that (13.5.6) can be replaced by

$$\dot{\xi}_j = \frac{\partial F}{\partial \eta_j} + X_j, \qquad \dot{\eta}_j = -\frac{\partial F}{\partial \xi_j} \qquad (j = 1, 2, 3) \tag{13.5.7}$$

where

$$F = -\frac{1}{2}\sum_{i=1}^{3} \xi_i^2 - V$$

Brouwer and Hori prove a theorem that states the conditions whereby a system of differential equations

$$\dot{x}_j = \frac{\partial F}{\partial y_j} + X_j, \qquad \dot{y}_j = -\frac{\partial F}{\partial x_j} - Y_j \qquad (j = 1, 2, \ldots, n)$$

in which X_j, Y_j are arbitrary functions of the variables x_j, y_j and t, can be transformed canonically. They showed that a transformation

$$(x_j\,;\, y_j) \rightarrow (x_j'; y_j')$$

which is canonical in the absence of both X_j and Y_j, leads to the transformed equations of motions

$$\dot{x}_j' = \frac{\partial F^*}{\partial y_j'} + X_j', \qquad \dot{y}_j' = -\frac{\partial F^*}{\partial x_j'} - Y_j' \qquad (j = 1, 2, \ldots, n)$$

with

$$F^*(x_j'; y_j') = F(x_j; y_j)$$

if the functions X_j' and Y_j' are obtained from the expressions

$$X_j' = \sum_{k=1}^{n} \left(X_k \frac{\partial y_k}{\partial y_j'} + Y_k \frac{\partial x_k}{\partial y_j'} \right)$$

$$Y_j' = \sum_{k=1}^{n} \left(X_k \frac{\partial y_k}{\partial x_j'} + Y_k \frac{\partial x_k}{\partial x_j'} \right)$$

The Brouwer–Hori theorem can be applied to (13.5.7). It is assumed that, with $X_j = 0$, $j = 1, 2, 3$, a canonical transformation from the variables $(\xi_j; \eta_j)$ to the Delaunay variables $(L_j; l_j)$ is known. Then Equations (13.5.7) may be replaced by the system

$$\dot{L}_j = \frac{\partial F}{\partial l_j} + P_j, \qquad \dot{l}_j = -\frac{\partial F}{\partial L_j} - Q_j \qquad (j = 1, 2, 3)$$
$$F = F(L, G, H, l, g, -) \tag{13.5.8}$$

where P_j and Q_j are functions of the Delaunay variables through the equations

$$P_j = \sum_{k=1}^{3} X_k \frac{\partial \eta_k}{\partial l_j}, \qquad Q_j = \sum_{k=1}^{3} X_k \frac{\partial \eta_k}{\partial L_j} \tag{13.5.9}$$

Equations (13.5.8) represent the extension of Equations (13.5.2) to include the effects of air drag.

A canonical transformation that relates $(L_j; l_j)$ to the double-primed variables $(L_j''; l_j'')$ has already been determined for the drag-free problem. Thus the Brouwer–Hori theorem can be applied again to replace the system (13.5.8) with the equations

$$\dot{L}_j'' = P_j'', \qquad \dot{l}_j'' = -\frac{\partial F^{**}}{\partial L_j''} - Q_j'' \qquad (j = 1, 2, 3) \tag{13.5.10}$$

where

$$P_j'' = \sum_{k=1}^{3} P_k \frac{\partial l_k}{\partial l_j''} + \sum_{k=1}^{3} Q_k \frac{\partial L_k}{\partial l_j''} \tag{13.5.11a}$$

$$Q_j'' = \sum_{k=1}^{3} P_k \frac{\partial l_k}{\partial L_j''} + \sum_{k=1}^{3} Q_k \frac{\partial L_k}{\partial L_j''} \tag{13.5.11b}$$

The perturbations arising from atmospheric resistance enter into the equations of motion (13.5.10) through P_j'' and Q_j''. Brouwer and Hori assumed a drag acceleration of the form (13.3.1) in which they neglect the rotation of the atmosphere so that $\mathbf{V} = \mathbf{v}$ [see (13.3.4)]. They further assume an exponential representation of air density given by (13.4.13) so that

$$X_j = -A^* v \exp(-\alpha r)\, \xi_j \tag{13.5.12}$$

where

$$A^* = \frac{1}{2} C_D \frac{A}{m} \rho_0 \exp(\alpha q_0), \qquad \alpha = \frac{1}{H} \tag{13.5.13}$$

By identifying the canonical variables with the solution to the drag-free problem, Brouwer and Hori, after some manipulation, rewrite (13.5.10) in the form

$$\begin{aligned} \dot{L}_j'' &= A^* v \exp(-\alpha r)\, p_j - A^* v \exp(-\alpha r)\, \delta p_j \qquad (j = 1, 2, 3) \\ \dot{l}_j'' &= -\frac{\partial F^{**}}{\partial L_j''} + A^* v \exp(-\alpha r)\, q_j + A^* v \exp(-\alpha r)\, \delta q_j \end{aligned} \tag{13.5.14}$$

where p_j, q_j are explicit functions of the Delaunay variables and δp_j, δq_j are explicit functions of the double-primed variables.

The integrals of (13.5.14) constitute an approximation to the solution of the problem of describing the motion of an artificial satellite under the combined influences of gravity and drag, for, in the coupled theory, we are essentially studying the perturbations of the drag-free elements L'', G'', H'', l'', g'', h'' due to air resistance. Brouwer and Hori obtained approximations to these integrals by first developing the density function in a Taylor series complete to the fifth power of the eccentricity. After expressing the Delaunay variables in terms of double-primed variables, they used the method of variation of parameters to effect the integration. For satellites with low perigee heights, the truncation of the series expansion of the density function does not provide sufficient convergence because the absolute value of the argument of the exponential increases as the perigee height decreases. Certain additional limitations appear in the theory: some of the formulas contain small divisors for small eccentricities, small inclination angles, and angles near the critical inclination case. For example, in one place in the final equations for l and g in the coupled theory, the factor $1/e$ appears because of its presence in l' and g' in the drag-free theory. In the same equations, the factor $1/e$ appears in l'' and g'' because of the effect of air resistance. Questions relating to some of the limitations cited above have been examined independently by Lyddane and Davenport [31, 32]. They give improved formulas to handle the small eccentricity and small inclination problems in the drag-free problem.

Lane [23] has also used the method of von Zeipel transformations as introduced by Brouwer and Hori to study the motion of an artificial satellite under the joint effects of gravity and drag. He uses, however, a power-law model atmosphere of the form (13.4.10), where n is restricted to positive integers, instead of the spherically symmetrical exponential model atmosphere used by Brouwer and Hori. This means that instead of the single parameter α, two disposable parameters n and s appear in the final solution. These parameters are to be chosen to fit the power function to the best available atmospheric data. Wherever possible, Lane has used the results obtained by Brouwer and Hori, but he has rederived and expressed them in a form suitable for use with the power-law density model.

The theory given by Lane is carried out to the same order as the Brouwer-Hori theory and has the same limitations as to small eccentricities, small inclination angles, and critical inclination case. Nevertheless, he has been able to avoid the convergence problem for low perigee heights that is present in the Brouwer–Hori theory, because the power density function does not have to be expanded in Taylor series in order to carry out the integration of the equations of motion for the double-primed variables.

Lane and Cranford have extended and improved the Lane theory (see Lane and Cranford [32a]). They have been able to reformulate Lane's equations in such a way as to avoid the limitations to the theory for small eccentricities and small inclination angles. In addition, they have been able to incorporate into the theory certain terms arising from drag which are not present in either the Brouwer–Hori equations or the original Lane development. These terms, which were previously neglected in both the Brouwer–Hori and Lane theories, because they contain the square of a small first-order parameter, have been found to be significant for Earth satellites with low perigee heights.

Exercises

1. Derive Equations (13.2.5) in the text.
2. Derive Equations (13.2.12), (13.2.13), and (13.2.14) in the text.
3. Show that the component along the z_q-axis of the orbital angular momentum per unit mass is a constant of the motion in the first-order gravitational theory.
4. Consider the problem of describing the motion of a particle of unit mass in the Earth's gravitational field, using the method of canonical

transformations. Show that the Hamilton–Jacobi partial differential equation for the generating function S may be written

$$(\nabla S)^2 - 2(\alpha - V) = 0 \tag{1}$$

where ∇ is the gradient operator, V is the potential energy of the particle, and α is the total energy (a constant of the motion).

5. If we designate a set of orthogonal curvilinear coordinates by (u_1, u_2, u_3), it can be shown (see [33]) that the Hamilton–Jacobi equation in Exercise 4 will separate if the potential energy has the form

$$V = \sum_{j=1}^{3} \frac{f_j(u_j)}{h_j^2}$$

where the h_j are the scale factors defined in (1.5.10) and $f_j(u_j)$ is a function of the single coordinate u_j. Assuming that V depends only on the radial distance r from geocenter and the codeclination θ, show, by solving Laplace's equation, that a solution consistent with Hamilton–Jacobi separability is

$$V = \frac{C_1}{r} + \frac{C_2 \cos\theta}{r^2}$$

where C_1 and C_2 are integration constants.

6. Show that, for the case of a satellite whose surface intersects a straight line no more than twice, the effective cross-sectional area of the satellite [see (13.3.1)] is equal to one-fourth its surface area.
7. Verify Equation (13.3.23) in the text.
8. Derive the power-law representation of air density given as (13.4.10) in the text. Show that if the "density scale height" is assumed to be constant, the power-law representation passes over to the exponential representation given by (13.4.13). Show, also, that if atmospheric oblateness is taken into account, (13.4.10) is to be replaced by (13.4.14).

Chapter 14

Gravity Effects on the Rotational Motion of an Artificial Earth Satellite

14.1. INTRODUCTION†

Since 1957 a considerable amount of work has been directed toward the development of a precise orbit theory for determining the geocentric motions of artificial Earth satellites. Consequently, significant advances have been made in our ability to establish the long-term ephemerides of such orbiting bodies. Considerably less progress has been made toward the development of an adequate theory for the rotational motions of artificial satellites, although the proper interpretation of much of the scientific data obtained from satellite experiments depends on information of the type that such a theory could conceivably supply. The Pegasus satellites (see Figure 14.1.1) offer a good example of the need for such a theory. These satellites are neither actively nor passively stabilized in attitude, and their size and shape cause them to be subject to significant gravity-gradient (gravitational) and damping torques.

† Much of the material discussed in this chapter is an outgrowth of work carried on by the author and several of his graduate students [34–36]. The work has been sponsored in part by the National Aeronautics and Space Agency (NASA), Marshall Space Flight Center, Huntsville, Alabama, under contract NAS8-20175, and in part by NASA, Washington, D.C., under Grant NGR-01-003-036.

The development of much of the material to be discussed in this chapter was motivated by the attempts of Holland and Sperling [37, 38] to explain the observed rotational motions of the Pegasus satellites. Beletsky [39, 40] and Colombo [41, 42] had previously developed approximate analytical methods for predicting the motion of the angular momentum of such artificial satellites if the satellite orbit plane does not regress.

FIGURE 14.1.1

Preliminary analyses of the relative magnitudes of the various torques acting on the Pegasus satellites indicated that the gravity-gradient torque should be the dominant one. However, the gravity-gradient theories for nonregressing orbits did not explain the observed variations in the angular momenta of the Pegasus satellites. Holland and Sperling showed that the discrepancy between these theories and the observations is primarily due to the precession of the satellite orbital plane.† By incorporating nodal precession effects in their theory, they have achieved excellent agreement with certain of the observed motions.

In essence, they derived new results that predict the principal effects of the gravitational and magnetic torques on the rotational motion of an asymmetrical rigid body in a regressing orbit about an oblate attracting mass. Using a method of averages, they developed an approximate theory in the form of three integrals that predict the motion of the angular momentum (due to the rotation about the center of mass) of the rigid body as viewed from the orbital plane. Their theory should be valid if the rotational kinetic energy of the satellite is large when compared to the effects of the perturbing torques.

In the references [37, 38] just cited, the method used was a method of averaging. Only the motion of the angular momentum vector was predicted; the motion of the rigid body relative to the angular momentum vector is not given. Also, by the very nature of the method, periodic motions were necessarily ignored.

The theoretical results described in the later sections of this chapter reflect beginning attempts to develop a general theory for predicting the motions, not just of the angular momenta of orbiting bodies, but their complete motions in space as well. The effects of orbital regression are taken into account.

Much of the analytical work on the problem of the rotation of rigid bodies appears in the classical literature, where the analyses have been limited essentially to the solutions of the free Eulerian motions of uniaxial and triaxial bodies. Free Eulerian motions are rotations of the rigid body in the absence of external torques; rigid bodies with an axis of dynamic symmetry through the center of mass will be referred to as uniaxial, while asymmetrical rigid bodies will be referred to as triaxial. In the more recent literature, the numerical approach, as well as the analytical, has been employed in some of the research effort in this area. The numerical techniques have certain serious disadvantages in such work, however, primarily because the integration step time must be kept small compared to the period of the most rapid motion of the system. From the observed motions of the Pegasus satellites, small perturbing torques are found to have profound long-term effects on the free Eulerian motion of a rigid body; effects that cannot be approximated meaningfully by the classical analytical treatments of the rigid-body problem. No adequate analytical theory for the accurate prediction of the rotational motions

† In later publications both Colombo [43] and Beletsky [44] take into account the effect of nodal regression for axisymmetric bodies.

of artificial Earth satellites now exists. By an adequate theory, we mean one that solves the "main" problem of satellite rotational motion, namely, describing the motion of a satellite rotating in the gravitational field of the Earth and being perturbed by damping torques.

For our purposes here, in discussing the motion of a small rigid satellite about its center of mass when it is in orbit about an oblate planet, we may neglect the coupling of the satellite's rotational and translational motions (see e.g., Holland [37]). Two basic types of motion about the center of mass are to be distinguished. A librational motion about a position of relative equilibrium is possible if the angular velocity of the satellite's rotation is less than roughly 0.1. deg/sec (see Beletsky [44]) A second type of motion occurs if the rotational kinetic energy of the satellite is large when compared with the effects of the perturbing forces (angular velocity of rotation greater than 1 deg/sec). Here the unperturbed motion is free Eulerian and the small perturbing forces introduce effects that accumulate with time and result in significant secular deviations. A familiar example is the motion of the Earth's axis of rotation under the gravitational perturbations of the Sun and Moon. In this discussion, we plan to limit ourselves to motions of the second type (that is, motions in the regime of relatively large rotational kinetic energy), of which the luni-solar precession just cited is a special case wherein certain simplifying assumptions not present in the general theory are made.

The Pegasus satellites have provided a good opportunity for observing the rotational motions of spacecraft. The satellites are instrumented with attitude sensors, and a considerable amount of data indicating the presence of at least three non-Eulerian effects in the high rotational energy regime has been obtained.

One important feature of the motion of the Pegasus satellites, which appears to be caused by gravitational influences, is that the angular momentum, which is constant in free Eulerian motion, undergoes a long-period ($\approx$25 days) change in its direction with respect to the satellite orbital plane. A second observed effect is a secular increase of the coning angle (i.e., the angle between what is approximately a symmetry axis and the angular momentum of the satellite), from some initially small value to a final value of 90 deg, at which time the satellite begins to spin about its axis of maximum moment of inertia. A third effect, which is observed after the satellite enters this final spin phase, is a secular decrease in the magnitude of the angular velocity, thus indicating the presence of dissipative or damping torques. Neither of these last two effects can be explained on the basis of gravitational perturbations.

The presence of gravitational and dissipative influences in the main problem of satellite rotational motion in the high rotational energy regime points up a striking parallel between this problem and the main problem of satellite orbital theory, which, as we have seen, is to describe the motion of the center of mass of a close-Earth satellite in an atmosphere. The approach used by Brouwer and Hori [22] in their development of an analytical theory for this latter problem suggests a parallel approach to the former one. They first solved the problem of the motion of the center of mass of the satellite in the Earth's aspherical gravitational field, and then used this "drag-free" solution as a point of departure for the development of a coupled theory that includes the effects of atmospheric drag. An analogy with orbital theory suggests that a reasonable approach to the main problem of satellite rotational motion in the high rotational energy regime is to consider first the problem of the rotational motion in the Earth's gravitational field. The solution of this "damping-free" problem may then be used as a point of departure for the derivation of a theory that includes the effects of dissipative torques.

Only the damping-free aspect of the satellite rotational motion, in the high rotational energy regime, is treated in this chapter. The method of canonical transformations, with the unperturbed problem taken to correspond to that of a free triaxial body, is used. An exact solution of this unperturbed problem is given. This exact solution is used as a starting point to study the case of a uniaxial rigid body in a circular orbit about an attracting point mass. The inclination of the orbital plane is assumed constant, but the plane is permitted to regress at a constant rate. Analytical expressions (to first order in a small parameter) that describe the motion are obtained.† A brief discussion of certain important relationships in the general theory of rigid-body motion, which will be drawn upon in developing the solution, are given in the next several sections.

14.2. BASIC CONCEPTS IN RIGID-BODY MECHANICS

With reference to Figure 14.2.1, let $Ox^*y^*z^*$ represent a rectangular coordinate system that has its origin O at the center of mass of a rigid body and its orientation fixed in space. The $Ox'y'z'$-system is fixed in the rigid body, which is rotating with angular velocity $\boldsymbol{\omega}$ about an instantaneous axis. Let G be the matrix D of (2.5.1) in which the notation Ω, i, ω is replaced by the notation ψ, θ, ϕ, respectively. It follows directly from (2.5.6) that a vector $\mathbf{A}$, represented in the body-fixed

† References to more recent results for the triaxial case are given in a note added in proof, p. 381.

system by $\mathbf{A}'$, can be transformed to a representation $\mathbf{A}^*$ in the space-fixed system, and conversely, by the matrix equations

$$\mathbf{A}^* = G\mathbf{A}', \qquad \mathbf{A}' = G^T\mathbf{A}^* \tag{14.2.1}$$

where

$$\mathbf{A}^* = \begin{pmatrix} A_{x^*} \\ A_{y^*} \\ A_{z^*} \end{pmatrix}, \qquad \mathbf{A}' = \begin{pmatrix} A_{x'} \\ A_{y'} \\ A_{z'} \end{pmatrix}$$

G^T is the transpose of the matrix G and the elements A_{x^*}, A_{y^*}, and A_{z^*} are the components of $\mathbf{A}$ in the $Ox^*y^*z^*$-system, while $A_{x'}$, $A_{y'}$, and $A_{z'}$ are the components of $\mathbf{A}$ in the $Ox'y'z'$-system. Since G represents an orthogonal transformation, $G^T = G^I$, where G^I is the inverse matrix of G. The angles ψ, θ, and ϕ are referred to as Euler angles.

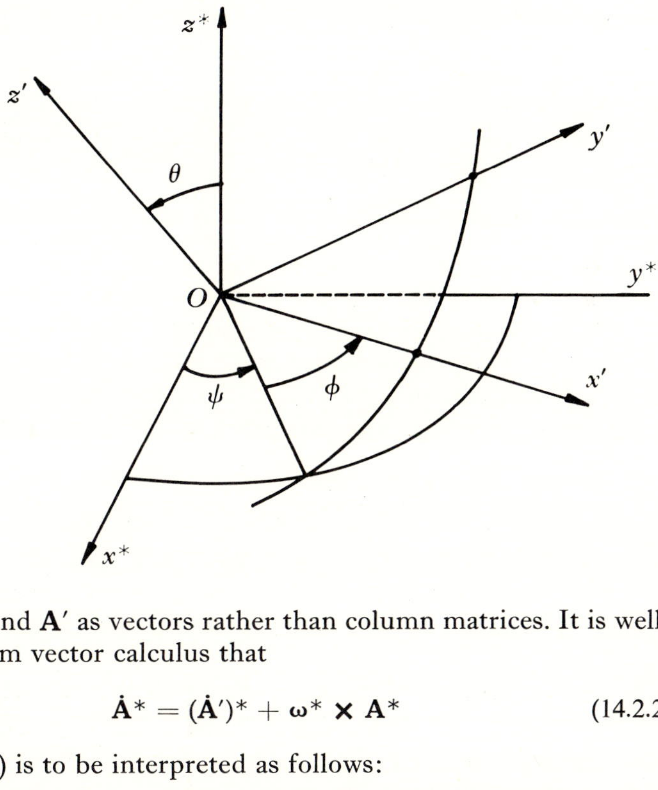

FIGURE 14.2.1

Consider $\mathbf{A}^*$ and $\mathbf{A}'$ as vectors rather than column matrices. It is well-known result from vector calculus that

$$\dot{\mathbf{A}}^* = (\dot{\mathbf{A}}')^* + \boldsymbol{\omega}^* \times \mathbf{A}^* \tag{14.2.2}$$

Equation (14.2.2) is to be interpreted as follows:

1. Thinking of $\mathbf{A}$ as being represented in the $Ox^*y^*z^*$-system by $\mathbf{A}^*$, we take the time derivative of $\mathbf{A}^*$ as viewed in the $Ox^*y^*z^*$-system. The result is $\dot{\mathbf{A}}^*$.

2. Thinking of $\mathbf{A}$ as being represented in the $Ox'y'z'$-system by $\mathbf{A}'$,

we take the time derivative of $\dot{\mathbf{A}}'$ as viewed in the $Ox'y'z'$-system. Call this time derivative $\dot{\mathbf{A}}'$. Then transform $\dot{\mathbf{A}}'$ to the $Ox^*y^*z^*$-system. The result is $(\dot{\mathbf{A}}')^*$.

3. Thinking of the vector $\boldsymbol{\omega}$ as being represented in the $Ox^*y^*z^*$-system by $\boldsymbol{\omega}^*$, we form the cross product $\boldsymbol{\omega}^* \times \mathbf{A}^*$.

It is readily verified that the vector equation (14.2.2) is equivalent to the matrix equation

$$\dot{\mathbf{A}}^* = G\dot{\mathbf{A}}' + \Omega\mathbf{A}^* \tag{14.2.3}$$

where

$$\dot{\mathbf{A}}^* = \begin{pmatrix} \dot{A}_{x^*} \\ \dot{A}_{y^*} \\ \dot{A}_{z^*} \end{pmatrix}, \qquad \dot{\mathbf{A}}' = \begin{pmatrix} \dot{A}_{x'} \\ \dot{A}_{y'} \\ \dot{A}_{z'} \end{pmatrix}$$

and Ω, a tensor, is given in matrix notation by

$$\Omega = \begin{pmatrix} 0 & -\omega_{z^*} & \omega_{y^*} \\ \omega_{z^*} & 0 & -\omega_{x^*} \\ -\omega_{y^*} & \omega_{x^*} & 0 \end{pmatrix}$$

The elements ω_{x^*}, ω_{y^*}, and ω_{z^*} are the components of $\boldsymbol{\omega}$ in the $Ox^*y^*z^*$-system.

Forming the time derivative of both sides of (14.2.1), we can write

$$\dot{\mathbf{A}}^* = G\dot{\mathbf{A}}' + \dot{G}\mathbf{A}' \tag{14.2.4}$$

where $\dot{G}$ is the matrix whose elements are the time derivatives of the corresponding elements of G. Equating (14.2.3) and (14.2.4) and using the second equation in (14.2.1), we see that we must have

$$\Omega = \dot{G}G^T \tag{14.2.5}$$

Expressed in matrix notation, the tensor Ω transforms according to the equation

$$\Omega = G\Omega'G^T \tag{14.2.6}$$

where

$$\Omega' = \begin{pmatrix} 0 & -\omega_{z'} & \omega_{y'} \\ \omega_{z'} & 0 & -\omega_{x'} \\ -\omega_{y'} & \omega_{x'} & 0 \end{pmatrix}$$

and the elements $\omega_{x'}$, $\omega_{y'}$, and $\omega_{z'}$ are components of $\boldsymbol{\omega}$ in the $Ox'y'z'$–system. Thus it follows that

$$\Omega' = G^T\Omega G = G^T\dot{G} \tag{14.2.7}$$

After the operation of performing the time differentiation of each of the elements in the matrix G has been carried out, we readily obtain, from (14.2.6) and (14.2.7), equations that express the components of $\boldsymbol{\omega}$ in both the $Ox^*y^*z^*$-system and the $Ox'y'z'$-system in terms of the Euler angles ψ, θ, ϕ and their time derivatives. These equations, known as Euler's geometrical equations, are given in matrix notation by

$$\boldsymbol{\omega}^* = M\xi, \qquad \boldsymbol{\omega}' = N\xi \tag{14.2.8}$$

where

$$\boldsymbol{\omega}^* = \begin{pmatrix} \omega_{x^*} \\ \omega_{y^*} \\ \omega_{z^*} \end{pmatrix}, \qquad \boldsymbol{\omega}' = \begin{pmatrix} \omega_x{}' \\ \omega_y{}' \\ \omega_z{}' \end{pmatrix}, \qquad \xi = \begin{pmatrix} \dot{\psi} \\ \dot{\theta} \\ \dot{\phi} \end{pmatrix}$$

and

$$M = \begin{pmatrix} 0 & \cos\psi & \sin\psi\sin\theta \\ 0 & \sin\psi & -\cos\psi\sin\theta \\ 1 & 0 & \cos\theta \end{pmatrix}, \qquad N = \begin{pmatrix} \sin\phi\sin\theta & \cos\phi & 0 \\ \cos\phi\sin\theta & -\sin\phi & 0 \\ \cos\theta & 0 & 1 \end{pmatrix}$$

Note that the matrices M and N are not orthogonal. Also, although it has three "components" ξ is not to be construed as a vector. The inverses of (14.2.8) are

$$\xi = M^I\boldsymbol{\omega}^*, \qquad \xi = N^I\boldsymbol{\omega}' \tag{14.2.9}$$

where

$$M^I = \begin{pmatrix} -\sin\psi\cot\theta & \cos\psi\cot\theta & 1 \\ \cos\psi & \sin\psi & 0 \\ \sin\psi\csc\theta & -\cos\psi\csc\theta & 0 \end{pmatrix} \tag{14.2.10}$$

$$N^I = \begin{pmatrix} \sin\phi\csc\theta & \cos\phi\csc\theta & 0 \\ \cos\phi & -\sin\phi & 0 \\ -\sin\phi\cot\theta & -\cos\phi\cot\theta & 1 \end{pmatrix} \tag{14.2.11}$$

14.3. ROTATIONAL KINETIC ENERGY OF A RIGID BODY ABOUT ITS CENTER OF MASS

We saw in (1.5.5) that the total kinetic energy T of a system of particles may be written as

$$T = \tfrac{1}{2}M\dot{R}^2 + T_c \qquad [1.5.5]$$

where the first term on the right-hand side is the kinetic energy of translation of the center of mass O of the system and T_c is the kinetic energy of the motion with respect to the center of mass. We suppose that the rigid body, composed of n mass elements m_i, is rotating with angular velocity $\boldsymbol{\omega}$ about an instantaneous axis passing through O (see Figure 14.3.1). The kinetic energy of rotation about the instantaneous

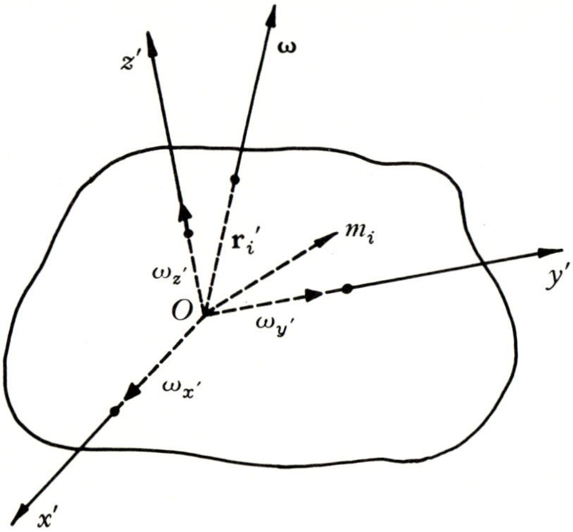

FIGURE 14.3.1

axis can be obtained by summing over the individual mass elements. Using (1.5.2), we write

$$
\begin{aligned}
T_c &= \frac{1}{2}\sum_{i=1}^{n}\frac{1}{m_i}\mathbf{p}_i'\cdot\mathbf{p}_i' \\
&= \frac{1}{2}\sum_{i=1}^{n} m_i(\boldsymbol{\omega}'\times\mathbf{r}_i')\cdot\mathbf{r}_i' \\
&= \frac{1}{2}\boldsymbol{\omega}'\cdot\sum_{i=1}^{n}(\mathbf{r}_i'\times m_i\dot{\mathbf{r}}_i') \\
&= \frac{1}{2}\boldsymbol{\omega}'\cdot\mathbf{h}'
\end{aligned}
\tag{14.3.1}
$$

where, in writing the last step, we have made use of (1.4.3), but we have replaced the notation $\mathbf{L}$ by $\mathbf{h}$ to avoid notational difficulties in the subsequent discussion. The vector $\mathbf{h}$ used here is not to be confused with the vector $\mathbf{h}$ used in earlier chapters. The prime notation is used to indicate that all the vectors appearing in (14.3.1) are referenced to the rectangular coordinate system $Ox'y'z'$ shown in Figure 14.3.1. The $Ox'y'z'$-system is considered to be fixed in the body.

Since

$$\mathbf{h}' = \sum_{i=1}^{n} m_i \mathbf{r}_i' \times \dot{\mathbf{r}}_i'$$

$$= \sum_{i=1}^{n} m_i \mathbf{r}_i' \times (\boldsymbol{\omega}' \times \mathbf{r}_i') \tag{14.3.2}$$

it is a straightforward matter to carry out the indicated operations (using the integral form of notation) and show that

$$\mathbf{h}' = I'\boldsymbol{\omega}' \tag{14.3.3}$$

where the moment of inertia tensor I' can be represented by the matrix

$$I' = \begin{pmatrix} I_{x'} & P_{x'y'} & P_{x'z'} \\ P_{x'y'} & I_{y'} & P_{y'z'} \\ P_{x'z'} & P_{y'z'} & I_{z'} \end{pmatrix} \tag{14.3.4}$$

The quantities $I_{x'}$, $I_{y'}$, $I_{z'}$, $P_{x'y'}$, $P_{x'z'}$, and $P_{y'z'}$ are the moments and products of inertia introduced in (12.5.2) but are here referred to $Ox'y'z'$-axes. When (14.3.3) is substituted into (14.3.1), we can represent T_c by the matrix equation

$$T_c = \tfrac{1}{2}\boldsymbol{\omega}'^T I' \boldsymbol{\omega}' = \tfrac{1}{2}\xi^T N^T I' N \xi \tag{14.3.5}$$

where the superscript T designates the transpose of a given matrix.

If $Ox'y'z'$ constitutes a set of principal axes for the rigid body, Equation (14.3.4) reduces to

$$I' = \begin{pmatrix} A & 0 & 0 \\ 0 & B & 0 \\ 0 & 0 & C \end{pmatrix} \tag{14.3.6}$$

where A, B, and C are the principal moments of inertia introduced in (12.6.1), but are here referred to the $Ox'y'z'$-axes. We define a new matrix S by the equation

$$S = N^T I' N \tag{14.3.7}$$

where I' is given by (14.3.6). Then S, its inverse S^I, and $T_c = \frac{1}{2}\xi^T S \xi$ are given explicitly by

$$S = \begin{pmatrix} (A\sin^2\phi + B\cos^2\phi)\sin^2\theta + C\cos^2\theta & (A - B)\sin\phi\cos\phi\sin\theta & C\cos\theta \\ (A - B)\sin\phi\cos\phi\sin\theta & A\cos^2\phi + B\sin^2\phi & 0 \\ C\cos\theta & 0 & C \end{pmatrix} \tag{14.3.8}$$

$$S^I =$$

$$\begin{bmatrix} \left(\frac{\sin^2\phi}{A} + \frac{\cos^2\phi}{B}\right)\frac{1}{\sin^2\theta} & \left(\frac{1}{A} - \frac{1}{B}\right)\frac{\sin\phi\cos\phi}{\sin\theta} & -\left(\frac{\sin^2\phi}{A} + \frac{\cos^2\phi}{B}\right)\frac{\cot\theta}{\sin\theta} \\ \left(\frac{1}{A} - \frac{1}{B}\right)\frac{\sin\phi\cos\phi}{\sin\theta} & \frac{\cos^2\phi}{A} + \frac{\sin^2\phi}{B} & \left(\frac{1}{B} - \frac{1}{A}\right)\frac{\sin\phi\cos\phi}{\tan\theta} \\ -\left(\frac{\sin^2\phi}{A} + \frac{\cos^2\phi}{B}\right)\frac{\cot\theta}{\sin\theta} & \left(\frac{1}{B} - \frac{1}{A}\right)\frac{\sin\phi\cos\phi}{\tan\theta} & \left(\frac{\sin^2\phi}{A} + \frac{\cos^2\phi}{B}\right)\cot^2\theta + \frac{1}{C} \end{bmatrix} \tag{14.3.9}$$

$$\begin{aligned} T_c = {} & \tfrac{1}{2}[\sin^2\theta(A\sin^2\phi + B\cos^2\phi) + C\cos^2\theta]\,\dot{\psi}^2 \\ & + (A - B)\sin\theta\sin\phi(\cos\phi)\,\dot{\psi}\dot{\theta} + \tfrac{1}{2}(A\cos^2\phi + B\sin^2\phi)\,\dot{\theta}^2 \\ & + C(\cos\theta)\,\dot{\psi}\dot{\phi} + \tfrac{1}{2}C\dot{\phi}^2 \end{aligned} \tag{14.3.10}$$

14.4. THE HAMILTONIAN FUNCTION FOR A SMALL RIGID BODY PERTURBED BY GRAVITY-GRADIENT TORQUE

Consider now a dynamical system for which the Lagrangian function L is of the form

$$L = T_c(q_j\,;\,\dot{q}_j\,;\,t) - V(q_j\,;\,t) \qquad (j = 1, 2, 3) \tag{14.4.1}$$

where the kinetic energy T_c is a function of the generalized coordinates q_j, the generalized velocities $\dot{q}_j$, and the time t, while the potential energy V depends, explicitly, only on the generalized coordinates and t. Using the definition of the generalized momenta p_j, we write

$$p_j = \frac{\partial L}{\partial \dot{q}_j} = \frac{\partial T}{\partial \dot{q}_j} \qquad (j = 1, 2, 3) \tag{14.4.2}$$

If we identify $(q_1, q_2, q_3, p_1, p_2, p_3)$ with $(\psi, \theta, \phi, p_\psi, p_\theta, p_\phi)$ respectively, and assume that $T_c = \frac{1}{2}\xi^T S \xi$, we have

$$p_j = \frac{1}{2}\frac{\partial}{\partial \dot{q}_j}(\xi^T S \xi) \qquad (j = 1, 2, 3) \tag{14.4.3}$$

One can readily carry out the partial derivatives indicated in (14.4.3). In this way, noting that $S = S^T$, it is a straightforward matter to deduce the matrix equations

$$P = S\xi \tag{14.4.4}$$

$$\xi = S^I P \tag{14.4.5}$$

where

$$P = \begin{pmatrix} p_\psi \\ p_\theta \\ p_\phi \end{pmatrix} \tag{14.4.6}$$

Thus T_c, expressed in terms of the Euler angles and their conjugate momenta, has the form

$$T_c = \tfrac{1}{2}P^T S^T P \tag{14.4.7}$$

where P^T is the transpose of P.

Relations between the matrix P and the angular momentum $\mathbf{h}$ can be obtained from (14.2.8) and (14.3.3). We find that

$$P = N^T\mathbf{h}' \tag{14.4.8}$$

$$P = M^T\mathbf{h}^* \tag{14.4.9}$$

where $\mathbf{h}^*$ is referenced to the $Ox^*y^*z^*$-system.

For the dynamical system that we are describing, the Hamiltonian function H is equal to $T_c + V$. Explicitly, we write

$$\begin{aligned}
H = {} & \frac{1}{2}\left(\frac{\sin^2\phi}{A} + \frac{\cos^2\phi}{B}\right)\frac{p_\psi{}^2}{\sin^2\theta} + \frac{1}{2}\left(\frac{\cos^2\phi}{A} + \frac{\sin^2\phi}{B}\right)p_\theta{}^2 \\
& + \frac{1}{2}\left[\left(\frac{\sin^2\phi}{A} + \frac{\cos^2\phi}{B}\right)\cot^2\theta + \frac{1}{C}\right]p_\phi{}^2 \\
& + \left(\frac{1}{A} - \frac{1}{B}\right)\frac{\sin\phi\cos\phi}{\sin\theta}p_\psi p_\theta \\
& + \left(\frac{1}{B} - \frac{1}{A}\right)\sin\phi\cos\phi\cot\theta\, p_\theta p_\phi - \left(\frac{\sin^2\phi}{A} + \frac{\cos^2\phi}{B}\right)\frac{\cot\theta}{\sin\theta}p_\psi p_\theta \\
& + V(\psi, \theta, \phi, t)
\end{aligned} \tag{14.4.10}$$

The canonical equations of motion are

$$\begin{aligned}
\dot\psi = {} & \left(\frac{\sin^2\phi}{A} + \frac{\cos^2\phi}{B}\right)\csc^2\theta(p_\psi - p_\phi\cos\theta) \\
& + \left(\frac{1}{A} - \frac{1}{B}\right)\sin\phi\cos\phi(\csc\theta)\,p_\theta + \frac{\partial V}{\partial p_\psi} \\
\dot\theta = {} & \left(\frac{1}{A} - \frac{1}{B}\right)\sin\phi\cos\phi\csc\theta(p_\psi - p_\phi\cos\theta) \\
& + \left(\frac{\cos^2\phi}{A} + \frac{\sin^2\phi}{B}\right)p_\theta + \frac{\partial V}{\partial p_\theta} \\
\dot\phi = {} & -\left(\frac{\sin^2\phi}{A} + \frac{\cos^2\phi}{B}\right)\csc\theta\cot\theta(p_\psi - p_\phi\cos\theta) \\
& - \left(\frac{1}{A} - \frac{1}{B}\right)\sin\phi\cos\phi(\cot\theta)\,p_\theta + \frac{p_\phi}{C} + \frac{\partial V}{\partial p_\phi}
\end{aligned} \tag{14.4.11}$$

$$\dot{p}_\psi = -\frac{\partial V}{\partial \psi}$$

$$\dot{p}_\theta = \left(\frac{\sin^2\phi}{A} + \frac{\cos^2\phi}{B}\right)\csc^3\theta[(p_\psi{}^2 + p_\phi{}^2)\cos\theta - (1+\cos^2\theta)\,p_\psi p_\phi]$$
$$+\left(\frac{1}{A} - \frac{1}{B}\right)\sin\phi\cos\phi\csc^2\theta\, p_\theta(p_\psi\cos\theta - p_\phi) - \frac{\partial V}{\partial\theta}$$

$$\dot{p}_\phi = \left(\frac{1}{A} - \frac{1}{B}\right)\{\sin\phi\cos\phi[p_\theta{}^2 - \csc^2\theta(p_\psi - p_\phi\cos\theta)^2]$$
$$-(\cos^2\phi - \sin^2\phi)\csc\theta\, p_\theta(p_\psi - p_\phi\cos\theta)\} - \frac{\partial V}{\partial\phi}$$

Let us assume that we have found a complete integral $W_0(q_j;\alpha_j;t)$, $j = 1, 2, 3$, of the Hamilton–Jacobi equation corresponding to a dynamical problem for which the Hamiltonian function H_0 is given by the right-hand side of (14.4.10), neglecting V. Further assume that W_0 has been used to generate a canonical transformation from the old variables ψ, θ, ϕ, p_ψ, p_θ, p_ϕ to the canonical parameters $(\alpha_k;\beta_k)$, $k = 1, 2, 3$. The equations

$$\dot{\alpha}_k = \frac{\partial V}{\partial\beta_k}, \qquad \dot{\beta}_k = -\frac{\partial V}{\partial\alpha_k} \qquad (k = 1, 2, 3) \tag{14.4.12}$$

give the time derivatives of the new variables. It may be convenient to replace the $(\alpha_k;\beta_k)$ with another complete set of independent variables (not necessarily canonical) through the use of Poisson bracket expressions. The procedure has already been introduced in Chapter 11. To effect the change, we note that if $\{x_1, x_2, x_3, x_4, x_5, x_6\}$ denotes an ordered set of independent functions of $(\alpha_k;\beta_k;t)$, Equation (11.6.7) can be written in the modified form

$$\dot{x}_i = \dot{x}_{i0} + \sum_{j=1}^{6}\{x_i, x_j\}\frac{\partial V}{\partial x_j} \qquad (i = 1, 2,\ldots, 6) \tag{14.4.13}$$

where

$$\{x_i, x_j\} = \sum_{k=1}^{3}\left(\frac{\partial x_i}{\partial\alpha_k}\frac{\partial x_j}{\partial\beta_k} - \frac{\partial x_i}{\partial\beta_k}\frac{\partial x_j}{\partial\alpha_k}\right)$$

is the Poisson bracket for x_i and x_j and $\dot{x}_{i0} = \partial x_i/\partial t$ is the time derivative of x_i in the absence of an external perturbing potential.

The axis about which a rotation R is executed will be referred to as the axis associated with the angle variable R. For example, Oz^* in Figure 14.2.1 will be called the axis associated with ψ. If x_j is an angle variable and $\mathbf{e}_{x_j}$ is a unit vector directed positively (in the sense of a right-hand screw) along the axis associated with x_j, it can be shown that

$$\frac{\partial V}{\partial x_j} = -N_{x_j} \qquad (N_{x_j} = \mathbf{N} \cdot \mathbf{e}_{x_j}) \tag{14.4.14}$$

where $\mathbf{N}$ is the external perturbing torque. For a suitable choice of independent variables, as we shall see, Equation (14.4.14) will permit us to express the right-hand sides of (14.4.13) in terms of the components of the external torque. Equations (14.4.13) are thus quite analogous in form to the Lagrange planetary equations given in (7.8.1) and (7.8.2).

If a small rigid body of unit mass is under the influence of an external (gravity-gradient) torque produced by an attracting point mass E, an expression for the associated gravitational potential V may be deduced readily from Equation (12.6.6) and the discussion leading to it. Consider the mass E to be at the point P in Figure 12.4.1. Let O be the center of mass of the rigid body and let $Ox'y'z'$ constitute a set of principal axes for the rigid body at O (see Figure 14.4.1). Clearly the potential V, which the rigid body possesses because of the presence of the attracting mass at P, is, through the second harmonic, given by

$$V = -\frac{GE}{r} - \frac{GE}{2r^3}[A + B + C - 3(A\lambda^2 + B\mu^2 + C\nu^2)] \tag{14.4.15}$$

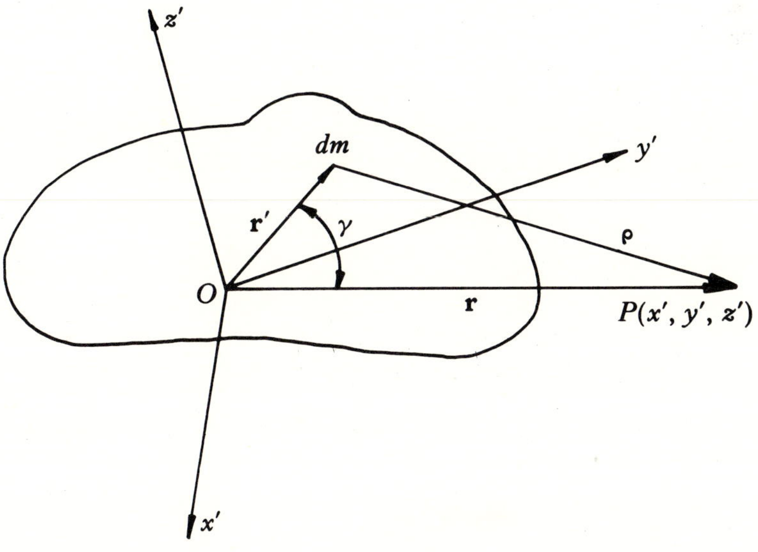

FIGURE 14.4.1

where λ, μ, and ν are direction cosines of the line segment from O to P, with respect to the $Ox'y'z'$–reference frame, and the sign of V has been chosen so that the force field per unit mass is $-\nabla V$.

Designate the direction angles of the line segment OP with respect to the $Ox'y'z'$-frame by α, β, and γ. Since $\cos^2\alpha + \cos^2\beta + \cos^2\gamma = 1$, Equation (14.4.15) can be written as

$$V = -\frac{GE}{r} - \frac{(A - 2B + C)\,GE}{r^3} + U \tag{14.4.16}$$

where

$$U = \frac{3}{2}\frac{GE}{r^3}\left[(C - B)\cos^2\gamma + (A - B)\cos^2\alpha\right]$$

The first term on the right-hand side of (14.4.16) is the familiar gravitational potential for two point masses. Since only that portion of V which produces rotational motion is of interest here, this term may be omitted. The second term is independent of orientation and will contribute nothing to the partial derivatives of V appearing in (14.4.12) or (14.4.13). Thus we can replace V by U in studying the perturbations of the motion of a free rigid body due to the gravity-gradient torque.

The torque $\mathbf{N}$, which the attracting mass E exerts on the rigid body, may be found from a development similiar to that in Chapter 12, Section 12.4. Again referring to Figure 14.4.1, we write for the increment of torque $d\mathbf{N}$, the relation

$$d\mathbf{N} = \frac{GE}{\rho^3}(\mathbf{r}' \times \boldsymbol{\rho})\,dm \qquad (\rho = |\,\boldsymbol{\rho}\,|)$$

Integrating over the mass elements of the rigid body, we have

$$\mathbf{N} = GE \iiint_M \frac{\mathbf{r}' \times \boldsymbol{\rho}}{\rho^3}\,dm$$

Using (12.4.11), we can express ρ^{-3} in powers of r'/r. If the development is carried out and terms of order higher than $(r'/r)^2$ are dropped, it can be shown that

$$\mathbf{N} = \frac{3GE}{r^3}\,\mathbf{e}_r \times (I'\mathbf{e}_r) \tag{14.4.17}$$

where I' is given in (14.3.6) and $\mathbf{e}_r = \mathbf{r}/r$.

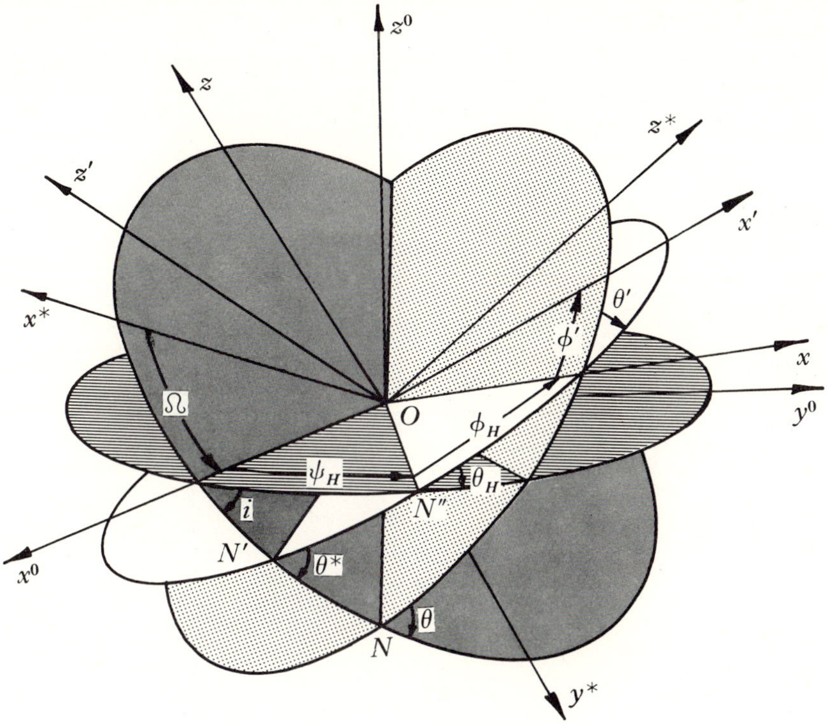

FIGURE 14.5.1

From J. W. Crenshaw and P. M. Fitzpatrick, Gravity effects on the rotational motion of a uniaxial artificial satellite *AIAA J.* **6** (11), p. 2141 (1968).

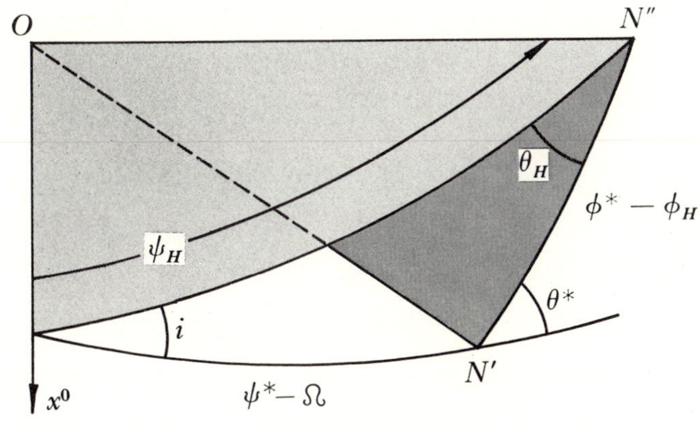

FIGURE 14.5.2a

Parts (a)–(c) are reprinted from J. W. Crenshaw and P. M. Fitzpatrick, Gravity effects on the rotational motion of a uniaxial artificial satellite *AIAA J.* **6** (11), p. 2142 (1968).

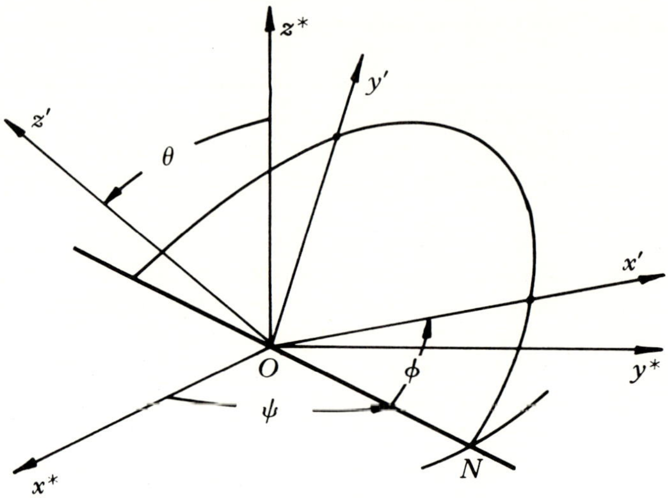

FIGURE 14.5.2b

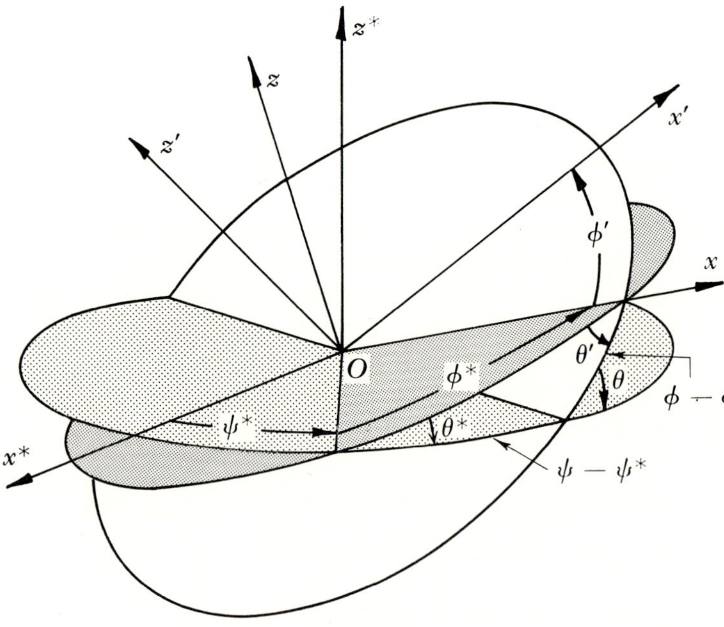

FIGURE 14.5.2c

14.5. COORDINATE SYSTEMS

In discussing the rotational motion of a rigid body in orbit about an attracting point mass E, four rectangular coordinate systems and five sets of Euler angles relating them are of particular interest. These four coordinate systems and the associated angles are depicted in Figures 14.5.1 and 14.5.2.

The $Ox^*y^*z^*$-rectangular system of axes has its origin at the center of mass of a triaxial rigid body and its orientation fixed in space. The $Ox'y'z'$-rectangular system is fixed in the body and constitutes a set of principal axes for the body at O. A typical set of Euler angles ψ, θ, ϕ is also depicted. The $Ox'y'z'$-system is obtained from the $Ox^*y^*z^*$-system by three successive rotations through the angles ψ, θ, ϕ in the senses shown.

Two additional rectangular reference frames, with origins at O, are defined with respect to Figures 14.5.1 and 14.5.2. The fundamental plane of the $Ox^0y^0z^0$-system is the orbital plane of the attracting point mass about the rigid body, the z^0-axis being taken positive in the sense of the angular momentum of the attracting point mass. The $Ox^0y^0z^0$-system, which is obtained from the $Ox^*y^*z^*$-system by two successive rotations through the angles Ω and i, has its positive x^0-axis directed toward the ascending node associated with the fundamental planes of the two systems. The $Oxyz$-system is related to the rotational angular momentum of the rigid body, the positive z-axis being taken in the sense of its angular momentum. The $Oxyz$-system, which is obtained from the $Ox^0y^0z^0$-system by three successive rotations through the angles ψ_H, θ_H, and ϕ_H, has its positive x-axis directed along the line of nodes associated with the xy- and $x'y'$-planes, in the sense shown.

The notation $\mathbf{i}$, $\mathbf{j}$, $\mathbf{k}$, with appropriate superscripts as required, will be used throughout to designate unit basis vectors along the corresponding superscripted axes x, y, and z respectively. The notation $\mathbf{e}_r$ will designate a unit vector from O to E making an angle l with the x^0-axis (see Figure 14.9.1).

14.6. THE UNPERTURBED HAMILTONIAN FUNCTION EXPRESSED IN A SIMPLIFIED FORM

We consider the problem of describing the free motion of a triaxial rigid body (of unit mass) as viewed in the $Ox^*y^*z^*$-system. Since no external torques are present, the angular momentum $\mathbf{h}$ is a constant of the motion. We assume only that the direction of $\mathbf{h}$ is fixed in space so that the

$Oxyz$-system will serve as a space-fixed reference system. Although our eventual goal is to give a description of the motion in the $Ox^*y^*z^*$-system, we first give a description of the motion in the $Oxyz$-system. To do this, we will use the generalized coordinates (ϕ^*, θ', ϕ') and their associated conjugate momenta $(p_{\phi^*}, p_{\theta'}, p_{\phi'})$. Note that the $Ox'y'z'$-system is obtained from the $Oxyz$-system by three successive rotations through the Euler angles ϕ^*, θ', ϕ'. The unperturbed Hamiltonian $H_0 = H - V$ and the associated canonical equations of motion are given in terms of these variables by (14.4.10) and (14.4.11), if we replace ψ, θ, ϕ, p_ψ, p_θ, p_ϕ by ϕ^*, θ', ϕ', p_{ϕ^*}, $p_{\theta'}$, $p_{\phi'}$, respectively.

In the $Oxyz$-system, the angular momentum $\mathbf{h}$ can be represented by a column vector as

$$\mathbf{h} = \begin{pmatrix} 0 \\ 0 \\ h \end{pmatrix} \qquad (h = |\,\mathbf{h}\,|) \tag{14.6.1}$$

It follows directly from (14.4.9) that, for Euler angles ϕ^*, θ', ϕ',

$$p_{\phi^*} = h \tag{14.6.2a}$$

$$p_{\theta'} = 0 \tag{14.6.2b}$$

$$p_{\phi'} = h \cos \theta' \tag{14.6.2c}$$

If Equations (14.6.2) are introduced into the Hamiltonian H_0 and the associated canonical equations mentioned above, we find that H_0 is given by

$$H_0 = \left(\frac{\sin^2 \phi'}{2A} + \frac{\cos^2 \phi'}{2B}\right)(p_{\phi^*}^2 - p_{\phi'}^2) + \frac{1}{2C} p_{\phi'}^2 \tag{14.6.3}$$

The associated canonical equations are

$$\begin{aligned}
\dot{\phi}^* &= \left(\frac{\sin^2 \phi'}{A} + \frac{\cos^2 \phi'}{B}\right) h \\
\dot{\theta}' &= \left(\frac{1}{A} - \frac{1}{B}\right)(\sin \phi' \cos \phi' \sin \theta')\, h \\
\dot{\phi}' &= -\left(\frac{\sin^2 \phi'}{A} + \frac{\cos^2 \phi'}{B}\right) h \cos \theta' + \frac{p_{\phi'}}{C} \\
\dot{p}_{\phi^*} &= 0, \qquad \dot{p}_{\theta'} = 0 \\
\dot{p}_{\phi'} &= \left(\frac{1}{B} - \frac{1}{A}\right)(\sin \phi' \cos \phi' \sin^2 \theta')\, h^2 = -h(\sin \theta')\, \dot{\theta}'
\end{aligned} \tag{14.6.4}$$

Since H_0, as given in (14.6.3), does not contain the time explicitly, it may be set equal to a constant of the motion, say α_1.

In view of the two constraints imposed by (14.6.2b) and (14.6.2c), only four of the six variables ϕ^*, θ', ϕ', p_{ϕ^*}, $p_{\theta'}$, $p_{\phi'}$ are independent. Thus a knowledge of the time dependence of ϕ^*, ϕ', p_{ϕ^*}, $p_{\phi'}$ coupled with the equations $p_{\theta'} = 0$, $h \cos \theta' = p_{\phi'}$, gives a complete description of the motion of the triaxial body in the $Oxyz$-system.

We now ask: How does a description of the motion in the $Oxyz$-system help us to describe the motion in the $Ox^*y^*z^*$-system, that is, in the terms of the variables $\psi, \theta, \phi, p_\psi, p_\theta, p_\phi$? To answer this question, we first note that, through the use of some elementary spherical trigonometry, we can relate the variables ψ, θ, ϕ to the variables $\phi^*, \theta', \phi', \theta^*$, and ψ^*. Through the use of (14.4.8), (14.4.9), and (14.2.1), we can relate the variables p_ψ, p_θ, p_ϕ to the variables $p_{\phi'}, p_{\phi^*}, \theta^*, \psi^*, \phi^*$. Explicitly, we can write these relationships as follows:

$$\begin{aligned} \cos \theta &= \cos \theta' \cos \theta^* - \sin \theta' \sin \theta^* \cos \phi^* \\ \sin \theta &= (1 - \cos^2 \theta)^{1/2} \end{aligned} \tag{14.6.5a}$$

$$\begin{aligned} \sin \theta^* \sin \theta \cos(\psi - \psi^*) &= \cos \theta' - \cos \theta^* \cos \theta \\ \sin \theta \sin(\psi - \psi^*) &= \sin \phi^* \sin \theta' \end{aligned} \tag{14.6.5b}$$

$$\begin{aligned} \sin \theta' \sin \theta \cos(\phi - \phi') &= \cos \theta^* - \cos \theta' \cos \theta \\ \sin \theta \sin(\phi - \phi') &= \sin \phi^* \sin \theta^* \end{aligned} \tag{14.6.5c}$$

$$p_\theta = -p_{\phi^*} \sin \theta' \sin(\phi - \phi') \tag{14.6.5d}$$

$$p_\psi = h \cos \theta^* \tag{14.6.5e}$$

$$p_\phi = p_{\phi'} \tag{14.6.5f}$$

Since θ^* and ψ^* are prescribed constants, independent of each other and independent of ϕ^*, ϕ', p_{ϕ^*}, $p_{\phi'}$, the independent quantities ϕ^*, ϕ', θ^*, ψ^*, p_{ϕ^*}, $p_{\phi'}$ serve to describe the motion of the triaxial body in the $Ox^*y^*z^*$-system.

If we wish to study perturbations of the force-free motion of the triaxial rigid body, using the canonical perturbation equations of Hamilton–Jacobi theory, it will be necessary to replace θ^* by an equivalent parameter p_{ψ^*}, the momentum conjugate to ψ^*. That either θ^* or p_{ψ^*} give an equivalent description of the motion follows from Equation (14.6.5e). It is clear from this equation that the momentum conjugate to any angle lying in the x^*y^*-plane is independent of the angle ψ and depends only on h and θ^*. Thus we see that

$$p_\psi = p_{\psi^*} = h \cos \theta^* = p_{\phi^*} \cos \theta^*$$

and the six independent quantities ϕ^*, ϕ', ψ^*, p_{ϕ^*}, $p_{\phi'}$, p_{ψ^*} will describe the motion of the triaxial rigid body with respect to the $Ox^*y^*z^*$-system. The unperturbed Hamiltonian function from which ϕ^*, ϕ', p_{ϕ^*} and $p_{\phi'}$ are to be obtained is of course, still given by (14.6.3). Furthermore, we can consider H_0, as given by (14.6.3), to be the Hamiltonian function of an extended system of variables ϕ^*, ϕ', ψ^*, p_{ϕ^*}, $p_{\phi'}$, p_{ψ^*}, which satisfy the canonical equations of motion

$$\dot{\phi}^* = \frac{\partial H_0}{\partial p_{\phi^*}} \tag{14.6.6a}$$

$$\dot{\phi}' = \frac{\partial H_0}{\partial p_{\phi'}} \tag{14.6.6b}$$

$$\dot{\psi}^* = \frac{\partial H_0}{\partial p_{\psi^*}} \tag{14.6.6c}$$

$$-\dot{p}_{\phi^*} = \frac{\partial H_0}{\partial \phi^*} \tag{14.6.6d}$$

$$-\dot{p}_{\phi'} = \frac{\partial H_0}{\partial \phi'} \tag{14.6.6e}$$

$$-\dot{p}_{\psi^*} = \frac{\partial H_0}{\partial \psi^*} \tag{14.6.6f}$$

subject to the two constraints

$$\psi^* = \text{constant} \tag{14.6.7a}$$

$$p_{\psi^*} = h \cos \theta^* = \text{constant} \tag{14.6.7b}$$

This follows from the fact that the two differential equations, (14.6.6c) and (14.6.6f), which have been added to the system, are entirely consistent with the equations of constraint (14.6.7).

14.7. THE UNPERTURBED PROBLEM

The unperturbed motion is taken to correspond to the case of a free triaxial body for which the perturbing potential U, due to gravity-gradient torque, is zero. In terms of the variables ϕ^*, ϕ', ψ^*, p_{ϕ^*}, $p_{\phi'}$, p_{ψ^*}, the unperturbed Hamiltonian H_0 has the form given in (14.6.3). The Hamilton–Jacobi equation associated with (14.6.3) is

$$\alpha_1 + \frac{\partial W_0}{\partial t} = 0 \qquad (H_0 = \alpha_1) \tag{14.7.1}$$

where if $W_0(\phi^*, \phi', \psi^*, \alpha_1, \alpha_2, \alpha_3, t)$ is chosen to satisfy (14.7.1), it will serve as the generator of a canonical transformation from the old variables $\phi^*, \phi', \psi^*, p_{\phi^*}, p_{\phi'}, p_{\psi^*}$ to a new set of independent variables $\alpha_1, \alpha_2, \alpha_3, \beta_1, \beta_2, \beta_3$, which are constants for the unperturbed motion.

The transformation equations are

$$
\begin{aligned}
p_{\phi^*} &= \frac{\partial W_0}{\partial \phi^*}, \qquad & \beta_1 &= -\frac{\partial W_0}{\partial \alpha_1} \\
p_{\phi'} &= \frac{\partial W_0}{\partial \phi'}, \qquad & \beta_2 &= -\frac{\partial W_0}{\partial \alpha_2} \\
p_{\psi^*} &= \frac{\partial W_0}{\partial \psi^*}, \qquad & \beta_3 &= -\frac{\partial W_0}{\partial \alpha_3}
\end{aligned}
\tag{14.7.2}
$$

Substitution of (14.7.2) into (14.7.1) gives the Hamilton–Jacobi equation for determining W_0. We find that

$$
\frac{1}{2}\left(\frac{\sin^2\phi'}{A} + \frac{\cos^2\phi'}{B}\right)\left[\left(\frac{\partial W_0}{\partial \phi^*}\right)^2 - \left(\frac{\partial W_0}{\partial \phi'}\right)^2\right] + \frac{1}{2C}\left(\frac{\partial W_0}{\partial \phi'}\right)^2 + \frac{\partial W_0}{\partial t} = 0
\tag{14.7.3}
$$

Using separation of variables, we obtain the complete integral

$$
W_0 = -\alpha_1 t + \alpha_2\phi^* + \alpha_3\psi^* + W_1(\phi')
\tag{14.7.4}
$$

where

$$
p_{\phi^*} = \alpha_2 = h, \qquad p_{\psi^*} = \alpha_3
$$

That $\alpha_2 = h$ follows from (14.6.2a). The symbol W_1 has the meaning

$$
W_1(\phi') = \int_{\phi_0'}^{\phi'} p_{\phi'}\, d\phi'
\tag{14.7.5}
$$

where ϕ_0' is understood to be an absolute constant, not yet prescribed, and

$$
p_{\phi'} = \pm\left[C\left(\frac{a' + b'\sin^2\phi'}{c' + d'\sin^2\phi'}\right)\right]^{1/2}
$$

$$
a' = A(2B\alpha_1 - \alpha_2^2), \qquad b' = \alpha_2^2(A - B)
$$

$$
c' = A(B - C), \qquad d' = C(A - B)
$$

The explicit form of the canonical transformation follows directly from (14.7.2) and (14.7.4). The equations of transformation are

$$\begin{aligned} \phi^* + \beta_2 &= M(\phi') \\ t - \beta_1 &= L(\phi') \\ \beta_3 &= -\psi^* \\ p_{\phi^*} &= \alpha_2 = h \\ p_{\phi'} &= \left[C\left(\frac{a' + b' \sin^2\phi'}{c' + d' \sin^2\phi'}\right)\right]^{1/2} \\ p_{\psi^*} &= \alpha_3 \end{aligned} \tag{14.7.6}$$

The symbols L and M have the meanings

$$L(\phi') = AB\sqrt{C}\, I_2(\phi') \tag{14.7.7a}$$

$$M(\phi') = -\sqrt{C}\, \alpha_2 I_3(\phi') \tag{14.7.7b}$$

where

$$I_2(\phi') = \int_{\phi_0'}^{\phi'} \frac{d\phi'}{[(a' + b' \sin^2\phi')(c' + d' \sin^2\phi')]^{1/2}} \tag{14.7.8}$$

$$I_3(\phi') = \int_{\phi_0'}^{\phi'} \frac{[(A - B)\sin^2\phi' - A]\, d\phi'}{[(a' + b' \sin^2\phi')(c' + d' \sin^2\phi')]^{1/2}} \tag{14.7.9}$$

The right-hand members of the two equations in (14.7.7) should be preceded by $\pm$ and $\mp$ symbols respectively. The choice of sign in these equations is determined by the choice of sign for $p_{\phi'}$. Since $p_{\phi'} = h\cos\theta'$, $p_{\phi'}$ is to be taken as positive or negative depending on whether $\cos\theta'$ is positive or negative. In writing (14.7.6) and (14.7.7), we assume that $0 < \theta' < \pi/2$. We also assume that $A > B > C$; hence the quantities b', c', and d' are all positive. The quantity a' may be positive, negative, or zero. We assume that $a' \geqslant 0$.

The canonical constants can be identified. The constant α_1 has already been identified with H_0, which is simply the kinetic energy of the body. Similarly, α_2 has been identified with h, the magnitude of the angular momentum $\mathbf{h}$. Further, α_3, from its identification with p_{ψ^*}, is the magnitude of the projection of $\mathbf{h}$ onto the z^*-axis. The constant β_1 is an epoch time corresponding to a value of $\phi' = \phi_0'$. The constant β_2 is a reference value of $-\phi^*$ when $\phi' = \phi_0'$ and β_3 is equal to $-\psi^*$.

The solution given in (14.7.6) and (14.7.7) can be inverted so as to express the variables ϕ^*, ϕ', ψ^*, p_{ϕ^*}, $p_{\phi'}$, p_{ψ^*} in terms of the canonical constants α_1, α_2, α_3, β_1, β_2, β_3 and the time. In carrying out the inversion, it is convenient to refer to numbered formulas given by Byrd and Friedman [45]. Such formulas will be indicated by prefixing the number designation by the symbols BF.

At this point, it is convenient to introduce certain parameters that we will use later. We define

$$n_1^2 = \frac{b'}{a' + b'} \tag{14.7.10a}$$

$$n_2^2 = \frac{d'}{c' + d'} \tag{14.7.10b}$$

It can readily be shown that

$$\frac{a'}{b'} < \frac{c'}{d'} \tag{14.7.11}$$

Hence it follows immediately that

$$1 \geqslant n_1^2 \geqslant n_2^2 > 0 \tag{14.7.12}$$

We also define parameters ξ, k, g, and k' by the equations

$$\xi = \frac{1}{[(a' + b')(c' + d')]^{1/2}} \tag{14.7.13}$$

$$k = \left(\frac{n_1^2 - n_2^2}{1 - n_2^2}\right)^{1/2} \tag{14.7.14}$$

$$g = \frac{1}{(1 - n_2^2)^{1/2}} \tag{14.7.15}$$

$$k' = (1 - k^2)^{1/2} \tag{14.7.16}$$

From (14.7.12), we see that $1 - n_2^2 > n_1^2 - n_2^2$. Thus it follows that $0 < k < 1$ and k' is real.

To express $I_2(\phi')$ in a more convenient form, we introduce a new variable χ by the equation

$$\chi = \phi' + \frac{\pi}{2} \tag{14.7.17}$$

We now write

$$I_2(\phi') = \xi \int_{\chi_0}^{\chi} r(\chi)\, d\chi \tag{14.7.18}$$

where

$$\begin{aligned} \chi_0 &= \phi_0' + \frac{\pi}{2} \\ r(\chi) &= [(1 - n_2^2 \sin^2 \chi)(1 - n_1^2 \sin^2 \chi)]^{-1/2} \end{aligned} \tag{14.7.19}$$

Since the lower limit of integration of (14.7.8) may be taken to be an absolute constant, we choose $\phi_0' = -\pi/2$ so that $\chi_0 = 0$.

Consider the integral in (14.7.18) with $\chi_0 = 0$. Using (BF-284.00), we write

$$\int_0^\chi r(\chi)\, d\chi = gu = \frac{1}{\xi AB\sqrt{C}}(t - \beta_1) \tag{14.7.20}$$

where

$$\sin(\operatorname{am} u) = \operatorname{sn} u = \frac{(1 - n_2^2)^{1/2} \sin \chi}{(1 - n_2^2 \sin^2 \chi)^{1/2}} \tag{14.7.21}$$

From (14.7.20) we have

$$u = \lambda t + \epsilon \tag{14.7.22}$$

where

$$\lambda = \frac{1}{g\xi AB\sqrt{C}} \tag{14.7.23}$$

$$\epsilon = -\lambda\beta_1 \tag{14.7.24}$$

We can solve (14.7.21) to express $\sin \chi$ in terms of u. We find that, since the sign of $\sin \chi$ depends only on $\operatorname{sn} u$,

$$\sin \chi = \frac{\operatorname{sn} u}{(1 - n_2^2 \operatorname{cn}^2 u)^{1/2}} \tag{14.7.25}$$

Similarly, we find that

$$\cos \chi = \frac{(1 - n_2^2)^{1/2} \operatorname{cn} u}{(1 - n_2^2 \operatorname{cn}^2 u)^{1/2}} \tag{14.7.26}$$

Since

$$\chi = \phi' + \frac{\pi}{2}, \qquad \cos \chi = -\sin \phi', \qquad \sin \chi = \cos \phi'$$

we have

$$\sin \phi' = \frac{(1 - n_2^2)^{1/2} \operatorname{cn} u}{(1 - n_2^2 \operatorname{cn}^2 u)} \tag{14.7.27a}$$

$$\cos \phi' = \frac{\operatorname{sn} u}{(1 - n_2^2 \operatorname{cn}^2 u)^{1/2}} \tag{14.7.27b}$$

$$\phi' = \tan^{-1}\left(-\frac{\operatorname{cn} u}{g \operatorname{sn} u}\right) \tag{14.7.27c}$$

The quadrant of ϕ' is uniquely determined by studying the signs of $\operatorname{cn} u$ and $\operatorname{sn} u$. If, for example, $\operatorname{cn} u > 0$, $\operatorname{sn} u > 0$, then, from (14.7.27a) and (14.7.27b), we see that $\sin \phi' < 0$, $\cos \phi' > 0$, and ϕ' lies in the fourth quadrant. A similar analysis holds for the other possibilities.

Equation (14.7.27c) is not in a convenient form for calculation, for powers of t appear in the expansions for cn u and sn u. This difficulty can be avoided by introducing theta functions. If we use (BF-1050.01), (BF-1052.01), (BF-1052.02), and (BF-901.01), we can write

$$\tan \phi' = -\frac{\sqrt{k'}}{g} \frac{(\cos v + q^2 \cos 3v + q^6 \cos 5v + \cdots)}{(\sin v - q^2 \sin 3v + q^6 \sin 5v - \cdots)} \tag{14.7.28}$$

where

$$v = \frac{\pi}{2K} u \tag{14.7.29a}$$

$$K = \frac{\pi}{2}\left(1 + \frac{1}{4} \sum_{m=1}^{\infty} \frac{q^m}{1 + q^{2m}}\right) \tag{14.7.29b}$$

$$q = \frac{1}{2} k_1 \left[1 + 2\left(\frac{k_1}{2}\right)^4 + 15\left(\frac{k_1}{2}\right)^8 + 150\left(\frac{k_1}{2}\right)^{12} + 1707\left(\frac{k_1}{2}\right)^{16} + \cdots\right] \tag{14.7.29c}$$

$$k_1 = \frac{1 - \sqrt{k'}}{1 + \sqrt{k'}} \qquad (k_1^2 < k^2 < 1) \tag{14.7.29d}$$

The series for computing q is rapidly convergent.

If (14.7.27a) is substituted into the right-hand side of the equation for $p_{\phi'}$ as it appears in (14.7.6) and the definitions of n_1^2 and n_2^2 are used, we obtain the equation

$$p_{\phi'} = \left[\frac{C}{c'}(a' + b')(1 - n_2^2)(k'^2 + k^2 \operatorname{cn}^2 u)\right]^{1/2}$$

Now, from (BF-121.00), we have

$$\operatorname{dn}^2 u = k'^2 + k^2 \operatorname{cn}^2 u$$

Hence we can express $p_{\phi'}$ in terms of canonical constants and the time through the equation

$$p_{\phi'} = \left[\frac{C}{c'}(a' + b')(1 - n_2^2)\right]^{1/2} \operatorname{dn} u = \alpha_2^* \operatorname{dn} u \tag{14.7.30}$$

where we have set

$$\alpha_2^* = \left[\frac{C(2A\alpha_1 - h^2)}{A - C}\right]^{1/2} \tag{14.7.31}$$

Since $p_{\phi'} = h \cos \theta'$ from (14.6.2c), it follows that

$$\cos \theta' = \frac{\alpha_2^*}{h} \operatorname{dn} u \tag{14.7.32}$$

To cast (14.7.30) and (14.7.32) into more convenient forms for calculation, we can use (BF-1052.01) and (BF-1050.01) and express dn u in the form

$$\mathrm{dn}\, u = \frac{\sqrt{k'}\,[1 + 2\sum_1^\infty q^{m^2}\cos(2mv)]}{1 + 2\sum_1^\infty (-1)^m q^{m^2}\cos(2mv)} \tag{14.7.33}$$

It remains to express ϕ^* in terms of the canonical constants and the time. The first equation of (14.7.6) can be rewritten as

$$\phi^* + \beta_2 = -h\sqrt{C}\,\xi \int_0^\chi [(A - B)\cos^2\chi - A]\, r(\chi)\, d\chi \tag{14.7.34}$$

Using (14.7.20), we obtain

$$\phi^* + \beta_2 = \frac{h}{B}(t - \beta_1) - h\sqrt{C}\,\xi(A - B)\int_0^\chi \cos^2\chi\, r(\chi)\, d\chi \tag{14.7.35}$$

It follows from (BF-284.08) and (BF-432.03) that

$$\phi^* + \beta_2 = \frac{h}{B}(t - \beta_1) + \frac{\pi}{2K}[u\Lambda_0(\beta, k) - \Omega_5] \tag{14.7.36a}$$

where

$$\Lambda_0(\beta, k) = \frac{2}{\pi}[EF(\beta, k') + KE(\beta, k') - KF(\beta, k')]$$

$$F(\beta, k') = \int_0^\beta \frac{d\theta}{[1 - (k')^2\sin^2\theta]^{1/2}}, \qquad E(\beta, k') = \int_0^\beta [1 - (k')^2\sin^2\theta]^{1/2}\, d\theta$$

$$E = E\left(\frac{\pi}{2}, k\right), \qquad K = F\left(\frac{\pi}{2}, k\right)$$

$$\beta = \sin^{-1}(1 - \gamma^2)^{-1/2}, \qquad \gamma^2 = \frac{-n_2^{\,2}}{1 - n_2^{\,2}}$$

$$\Omega_5 = u + \tan^{-1}\left[\frac{2\sum_1^\infty (-1)^{m+1} q^{m^2}\sin(2mv)\sinh[2m(p - w)]}{1 + 2\sum_1^\infty (-1)^m q^{m^2}\cos(2mv)\cosh[2m(p - w)]}\right]$$

$$p = \frac{\pi K'}{2K}, \qquad K' = F\left(\frac{\pi}{2}, k'\right), \qquad w = \frac{\pi F(\beta, k')}{2K}$$

if $k < -\gamma^2 < \infty$. If $0 < -\gamma^2 < k$, we use (BF-431.03) and write

$$\phi^* + \beta_2 = \frac{h}{B}(t - \beta_1)\left[1 - \frac{(A - B)\,k'^2}{A(\gamma^2 - k^2)}\right] - \frac{\pi}{2K}[u\Lambda_0(\psi, k) + \Omega_1] \tag{14.7.36b}$$

where

$$\psi = \sin^{-1} [\gamma^2(\gamma^2 - k^2)^{-1/2}]$$

$$\Omega_1 = \frac{2K}{\pi} \tan^{-1} \left[\frac{2 \sum_1^\infty (-1)^{m+1} q^{m^2} \sin(2mv) \sinh(2mw)}{1 + 2 \sum_1^\infty (-1)^m q^{m^2} \cos(2mv) \cosh(2mw)}\right]$$

This completes the inversion of the transformation equations (14.7.6).

For convenience, the equations representing the solution of the free motions of a triaxial rigid body are recapitulated. The six formulas expressing ϕ^*, ϕ', ψ^*, p_{ϕ^*}, $p_{\phi'}$, p_{ψ^*} in terms of the canonical constants α_1, h, α_3, β_1, β_2, β_3 and the time are

$$\phi^* = -\beta_2 + \frac{h}{B}(t - \beta_1) - \frac{\pi}{2K}[\Omega_5 - u\Lambda_0(\beta, k)] \qquad (k < -\gamma^2 < \infty)$$

$$\phi^* = -\beta_2 + \frac{h}{B}(t - \beta_1) \left[1 - \frac{(A - B) k'^2}{A(\gamma^2 - k^2)}\right]$$

$$- \frac{\pi}{2K}[u\Lambda_0(\psi, k) + \Omega_1] \qquad (0 < -\gamma^2 < k)$$

$$\phi' = \tan^{-1}\left(-\frac{\operatorname{cn} u}{g \operatorname{sn} u}\right), \qquad \psi^* = -\beta_3 \tag{14.7.37}$$

$$p_{\phi^*} = \alpha_2 = h, \qquad p_{\phi'} = \alpha_2^* \operatorname{dn} u, \qquad p_{\psi^*} = \alpha_3$$

For the special case of a uniaxial body, let C be the moment of inertia referred to the symmetry axis of the body and take $B = A$. We then note that Equations (14.7.37) reduce to

$$\phi^* = -\beta_2 + \frac{h}{A}(t - \beta_1)$$

$$\phi' = \phi_0' + \frac{A - C}{AC}\alpha_2^*(t - \beta_1)$$

$$\psi^* = -\beta_3 \tag{14.7.38}$$

$$p_{\phi^*} = h, \qquad p_{\phi'} = \alpha_2^*, \qquad p_{\psi^*} = \alpha_3$$

Alternatively, by setting $B = A$ in (14.7.4), we obtain

$$W_0 = -\alpha_1 t + h\phi^* + \alpha_3\psi^* + \alpha_2^*(\phi' - \phi_0') \tag{14.7.39}$$

Only three of the four constants α_1, h, α_3, and α_2^* appearing in (14.7.39) are independent, for α_2^* is related to h and α_1 through (14.7.31). If we consider α_2^* to be a function of h and α_1 in forming (14.7.2), we obtain (14.7.38) after inversion. On the other hand, if we consider h to be a function of α_1 and α_2^* in forming (14.7.2), we obtain, after inversion, the equations

$$
\begin{aligned}
\phi^* &= \frac{h}{A}(t - \beta_1^*) \\
\phi' &= \phi_0' - \beta_2^* - \frac{\alpha_2^*}{A}\left(\frac{C - A}{C}\right)(t - \beta_1^*) \\
\psi^* &= -\beta_3 \\
p_{\phi^*} &= h, \qquad p_{\phi'} = \alpha_2^*, \qquad p_{\psi^*} = \alpha_3
\end{aligned}
\tag{14.7.40}
$$

where

$$h = \left[2A\alpha_1 + \left(\frac{C - A}{C}\right)\alpha_2^{*2}\right]^{1/2} \tag{14.7.41}$$

$$\beta_1^* = \frac{\alpha_2^*}{h}\left(\frac{C - A}{C}\right)\beta_2 \tag{14.7.42}$$

$$\beta_2^* = \beta_1 + \frac{A}{h}\beta_2 \tag{14.7.43}$$

Equations (14.7.38) and (14.7.40) comprise the equations of transformation from the variables $\psi, \theta, \phi, p_\psi, p_\theta, p_\phi$ to the two sets of canonical constants $\alpha_1, h, \alpha_3, \beta_1, \beta_2, \beta_3$ and $\alpha_1, \alpha_2^*, \alpha_3, \beta_1^*, \beta_2^*, \beta_3$. They are supplemented by auxiliary relations such as (14.6.2) and (14.6.7) and the trigonometric equations (14.6.5) that relate intermediate variables like ψ^*, θ^*, and ϕ^* to the original Euler angles ψ, θ, ϕ. The explicit equations of canonical transformation, which relate the original variables $\psi, \theta, \phi, p_\psi, p_\theta, p_\phi$ to the canonical constants $\alpha_1, \alpha_2^*, \alpha_3, \beta_1^*, \beta_2^*, \beta_3$, are

$$
\begin{aligned}
\psi &= -\beta_3 + \cot^{-1}\left\{\alpha_2^*\left(\frac{h^2 - \alpha_3^2}{h^2 - \alpha_2^{*2}}\right)^{1/2} \csc\left[\frac{h}{A}(t - \beta_1^*)\right]\right. \\
&\qquad \left. + \frac{\alpha_3}{h}\cot\left[\frac{h}{A}(t - \beta_1^*)\right]\right\} \\
\theta &= \cos^{-1}\left\{\frac{\alpha_2^*\alpha_3}{h_2} - \frac{[(h^2 - \alpha_2^{*2})(h^2 - \alpha_3^2)]^{1/2}}{h^2}\cos\left[\frac{h}{A}(t - \beta_1^*)\right]\right\} \\
\phi &= -\beta_2^* - \alpha_2^*\frac{(C - A)(t - \beta_1^*)}{AC} \\
&\qquad + \cot^{-1}\left\{\alpha_3\left(\frac{h^2 - \alpha_2^{*2}}{h_2 - \alpha_3^2}\right)^{1/2}\csc\left[\frac{h}{A}(t - \beta_1^*)\right]\right. \\
&\qquad \left. + \frac{\alpha_2^*}{h}\cot\left[\frac{h}{A}(t - \beta_1^*)\right]\right\} \\
p_\psi &= \alpha_3, \qquad p_\theta = -(h^2 - \alpha_3^2)^{1/2}\sin(\psi + \beta_3), \qquad p_\phi = \alpha_2^*
\end{aligned}
\tag{14.7.44}
$$

Equations (14.7.44) can be readily expressed in terms of the constants α_1, h, α_3, β_1, β_2, β_3 through the use of (14.7.31), (14.7.42), and (14.7.43).

14.8. FIRST-ORDER SOLUTION FOR GRAVITY-INDUCED PERTURBATIONS OF AN ORBITING UNIAXIAL BODY

Either (14.7.38) or (14.7.40) will serve as a basis for studying the perturbations of the free rotational motions of a uniaxial satellite in orbit about a point attracting mass. For this problem, the function U, appearing in (14.4.16), reduces to

$$U = K \cos^2 \gamma \qquad \left[K = \frac{3GE}{2r^3}(C - A)\right] \tag{14.8.1}$$

It can also be seen that the associated gravity-gradient torque (see 14.4.17) may be expressed in the form

$$\mathbf{N} = 2K(\mathbf{e}_r \cdot \mathbf{k}')(\mathbf{e}_r \times \mathbf{k}') \tag{14.8.2}$$

where $\mathbf{k}'$ is a unit vector directed along the positive z'-axis. Observe that in formulas (14.4.16) and (14.4.17) [and hence (14.8.1) and (14.8.2)], it does not matter whether $\mathbf{e}_r$ is regarded as a vector directed from the attracting point mass E to the center of mass of the rigid body or vice versa.

We can now study the gravity-induced variations of either the set of six canonical constants α_1, h, α_3, β_1, β_2, β_3 appearing in (14.7.38) or the set of six canonical constants α_1, α_2^*, α_3, β_1^*, β_2^*, β_3 appearing in (14.7.40). We have elected here to work with the latter set.

Equations (14.4.12), in which V is replaced by U as given in (14.8.1), can be used to study the perturbations. In terms of the variables α_1, α_2^*, α_3, β_1^*, β_2^*, β_3, the time t, and the angles i, Ω, and l which are assumed to be known functions of the time, the function $\cos \gamma$ may be written, after some tedious spherical trigonometry, as

$$\begin{aligned}\cos \gamma = {} & \sin i \sin l(\cos \theta' \cos \theta^* - \sin \theta' \sin \theta^* \cos \phi^*) \\ & -(\cos \theta' \sin \theta^* + \sin \theta' \cos \theta^* \cos \phi^*)[\cos i \sin l \cos(\Omega + \beta_3) \\ & + \cos l \sin(\Omega + \beta_3)] \\ & - \sin \theta' \sin \phi^* [\cos i \sin l \sin(\Omega + \beta_3) \\ & - \cos l \cos(\Omega + \beta_3)]\end{aligned} \tag{14.8.3}$$

The variational equations may be obtained by forming the appropriate partial derivatives of the potential function U using the expression for $\cos \gamma$ given in (14.8.3). The resulting expressions are rather lengthy and will not be reproduced here. Using the perturbation techniques described in Chapter 7, we can readily integrate the variational equations in closed form under the simplifying assumptions that (1) the inclination angle i is constant, (2) the node Ω regresses at a uniform rate, (3) the perturbing body moves in a circular orbit at a uniform rate, and (4) the rotational rate of the body about its center of mass is large compared to its orbital rate. This last criterion is satisfied essentially if the ratio $n/\omega \ll 1$, where $n = (GE/r^3)^{1/2}$ is the mean angular speed of the center of mass of the satellite in its orbit and ω is the angular speed of rotation of the body about its center of mass (see Beletsky [44]). The development is carried out to first order in K/h. The parameter K/h will be small if criterion (4) is met.

Secular terms, as well as long- and short-period terms, arise in the integration. The long-period terms have periods of the order of $2\pi/\dot{\Omega}$ (roughly, a 60-day period) while the short-period terms have periods of order $2\pi/\dot{l}$ and $2\pi/\dot{\phi}^*$ (roughly, periods that are less than or at most equal to the orbital period). Typically, the amplitudes of the short-period terms are several orders of magnitude less than the amplitudes of the long-period terms. Poisson terms appear in the integrated results, but only in the expressions for $\beta_1{}^*$ and $\beta_2{}^*$. These Poisson terms oscillate rapidly with "frequencies" that are comparable to the frequencies of the short-period terms. Except for the short-period and Poisson terms, the first-order solutions for the canonical variables are

$$\alpha_1 = \text{constant}, \qquad \alpha_2{}^* = \text{constant}$$

$$\alpha_3 = \frac{K}{8} \sin i\,(1 - 3\cos^2\theta')\sin\theta^*\,(4C_1\cos i\cos\theta^* + C_2\sin i\sin\theta^*) + \text{constant}$$

$$\begin{aligned}\beta_1{}^* = -\frac{KA}{8h^2}\Big[&4\left(1-\frac{3}{2}\sin^2 i\right)(6\cos^2\theta'\cos^2\theta^* - \cos^2\theta' - \cos^2\theta^*)\,t \\ &- 4S_1\sin i\cos i\cot\theta^*\,(1 - 2\cos^2\theta^* - 9\cos^2\theta' + 12\cos^2\theta'\cos^2\theta^*) \\ &- S_2\sin^2 i\,(6\cos^2\theta'\cos^2\theta^* - 3\cos^2\theta' - \cos^2\theta^*)\Big] + \text{constant}\end{aligned}$$

$$\begin{aligned}\beta_2{}^* = -\frac{K\cos\theta'}{8h}\Big[&4\left(1-\frac{3}{2}\sin^2 i\right)(1 - D\cos^2\theta' - D\cos^2\theta^* - 3\cos^2\theta^* \\ &+ 6D\cos^2\theta'\cos^2\theta^*)\,t + 4S_1\sin i\cos i\cot\theta^*(9D\cos^2\theta' - 6 - D \\ &+ 6\cos^2\theta^* + 2D\cos^2\theta^* - 12D\cos^2\theta'\cos^2\theta^*) + S_2\sin^2 i \\ &\times(-3 + 3D\cos^2\theta' + 3\cos^2\theta^* + D\cos^2\theta^* - 6D\cos^2\theta'\cos^2\theta^*)\Big] \\ &+ \text{constant}\end{aligned}$$

$$\beta_3 = -\frac{K}{8h}(1 - 3\cos^2\theta')\Big[4\cos\theta^*\Big(1 - \frac{3}{2}\sin^2 i\Big)t + 4S_1 \sin i \cos i$$

$$\times \csc\theta^*(1 - 2\cos^2\theta^*) - S_2 \sin^2 i \cos\theta^*\Big] + \text{constant} \tag{14.8.4}$$

where the long-period factors C_1, C_2, S_1, and S_2 are given by

$$C_1 = \frac{1}{\dot{\Omega}}[\cos(\Omega + \beta_3) - \cos(\Omega_0 + \beta_3)]$$

$$C_2 = \frac{1}{\dot{\Omega}}[\cos 2(\Omega + \beta_3) - \cos 2(\Omega_0 + \beta_3)]$$

$$S_1 = \frac{1}{\dot{\Omega}}[\sin(\Omega + \beta_3) - \sin(\Omega_0 + \beta_3)] \tag{14.8.5}$$

$$S_2 = \frac{1}{\dot{\Omega}}[\sin 2(\Omega + \beta_3) - \sin 2(\Omega_0 + \beta_3)]$$

$$D = \frac{C - A}{C}$$

and θ', θ^*, i, and Ω_0 are reference values of the respective quantities.

The Euler angles ψ, θ, ϕ and their conjugate momenta p_ψ, p_θ, p_ϕ for the perturbed problem are to be obtained by replacing the canonical constants appearing in the unperturbed solution by the canonical variables whose time dependence is given by (14.8.4). Even though there is a singularity in $\beta_1{}^*$, $\beta_2{}^*$, and β_3 at $\theta^* = 0$, the limits of the functions ψ, θ, ϕ, p_ψ, p_θ, and p_ϕ exist as $\theta^* \to 0$. Taking the values of these functions at $\theta^* = 0$ to be the values of the limits as $\theta^* \to 0$, we find that

$$\psi(\theta^* = 0) = \lim_{\theta^*\to 0}\psi = \Big[\frac{h}{A} + K\cos^2\theta'\Big(\frac{1 - \frac{3}{2}\sin^2 i}{h}\Big)\Big]t - \beta_{30} - \frac{h}{A}\beta_{10}^*$$

$$\theta(\theta^* = 0) = \lim_{\theta^*\to 0}\theta = \theta'$$

$$\phi(\theta^* = 0) = \lim_{\theta^*\to 0}\phi = -\Big[\alpha_2{}^*\Big(\frac{D}{A}\Big) + K\cos\theta'\Big(\frac{1 - \frac{3}{2}\sin^2 i}{h}\Big)\Big]t + \frac{\alpha_2{}^* D\beta_{10}^*}{A} - \beta_{20}^*$$

$$p_\psi(\theta^* = 0) = \lim_{\theta^*\to 0} p_\psi = h \qquad p_\theta(\theta^* = 0) = \lim_{\theta^*\to 0} p_\theta = 0 \qquad p_\phi = \alpha_2{}^* \tag{14.8.6}$$

where β_{10}^*, β_{20}^*, and β_{30} refer to initial values of these variables.

As $\dot{\Omega}$ approaches zero, the long period factors C_1, C_2, S_1, and S_2 approach functions that are proportional to the time t. Thus, for non-regressing orbits, the long-period terms containing C_1, C_2, S_1, and S_2 are replaced by secular terms.

The factor $\cos \gamma$ which appears in the potential U is free of $\beta_2{}^*$. Thus $\dot{\alpha}_2{}^* = 0$ and $\alpha_2{}^*$ is a constant of the perturbed motion. Since α_1 does not contain secular or long-period variations to first order, the magnitude h of the angular momentum vector, which is related to α_1 and $\alpha_2{}^*$ through (14.7.41), does not contain secular or long-period terms. Similarly, the coning angle $\theta' = \cos^{-1}(\alpha_2{}^*/h)$ does not contain secular or long-period terms. This means that the opening of the coning angle observed in the case of the Pegasus satellites cannot be explained on the basis of a gravitational theory. Neither can the observed decrease in the angular speed ω be explained on the basis of a gravitational theory, for the angular speed depends only on h and $\alpha_2{}^*$ through the equation

$$\omega^2 = \frac{h^2}{A^2} + (A^2 - C^2)\frac{\alpha_2^{*2}}{A^2C^2} \tag{14.8.7}$$

14.9. ALTERNATIVE FIRST-ORDER TREATMENT OF GRAVITY-INDUCED PERTURBATIONS OF AN ORBITING UNIAXIAL BODY

An alternative and more useful development of a first-order theory for the perturbed axisymmetric body can be made by introducing a new set of independent variables. We will use (14.4.13) and (14.7.40) to effect the transformation from the old canonical variables to the new variables $\psi_H, \theta_H, \phi_H, \theta', \phi', h$, where, for economy of notation, we let ϕ' stand for $\phi' - \phi_0{}'$. This new set is particularly convenient because ψ_H, θ_H, and ϕ_H relate directly to the orbital plane rather than to the xy–plane and Poisson terms do not appear in the first-order development. The angles ψ_H, θ_H, and ϕ_H are related to the variables that we have already been using through the equations

$$\cos \theta_H = \cos i \cos \theta^* + \sin i \sin \theta^* \cos(\psi^* - \Omega) \tag{14.9.1a}$$

$$\sin \theta^* \sin(\psi^* - \Omega) \cot \psi_H = -\cos \theta^* \sin i + \sin \theta^* \cos i \cos(\psi^* - \Omega) \tag{14.9.1b}$$

$$\sin i \sin(\psi^* - \Omega) \cot(\phi^* - \phi_H) = \cos i \sin \theta^* - \sin i \cos \theta^* \cos(\psi^* - \Omega) \tag{14.9.1c}$$

which are readily deducible from the spherical triangle in Figure 14.5.2a.

For the ordered set $\{\psi_H, \theta_H, \phi_H, \theta', \phi', h\}$, after introducing (14.4.14) into (14.4.13), we can write

$$\dot{x}_i = \dot{x}_{i0} - \sum_{j=1}^{5} \{x_i, x_j\} N_{x_j} \qquad (i = 1, 2, ..., 6) \tag{14.9.2}$$

where we have used the fact that $\partial V/\partial h = 0$, for any conservative perturbing torque. Actually, it may be shown more generally that Equations (14.9.2) hold even for nonconservative perturbing torques (see Exercise 10 at end of chapter).

The unperturbed rates of change [that is, the $\dot{x}_{i0}$ in (14.9.2)] can be obtained by differentiating (14.9.1) with respect to the time. The Poisson brackets associated with the set $\{\psi_H, \theta_H, \phi_H, \theta', \phi', h\}$ may be obtained in a straightforward manner from the transformation equations (14.7.40) together with (14.9.1). The nonzero brackets are found to be

$$\begin{aligned}
\{\psi_H, \theta_H\} &= -\{\theta_H, \psi_H\} = -\frac{1}{h \sin \theta_H} \\
\{\theta_H, \phi_H\} &= -\{\phi_H, \theta_H\} = -\frac{\cot \theta_H}{h} \\
\{\phi_H, \theta'\} &= -\{\theta', \phi_H\} = \frac{\cot \theta'}{h} \\
\{\phi_H, h\} &= -\{h, \phi_H\} = 1 \\
\{\theta', \phi'\} &= -\{\phi', \theta'\} = \frac{1}{h \sin \theta'}
\end{aligned} \tag{14.9.3}$$

If the orbital plane is assumed to precess at a constant rate $\dot{\Omega}$, with constant inclination i, the variational equations are

$$\begin{aligned}
\dot{\psi}_H &= -(\cos i + \sin i \cot \theta_H \cos \psi_H)\,\dot{\Omega} + \frac{1}{h \sin \theta_H} N_{\theta_H} \\
\dot{\theta}_H &= -\sin i \sin \psi_H \dot{\Omega} - \frac{1}{h \sin \theta_H} N_{\psi} + \frac{\cot \theta_H}{h} N_{\phi_H} \\
\dot{\phi}_H &= \frac{h}{A} + \frac{\sin i \cos \psi_H}{\sin \theta_H}\,\dot{\Omega} - \frac{\cot \theta_H}{h} N_{\theta_H} - \frac{\cot \theta'}{h} N_{\theta'} \\
\dot{\theta}' &= \frac{\cot \theta'}{h} N_{\phi_H} - \frac{1}{h \sin \theta'} N_{\phi'} \\
\dot{\phi}' &= -\frac{h}{A}\left(\frac{C - A}{C}\right) \cos \theta' + \frac{1}{h \sin \theta'} N_{\theta'} \\
\dot{h} &= N_{\phi_H}
\end{aligned} \tag{14.9.4}$$

The set of differential equations (14.9.4) is suitable for numerical integration by high-speed computer. Their use is completely analogous to the use of the method of variation of parameters for orbit prediction. They are general, and valid for dissipative as well as conservative torques. Finally, they serve as a convenient basis for developing analytical solutions for problems involving perturbations of a uniaxial body.

From the origin O, the attracting point mass E appears to move in the x^0y^0-plane, as shown in Figure 14.9.1. A unit vector $\mathbf{e}_r$ directed from 0 toward E, making an angle l with the x^0-axis, may be written

$$\mathbf{e}_r = \cos l\, \mathbf{i}^0 + \sin l\, \mathbf{j}^0 \tag{14.9.5}$$

The basis vectors $(\mathbf{i}^0, \mathbf{j}^0, \mathbf{k}^0)$ and $(\mathbf{i}', \mathbf{j}', \mathbf{k}')$ are readily expressed in terms of the $(\mathbf{i}, \mathbf{j}, \mathbf{k})$ basis (see Exercise 7 at end of chapter). Thus, using (14.8.2)

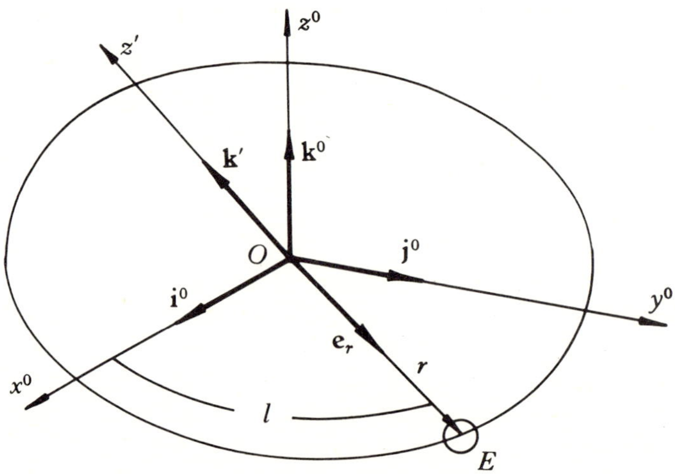

FIGURE 14.9.1

From J. W. Crenshaw and P. M. Fitzpatrick, Gravity effects on the rotational motion of a uniaxial artificial satellite *AIAA J.* **6** (11), p. 2143 (1968).

in (14.9.4), the following perturbation equations are obtained for this gravity-gradient torque:

$$\begin{aligned}
\dot{\psi}_H = &-(\cos i + \sin i \cot \theta_H \cos \psi_H)\, \dot{\Omega} \\
&- \left[\frac{2K \sin(\psi_H - l)}{h}\right] (\cot \theta_H \cos \theta' - \sin \theta' \cos \phi_H) \\
&\times [\sin(\psi_H - l)(\sin \theta_H \cos \theta' + \cos \theta_H \sin \theta' \cos \phi_H) \\
&+ \cos(\psi_H - l) \sin \theta' \sin \phi_H]
\end{aligned}$$

$$
\begin{aligned}
\dot{\theta}_H &= -\sin i \sin \psi_H \dot{\Omega} + \frac{2K}{h} [\cos(\psi_H - l) \cos \theta' \\
&\quad - \sin(\psi_H - l) \sin \theta' \sin \theta_H \sin \phi_H] \\
&\quad \times [\sin(\psi_H - l)(\sin \theta_H \cos \theta' + \cos \theta_H \sin \theta' \cos \phi_H) \\
&\quad + \cos(\psi_H - l) \sin \theta' \sin \phi_H] \\
\dot{\phi}_H &= \frac{h}{A} + \frac{\sin i \cos \psi_H}{\sin \theta_H} \dot{\Omega} \\
&\quad - \frac{2K}{h} [\sin(\psi_H - l)(\sin \theta_H \cos \theta' + \cos \theta_H \sin \theta' \cos \phi_H) \\
&\quad + \cos(\psi_H - l) \sin \theta' \sin \phi_H][-\cos(\psi_H - l) \sin \phi_H \cos \theta' \cot \theta' \\
&\quad + \sin(\psi_H - l) \cos \theta'(\sin \theta_H - \cos \theta_H \cot \theta' \cos \phi_H) \\
&\quad - \sin(\psi_H - l) \cos \theta_H(\cot \theta_H \cos \theta' - \sin \theta' \cos \phi_H)] \qquad (14.9.6) \\
\dot{\theta}' &= \frac{\cot \theta'}{h} \dot{h} \\
\dot{\phi}' &= -\frac{h}{A} \left(\frac{C - A}{C}\right) \cos \theta' - \frac{2K}{h} \\
&\quad \times [\cos(\psi_H - l) \sin \phi_H \cot \theta' + \sin(\psi_H - l) \\
&\quad \times (\sin \theta_H + \cos \theta_H \cos \phi_H \cot \theta')] \\
&\quad \times [\sin(\psi_H - l)(\sin \theta_H \sin \theta' - \cos \theta_H \cos \theta' \cos \phi_H) \\
&\quad - \cos(\psi_H - l) \cos \theta' \sin \phi_H] \\
\dot{h} &= 2K \sin \theta'[\sin(\psi_H - l) \cos \theta_H \sin \phi_H - \cos(\psi_H - l) \cos \phi_H] \\
&\quad \times [\sin(\psi_H - l)(\sin \theta_H \cos \theta' + \cos \theta_H \sin \theta' \cos \phi_H) \\
&\quad + \cos(\psi_H - l) \sin \theta' \sin \phi_H]
\end{aligned}
$$

A first approximation to the solution of (14.9.6) will be obtained for the important special case considered in Section 14.8. The same four simplifying assumptions stated there will apply here. Thus the angle $l = l_0 + nt$ will increase linearly with time from some reference value l_0. Since $\dot{\Omega}$ is typically of the order of 0.004 deg/min, we will treat both the terms containing $\dot{\Omega}$ and those containing K as perturbations.

The solutions of (14.9.6) with $\dot{\Omega}$ and K both zero will be referred to as the unperturbed solutions for the variables ψ_H, θ_H, ϕ_H, θ', ϕ', h.

These unperturbed solutions will be designated by the use of a circumflex (^). They are

$$
\begin{aligned}
&\hat{\psi}_H = \text{constant}, \qquad \hat{\theta}_H = \text{constant} \\
&\hat{\phi}_H = \hat{\phi}_H(t_1) + \frac{\hat{h}}{A}(t - t_1), \qquad \hat{\theta}' = \text{constant} \\
&\hat{\phi}' = \hat{\phi}'(t_1) - \frac{\hat{h}}{A}(t - t_1)\left(\frac{C - A}{C}\right)\cos\hat{\theta}', \qquad \hat{h} = \text{constant}
\end{aligned} \tag{14.9.7}
$$

where t_1 is an initial value of t.

When Equations (14.9.7) are substituted into (14.9.6), since the two nonconstant expressions for the unperturbed variables are linear functions of the time, only elementary integrals will appear in the right-hand members when they are interpreted in terms of the unperturbed values. Thus the first-order solutions of (14.9.6) may be obtained readily, using the techniques described in Chapter 7. The resulting expressions are found to have both secular and periodic terms, but Poisson terms do not appear. Only the secular terms are given below:

$$
\begin{aligned}
\psi_H(t) &= \psi_H(t_1) - (\cos i + \sin i \cot\hat{\theta}_H \cos\hat{\psi}_H)\,\dot{\Omega}(t - t_1) \\
&\quad + \frac{K}{2\hat{h}}(1 - 3\cos^2\hat{\theta}')\cos\hat{\theta}_H(t - t_1) \\
\theta_H(t) &= \theta_H(t_1) - \sin i \sin\hat{\psi}_H\dot{\Omega}(t - t_1) \\
\phi_H(t) &= \phi_H(t_1) + \frac{\hat{h}}{A}(t - t_1) + \sin i \cos\hat{\phi}_H \csc\hat{\theta}_H\dot{\Omega}(t - t_1) \\
&\quad - \frac{K}{2\hat{h}}[\cos^2\hat{\theta}' + \cos^2\hat{\theta}_H(1 - 6\cos^2\hat{\theta}')](t - t_1) \\
\theta'(t) &= \theta'(t_1) \\
\phi'(t) &= \phi'(t_1) - \frac{\hat{h}}{A}\left(\frac{C - A}{C}\right)\cos\hat{\theta}' \\
&\quad + \frac{K}{2\hat{h}}(1 - 3\cos^2\hat{\theta}_H)\cos\hat{\theta}'(t - t_1) \\
h(t) &= h(t_1)
\end{aligned} \tag{14.9.8}
$$

To first order, there are no secular changes in θ' or h. The variables ψ_H, θ_H, ϕ_H, θ', ϕ', h can be regarded as undergoing secular rates of change with t, which for small intervals of time (≈ 1 day) are given by (14.9.8). The results are further restricted to values of $\dot{\phi}_H$ not close to $\pm n$ or $\pm 2n$ due to the appearance of resonance conditions in the coefficients of certain of the periodic terms. The complete integrals, along with a discussion of the resonance conditions, may be found in the work of Crenshaw [46].

The secular rates of change of the variables ψ_H, θ_H, ϕ_H, θ', ϕ', h, which are directly deducible from (14.9.8), can be interpreted in terms of osculating elements by simply dropping the circumflex notation. Thus we write

$$\begin{aligned}
\langle\dot{\psi}_H\rangle &= -(\cos i + \sin i \cot \theta_H \cos \psi_H)\, \dot{\Omega} + \frac{K}{2h}(1 - 3\cos^2\theta')\cos\theta_H \\
\langle\dot{\theta}_H\rangle &= -\sin i \sin \psi_H \dot{\Omega} \\
\langle\dot{\phi}_H\rangle &= \frac{h}{A} + \frac{\sin i \cos \psi_H}{\sin \theta_H}\, \dot{\Omega} - \frac{K}{2h}[\cos^2\theta' + \cos^2\theta_H(1 - 6\cos^2\theta')] \\
\langle\dot{\theta}'\rangle &= 0 \\
\langle\dot{\phi}'\rangle &= -\frac{h}{A}\left(\frac{C-A}{C}\right)\cos\theta' + \frac{K}{2h}(1 - 3\cos^2\theta_H)\cos\theta' \\
\langle\dot{h}\rangle &= 0
\end{aligned} \tag{14.9.9}$$

The "average sign"—$\langle\ \rangle$—is used to stress the fact that these quantities have been averaged over the short periods of the motion. Equations (14.9.9) are suitable for numerical integration with a much longer integration step time than may be used with (14.9.4). Three of the equations in (14.9.9) correspond to the solution of the motion of the angular momentum obtained by Holland, and agree exactly when $B = A$. The remaining three equations, not given by Holland, give the motion of the body relative to its angular momentum. Together, Equations (14.9.9) constitute a complete first-order solution. Equations (14.9.9) have been verified by direct numerical integration, and excellent agreement with the computer results has been obtained (see Crenshaw [46]).

Equations (14.9.9) may be used to obtain an interpretation of the behavior of the angular momentum following a scheme used by Holland [37]. The components of the angular momentum **h**, expressed in the $Ox^0y^0z^0$-system are (see Figure 14.5.1)

$$h_{x^0} = h \sin\psi_H \sin\theta_H, \quad h_{y^0} = -h\cos\psi_H \sin\theta_H, \quad h_{z^0} = h\cos\theta_H \tag{14.9.10}$$

If the time derivatives of (14.9.10) are formed and $\dot{\psi}_H$, $\dot{\theta}_H$, and $\dot{h}$ are replaced with their secular values as given by (14.9.9), we find that

$$\begin{aligned}
\dot{h}_{x^0} &= ah_{y^0} - bh_{z^0} - ch_{y^0}h_{z^0} \\
\dot{h}_{y^0} &= -ah_{x^0} + ch_{z^0}h_{x^0}, \qquad \dot{h}_{z^0} = bh_{x^0}
\end{aligned} \tag{14.9.11}$$

where

$$a = \cos i\dot{\Omega}, \qquad b = \sin i\dot{\Omega}, \qquad c = \frac{K}{2h^2}(1 - 3\cos^2\theta') \tag{14.9.12}$$

A first integral of (14.9.11) is

$$h_{y^0} = \frac{c}{2b}h_{z^0}^2 - \frac{a}{b}h_{z^0} + \text{constant} \tag{14.9.13}$$

This equation represents a parabolic cylinder whose elements are parallel to the x^0-axis and which opens in the y^0-direction. The intersection of this cylinder with a sphere of radius h, centered at O, is a closed curve along which the tip of **h** appears to move as viewed in the $Ox^0y^0z^0$-frame of reference. A typical motion of the angular momentum **h**, such as has been observed for the Pegasus satellites, is sketched in Figure 14.9.2.

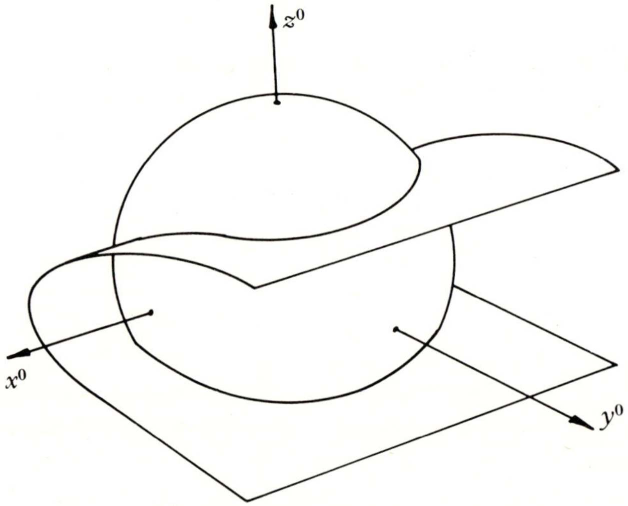

FIGURE 14.9.2

From J. W. Crenshaw and P. M. Fitzpatrick, Gravity effects on the rotational motion of a uniaxial artificial satellite *AIAA J.* **6** (11), p. 2144 (1968).

For the special case in which $K = 0$ (no gravity-gradient torque), Equation (14.9.13) becomes

$$h_{y^0} = -\frac{a}{b}h_{z^0} + \text{constant}$$

which represents a plane normal to the z^*-axis. For this special case, it is known that **h** is a constant vector in a nonrotating frame of reference.

However, the orbital plane, and consequently the $Ox^0y^0z^0$-coordinate system, precesses about the z^*-axis. For this reason the **h**-vector, when viewed from this system, appears to precess at a constant rate, $-\dot{\Omega}$, about the same axis.

For the special case in which $\dot{\Omega} = 0$ (no precession), then $a = b = 0$. A complete solution of (14.9.11) for this special case is

$$h_{x^0} = c_1 \cos(ch_{z^0}t), \qquad h_{y^0} = c_1 \sin(ch_{z^0}t), \qquad h_{z^0} = (h^2 - c_1^2)^{1/2}$$

where c_1 is a constant of integration. Thus in the absence of orbital precession the angular momentum itself precesses about the z^0-axis, with constant angular speed ch_{z^0}. This motion was predicted by Beletsky [39].†

Exercises

1. Derive Equations (14.4.4) and (14.4.5) in the text.
2. Verify Equations (14.4.8) and (14.4.9) in the text.
3. Verify Equation (14.8.2) in the text.
4. Verify Equations (14.9.1) in the text.
5. The components Q_j of generalized force associated with a generalized coordinate q_j has been defined in Equation (9.2.10) by the expression

$$Q_j = \sum_{i=1}^{n} \mathbf{F}_i \cdot \frac{\partial \mathbf{r}_i}{\partial q_j} \qquad (j = 1, 2, \ldots, r)$$

If q_j is an angle variable and $\mathbf{e}_{q_j}$ is a unit vector directed positively (in the sense of a right-hand screw) along the axis associated with q_j, show that

$$Q_j = N_{q_j} \qquad (N_{q_j} = \mathbf{N} \cdot \mathbf{e}_{q_j})$$

where

$$\mathbf{N} = \sum_{i=1}^{n} (\mathbf{r}_i \times \mathbf{F}_i)$$

is the external torque. For conservative forces that are derivable from the potential energy V of the system,

$$Q_j = -\frac{\partial V}{\partial q_j} \qquad (j = 1, 2, \ldots, r)$$

† References to more recent results for the triaxial case are given in a note added in proof, p. 381.

by (9.2.12). Hence

$$\frac{\partial V}{\partial q_j} = -N_{q_j}$$

6. For the special case that $B = A$ and $V = 0$, Equation (14.4.10) reduces to

$$H_0 = \frac{1}{2A} p_\theta^2 + \frac{1}{2C} p_\phi^2 + \frac{\csc^2\theta}{2A}(p_\psi - p_\phi \cos\theta)^2 \tag{1}$$

A function $W_0(\psi, \theta, \phi, \alpha_1, \alpha_2^*, \alpha_3, t)$ that generates a canonical transformation

$$\begin{aligned} p_\psi &= \frac{\partial W_0}{\partial \psi}, & \beta_1^* &= -\frac{\partial W_0}{\partial \alpha_1} \\ p_\theta &= \frac{\partial W_0}{\partial \theta}, & \beta_2^* &= -\frac{\partial W_0}{\partial \alpha_2^*} \\ p_\phi &= \frac{\partial W_0}{\partial \phi}, & \beta_3 &= -\frac{\partial W_0}{\partial \alpha_3} \end{aligned} \tag{2}$$

from the old variables $\psi, \theta, \phi, p_\psi, p_\theta, p_\phi$ to a new set of independent variables $\alpha_1, \alpha_2^*, \alpha_3, \beta_1^*, \beta_2^*, \beta_3$ in such a way that each of the new variables is a constant of the unperturbed motion may be found by obtaining a complete integral of the Hamilton–Jacobi partial differential equation

$$\frac{\partial W_0}{\partial t} + \frac{1}{2A}\left(\frac{\partial W_0}{\partial \theta}\right)^2 + \frac{1}{2C}\left(\frac{\partial W_0}{\partial \phi}\right)^2 + \frac{\csc^2\theta}{2A}\left(\frac{\partial W_0}{\partial \psi} - \cos\theta \frac{\partial W_0}{\partial \phi}\right)^2 = 0 \tag{3}$$

associated with H_0. Show that

$$\begin{aligned} W_0 = {} & -\alpha_1 t + \alpha_2^*\phi + \alpha_3\psi - \alpha_2^* \cos^{-1}\left[\frac{\alpha_3 - \alpha_2^*\cos\theta}{(h^2 - \alpha_2^{*2})^{1/2}\sin\theta}\right] \\ & -\alpha_3 \cos^{-1}\left[\frac{\alpha_2^* - \alpha_3\cos\theta}{(h^2 - \alpha_3^2)^{1/2}\sin\theta}\right] + h\cos^{-1}\left[\frac{\alpha_2^*\alpha_3 - h^2\cos\theta}{[(h^2 - \alpha_2^{*2})(h^2 - \alpha_3^2)]^{1/2}}\right] \end{aligned} \tag{4}$$

where the positive constant h is defined by

$$h^2 = 2A\alpha_1 + \alpha_2^{*2}\frac{C - A}{C} \tag{5}$$

is a complete integral of (3). Use this result to derive Equations (14.7.44) in the text.

7. Show that the unit vectors $\mathbf{i}^0$, $\mathbf{j}^0$, $\mathbf{k}^0$, $\mathbf{i}'$, $\mathbf{j}'$, $\mathbf{k}'$ are related to the unit vectors $\mathbf{i}$, $\mathbf{j}$, and $\mathbf{k}$ through the equations

$$\begin{aligned}\mathbf{i}^0 = {} & (\cos\psi_H \cos\phi_H - \sin\psi_H \sin\phi_H \cos\theta_H)\mathbf{i} \\ & - (\cos\psi_H \sin\phi_H + \sin\psi_H \cos\phi_H \cos\theta_H)\mathbf{j} \\ & + \sin\psi_H \sin\theta_H \,\mathbf{k}\end{aligned}$$

$$\begin{aligned}\mathbf{j}^0 = {} & (\sin\psi_H \cos\phi_H - \cos\psi_H \sin\phi_H \cos\theta_H)\mathbf{i} \\ & + (-\sin\psi_H \sin\phi_H + \cos\psi_H \cos\phi_H \cos\theta_H)\mathbf{j} \\ & - \cos\psi_H \sin\theta_H \,\mathbf{k}\end{aligned}$$

$$\mathbf{k}^0 = \sin\phi_H \sin\theta_H \,\mathbf{i} + \cos\phi_H \sin\theta_H \,\mathbf{j} + \cos\theta_H \,\mathbf{k}$$

$$\mathbf{i}' = \cos\phi' \,\mathbf{i} - \sin\phi' \cos\theta' \,\mathbf{j} + \sin\phi' \sin\theta' \,\mathbf{k}$$

$$\mathbf{j}' = -\sin\phi' \,\mathbf{i} + \cos\phi' \cos\theta' \,\mathbf{j} + \cos\phi' \sin\theta' \,\mathbf{k}$$

$$\mathbf{k}' = -\sin\theta' \,\mathbf{j} + \cos\theta' \,\mathbf{k}$$

8. Show that the nonzero Poisson brackets associated with the variables ψ^*, θ^*, ϕ^*, θ', ϕ', h, which are obtained from the transformation equations (14.6.5), (14.6.7), and (14.7.40), are

$$\{\psi^*, \theta^*\} = -\{\theta^*, \psi^*\} = -\frac{1}{h \sin\theta^*}$$

$$\{\theta^*, \phi^*\} = -\{\phi^*, \theta^*\} = -\frac{\cot\theta^*}{h}$$

$$\{\phi^*, \theta'\} = -\{\theta', \phi^*\} = \frac{\cot\theta'}{h}$$

$$\{\phi^*, h\} = -\{h, \phi^*\} = 1$$

$$\{\theta', \phi'\} - \{\phi', \theta'\} = \frac{1}{h \sin\theta'}$$

Hence show that the variational equations for the variables ψ^*, θ^*, ϕ^*, θ', ϕ', h are

$$\dot{\psi}^* = \frac{1}{h \sin \theta^*} N_{\theta^*}$$

$$\dot{\theta}^* = -\frac{1}{h \sin \theta^*} N_{\psi^*} + \frac{\cot \theta^*}{h} N_{\phi^*}$$

$$\dot{\phi}^* = \frac{h}{A} - \frac{\cot \theta^*}{h} N_{\theta^*} - \frac{\cot \theta'}{h} N_{\theta'}$$

$$\dot{\theta}' = \frac{\cot \theta'}{h} N_{\phi^*} - \frac{1}{h \sin \theta'} N_{\phi'}$$

$$\dot{\phi}' = -\frac{h}{A}\left(\frac{C-A}{C}\right) \cos \theta' + \frac{1}{h \sin \theta'} N_{\theta'}$$

$$\dot{h} = N_{\phi^*}$$

9. Show that the components h_{x^0}, h_{y^0}, and h_{z^0} of the angular momentum $\mathbf{h}$ in the orbital coordinate system $Ox^0y^0z^0$ are related to the canonical variables α_1, α_2^*, α_3, β_3 by

$$\begin{aligned} h_{x^0} &= -(h^2 - \alpha_3^2)^{1/2} \sin(\Omega + \beta_3) \\ h_{y^0} &= \alpha_3 \sin i - (h^2 - \alpha_3^2)^{1/2} \cos i \cos(\Omega + \beta_3) \\ h_{z^0} &= \alpha_3 \cos i + (h^2 - \alpha_3^2)^{1/2} \sin i \cos(\Omega + \beta_3) \end{aligned} \tag{1}$$

and thus

$$h_{y^0} = -h_{z^0} \cot i + \alpha_3 \csc i \tag{2}$$

By comparing the solution for α_3 in (14.8.4) with the square of the last equation of (1), show that

$$\alpha_3 = \frac{K}{4h^2\dot{\Omega}} (1 - 3 \cos^2 \theta')\, h_{z_0}^2 + \text{constant} \tag{3}$$

Since the substitution of (3) into (2) leads to Equation (14.9.13) in the text, it is clear that the motion of the angular momentum predicted by the perturbation theory developed in Section 14.8 agrees exactly with the motion predicted in Section 14.9.

10. If the component Q_j of the generalized force, associated with the generalized coordinate q_j, has the form

$$Q_j = -\frac{\partial V}{\partial q_j} + Q_j' \qquad (j = 1, 2, \ldots, r)$$

where Q_j' represents the nonconservative part of Q_j and V is a function of the generalized coordinates alone, Equation (9.2.10) can be written

$$\frac{dp_j}{dt} - \frac{\partial L}{\partial q_j} = Q_j' \qquad \left(p_j = \frac{\partial L}{\partial \dot{q}_j}\right)$$

where $L = T - V$. Show that the corresponding Hamilton equations of motion become

$$\dot{q}_j = \frac{\partial H}{\partial p_j}, \qquad \dot{p}_j = -\frac{\partial H}{\partial q_j} + Q_j'$$

where

$$H = \sum_{j=1}^{r} p_j \dot{q}_j - L$$

If H has the form given in (11.2.2) and the solution of the H_0-problem has been obtained by expressing $(q_j; p_j)$ in terms of the canonical constants $(\alpha_j; \beta_j)$ and the time t, show that the equations of motion (14.4.13) are modified to read

$$\dot{x}_i = \dot{x}_{i0} + \sum_{k=1}^{2r} \{x_i, x_k\} \left(\frac{\partial H_1}{\partial x_k} - Q'_{x_k}\right) \qquad (i = 1, 2, ..., 2r)$$

where

$$Q'_{x_k} = \sum_{j=1}^{r} Q_j' \frac{\partial q_j}{\partial x_k}$$

Note Added in Proof

Although we discussed the free motions of a triaxial body in Sections 14.6 and 14.7, our subsequent remarks were confined to the effects of gravity-gradient torque on an orbiting uniaxial body. In more recent studies [48, 49] gravity effects on orbiting triaxial satellites have been considered. Hitzl and Breakwell [48] applied canonical methods similar to those described in this chapter, Cochran *et al.* [49] used the method of general perturbations introduced by Hori [12a].

Appendix 1

Velocity and Acceleration in Polar Coordinates

Let us consider briefly the derivation of expressions for the velocity $\dot{\mathbf{r}}$ and the acceleration $\ddot{\mathbf{r}}$ in the case of polar coordinates (r, θ) in the plane (see Figure A.1.1). The transformation equations between rectangular coordinates (x, y) and polar coordinates (r, θ) are

$$x = r \cos \theta, \qquad y = r \sin \theta \tag{A.1.1}$$

The unit vectors $\mathbf{u}_r$ and $\mathbf{u}_\theta$ are chosen in the senses of increasing r and θ

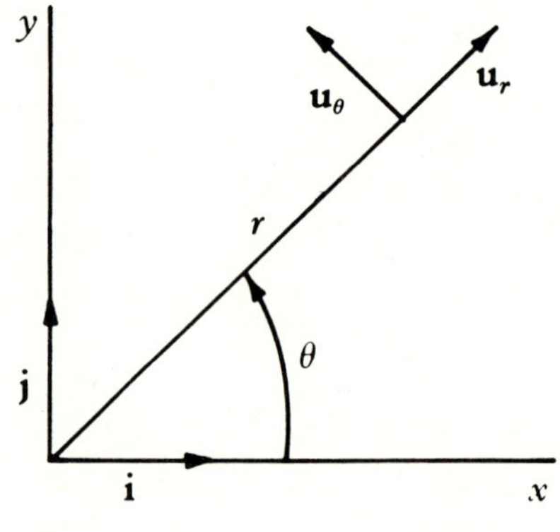

FIGURE A.1.1

respectively. They are related to the unit vectors **i** and **j** through the matrix equations

$$\begin{pmatrix}\mathbf{u}_r\\ \mathbf{u}_\theta\end{pmatrix} = R\begin{pmatrix}\mathbf{i}\\ \mathbf{j}\end{pmatrix}, \qquad \begin{pmatrix}\mathbf{i}\\ \mathbf{j}\end{pmatrix} = R^T\begin{pmatrix}\mathbf{u}_r\\ \mathbf{u}_\theta\end{pmatrix} \tag{A.1.2}$$

where

$$R = \begin{pmatrix}\cos\theta & \sin\theta\\ -\sin\theta & \cos\theta\end{pmatrix}, \qquad R^T = \begin{pmatrix}\cos\theta & -\sin\theta\\ \sin\theta & \cos\theta\end{pmatrix}$$

The time derivatives $\dot{\mathbf{u}}_r$ and $\dot{\mathbf{u}}_\theta$ may be expressed in terms of $\mathbf{u}_r$ and $\mathbf{u}_\theta$ by means of (A.1.2). We can write

$$\begin{aligned}\begin{pmatrix}\mathbf{u}_r\\ \mathbf{u}_\theta\end{pmatrix}^{\cdot} &= \dot{R}\begin{pmatrix}\mathbf{i}\\ \mathbf{j}\end{pmatrix} = \dot{R}R^T\begin{pmatrix}\mathbf{u}_r\\ \mathbf{u}_\theta\end{pmatrix}\\ &= \begin{pmatrix}-\sin\theta\dot{\theta} & \cos\theta\dot{\theta}\\ -\cos\theta\dot{\theta} & -\sin\theta\dot{\theta}\end{pmatrix}\begin{pmatrix}\cos\theta & -\sin\theta\\ \sin\theta & \cos\theta\end{pmatrix}\begin{pmatrix}\mathbf{u}_r\\ \mathbf{u}_\theta\end{pmatrix}\\ &= \begin{pmatrix}0 & \dot{\theta}\\ -\dot{\theta} & 0\end{pmatrix}\begin{pmatrix}\mathbf{u}_r\\ \mathbf{u}_\theta\end{pmatrix}\end{aligned} \tag{A.1.3}$$

The velocity and acceleration can then be written

$$\begin{aligned}\dot{\mathbf{r}} &= \dot{r}\mathbf{u}_r + r\dot{\mathbf{u}}_r = \dot{r}\mathbf{u}_r + r\dot{\theta}\mathbf{u}_\theta\\ \ddot{\mathbf{r}} &= \ddot{r}\mathbf{u}_r + \dot{r}\dot{\mathbf{u}}_r + \dot{r}\dot{\theta}\mathbf{u}_\theta + r\ddot{\theta}\mathbf{u}_\theta + r\dot{\theta}\dot{\mathbf{u}}_\theta\\ &= (\ddot{r} - r\dot{\theta}^2)\,\mathbf{u}_r + (2\dot{r}\dot{\theta} + r\ddot{\theta})\mathbf{u}_\theta\end{aligned} \tag{A.1.4}$$

Appendix 2

Conic Sections

A brief discussion of those properties of a conic section that are useful in the study of celestial mechanics is given here. To this end, we begin with the following definition:

Definition. Let e be a positive number, select a point F in the plane and a line l, not passing through F, and consider the point set S:

$$S = \{P \mid P \text{ is a point in the plane such that the distance from } P \text{ to } F \text{ is } e \text{ times the perpendicular distance from } P \text{ to } l\}$$

S is called a conic section with *eccentricity* e, *focus* F, and *directrix* l. If

1. $e = 1$, the conic is called a parabola.
2. $e > 1$, the conic is called a hyperbola.
3. $0 < e < 1$, the conic is called an ellipse.

It is a consequence of the definition that P does not coincide with F. This follows from the facts that e is positive and l does not pass through F; hence the distance PF is greater than zero.

By a judicious choice of F and l it is possible to obtain fairly simple equations for S in plane polar coordinates (r, θ) for each of the three cases cited in the definition. First, however, it should be noted that for any real numbers r, θ, and any integer n, the vector

$$\mathbf{r} = (-1)^n r[\cos(\theta + n\pi)\mathbf{i} + \sin(\theta + n\pi)\mathbf{j}]$$

drawn from the origin of the (r, θ) system, has the same terminal point in the associated xy-plane as does the vector

$$\mathbf{r} = r(\cos\theta\, \mathbf{i} + \sin\theta\, \mathbf{j})$$

where $\mathbf{i}$ and $\mathbf{j}$ are the usual unit vectors in the xy-rectangular reference frame. The terminal point of $\mathbf{r}$ is said to have polar coordinates $[(-1)^n r, \theta + n\pi]$. Clearly any point (x, y) in the plane has indefinitely many pairs of polar coordinates associated with it. Thus for any integer n, the points $[(-1)^n r, \theta + n\pi]$ all correspond to the same point (x, y) for prescribed values of r and θ.

The pole will be chosen at the point $(0, 0)$ and the directrix will be taken to be perpendicular to the polar axis at the point $(2p, 0)$. See Figure A.2.1. From the definition, the point P is on the conic if and only if

$$|r| = e|2p - r\cos\theta|, \qquad |r| > 0 \tag{A.2.1}$$

FIGURE A.2.1

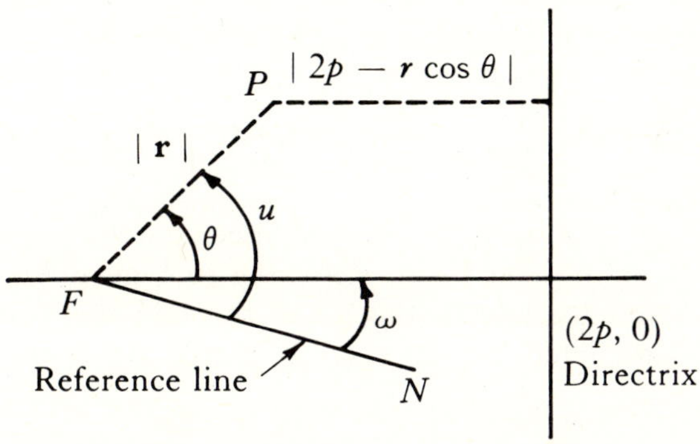

Now $r\cos\theta$ is never equal to $2p$ (i.e., the point P does not lie on the directrix), for then $|2p - r\cos\theta| = 0$ and $|r|$ could not be positive for e a positive real number.

The four possible cases are considered below:

$$\begin{aligned} r > 0: &\quad r\cos\theta < 2p, \quad r = e(2p - r\cos\theta) \\ &\quad r\cos\theta > 2p, \quad r = -e(2p - r\cos\theta) \\ r < 0: &\quad r\cos\theta < 2p, \quad r = -e(2p - r\cos\theta) \\ &\quad r\cos\theta > 2p, \quad r = e(2p - r\cos\theta) \end{aligned}$$

All possibilities are summed up in the two equations

$$r = e(2p - r\cos\theta), \qquad r = -e(2p - r\cos\theta)$$

which may be written in the forms

$$r = \frac{2ep}{1 + e\cos\theta} \qquad (e\cos\theta \neq -1) \tag{A.2.2a}$$

$$r = \frac{-2ep}{1 - e\cos\theta} \qquad (e\cos\theta \neq 1) \tag{A.2.2b}$$

It is readily seen that (A.2.2a) and (A.2.2b) have the same graph. Every point (x, y) that is on the graph of (a) is on the graph of (b), and conversely. To see this, let $[(-1)^n r, \theta + n\pi]$ be any representation of a point (x, y) which is on the graph of (a). Then the representation $[(-1)^n r, \theta + n\pi]$ satisfies (a). That is,

$$(-1)^n r = \frac{2ep}{1 + e\cos(\theta + n\pi)} \tag{A.2.3}$$

The same point (x, y) also has the representation $[(-1)^{n+1} r, \theta + (n + 1)\pi]$. This latter representation, when substituted into (b), leads to (A.2.3). Thus the representation $[(-1)^{n+1} r, \theta + (n + 1)\pi]$ of the point (x, y) satisfies (b) and hence (x, y) is on the graph of (b). A similar argument will establish that every point (x, y) on the graph of (b) is on the graph of (a).

It is customary in celestial mechanics to use (A.2.2a) to represent the conic in polar form. Referring to Figure A.2.1, we note that if FN is a reference line in the plane through the pole and u and ω are the angles depicted, Equation (A.2.2a) may be written as

$$r = \frac{2ep}{1 + e\cos(u - \omega)} \tag{A.2.4}$$

For the purpose of deducing the usual properties of conics, such as center, vertices, and semimajor axis, it is convenient to express the conic in rectangular form through the substitutions

$$x = r\cos\theta, \qquad r^2 = x^2 + y^2$$

Setting $e = 1$, we obtain from (A.2.1)

$$|r| = |2p - x| \tag{A.2.5}$$

Squaring (A.2.5) and simplifying, we obtain the expression

$$y^2 = -4p(x - p) \tag{A.2.6}$$

which is the rectangular form of a parabola with focus at (0, 0) vertex at $(p, 0)$, directrix $\{(x, y) \mid x = 2p\}$, and semilatus rectum of length $2|p|$.

If $e \neq 1$, we have

$$|r| = e|2p - x| \tag{A.2.7}$$

Squaring (A.2.7) and rearranging, we obtain

$$x^2 + \frac{4e^2px}{1-e^2} + \frac{y^2}{1-e^2} = \frac{4e^2p^2}{1-e^2}$$

After completing the square, we have

$$\left(x + \frac{2e^2p}{1-e^2}\right)^2 + \frac{y^2}{1-e^2} = \frac{4e^2p^2}{(1-e^2)^2}$$

whence

$$\frac{[x + 2e^2p/(1-e^2)]^2}{4e^2p^2/(1-e^2)^2} + \frac{y^2}{4e^2p^2/(1-e^2)} = 1 \tag{A.2.8}$$

If $0 < e < 1$, then (A.2.8) is the equation of an ellipse with the properties:

semilatus rectum length: $2e|p|$

semimajor axis: $a = \dfrac{2e\,|p|}{1-e^2}$

semiminor axis: $b = \dfrac{2e\,|p|}{(1-e^2)^{1/2}}$

foci: $(0, 0)$, $\left(-\dfrac{4e^2p}{1-e^2}, 0\right)$

center: $\left(-\dfrac{2e^2p}{1-e^2}, 0\right)$

vertices: $\left(-\dfrac{2e^2p}{1-e^2} \pm a, 0\right)$

directrices: $\left\{(x, y) \,\middle|\, x = 2p,\ x = -\dfrac{4e^2p}{1-e^2} - 2p\right\}$

If $e > 1$, then (A.2.8) is the equation of a hyperbola with the properties:

semilatus rectum length: $2e\,|\,p\,|$

semitransverse axis: $a = \dfrac{2e\,|\,p\,|}{e^2 - 1}$

semiconjugate axis: $b = 2e\,|\,p\,|(e^2 - 1)^{1/2}$

foci: $(0, 0), \quad \left(\dfrac{4e^2p}{e^2 - 1}, 0\right)$

center: $\left(\dfrac{2e^2p}{e^2 - 1}, 0\right)$

vertices: $\left(\dfrac{2e^2p}{e^2 - 1} \pm a, 0\right)$

directrices: $\left\{(x, y) \,\middle|\, x = 2p,\ x = \dfrac{4e^2p}{e^2 - 1} - 2p\right\}$.

asymptotes: $\left\{(x, y) \,\middle|\, y = \pm \dfrac{b}{a}\left(x + \dfrac{2e^2p}{1 - e^2}\right)\right\}$

Example A.2.1. State the usual properties of the conic section given by $r = 6(2 + \cos\theta)^{-1}$.

If the equation is written in the form

$$r = \frac{6}{2 + \cos\theta} = \frac{3}{1 + \frac{1}{2}\cos\theta}$$

we see that it represents an ellipse of eccentricity $e = \frac{1}{2}$ for which

the semilatus rectum $= 2e\,|\,p\,| = 3$

the semimajor axis $= \dfrac{2e\,|\,p\,|}{1 - e^2} = 4$

the semiminor axis $= \dfrac{2e\,|\,p\,|}{(1 - e^2)^{1/2}} = 2\sqrt{3}$

the foci are at $(0, 0), \quad \left(-\dfrac{4e^2p}{1 - e^2} = -4, 0\right)$

the vertices are at $\left(-\dfrac{2e^2p}{1 - e^2} \pm a, 0\right)$, that is, $(2, 0), \quad (-6, 0)$

the directrices are $x = 2p = 6; \quad x = -\dfrac{4e^2p}{1 - e^2} - 2p = -10$

Appendix 3

The Lagrange Expansion Theorem

We first state two theorems from complex function theory.[†] Concepts related to these theorems will be drawn upon in giving a proof of the Lagrange expansion theorem. The proof is included here because it provides a deep insight into the content of the Lagrange expansion theorem and the meaning of a convergence criterion for its use.

Theorem A.3.1. If $f(z)$ is analytic within a closed curve C, has no zeros on the boundary, and $f'(z)$ is continuous on C, then the number N of roots of $f(z)$ within C is determined by the number of complete sweeps through 2π radians of the angle of $f(z)$ as the complex variable z describes the boundary in the positive sense.

Theorem A.3.2 (Rouchè). If $f(z)$ and $g(z)$ are two functions, each of which is analytic in the interior and continuous on the boundary of a closed curve C, and $|g(z)| < |f(z)|$ everywhere on C, then the functions $f(z)$ and $f(z) + g(z)$ have the same number of zeros in C.

Suppose that there is a relation between the complex variables z and a given by

$$z = a + \alpha h(z) \tag{A.3.1}$$

[†] These theorems can be found in Volume 1 of Danese [47] and elsewhere.

where α is a variable parameter ($\alpha \neq 0$), $h(z)$ is a function analytic in, and continuous on, the boundary C of a circle of radius $|z - a|$, and $\zeta \neq a$ is a value of z within the circle which satisfies (A.3.1).

Theorem A.3.3 (Lagrange expansion theorem). If $\phi(z)$ is a function that is analytic in, and continuous on, a circle C of radius $|z - a|$ centered at a, then $\phi(\zeta)$ has a series expansion in α given by

$$\phi(\zeta) = \phi(a) + \sum_{n=1}^{\infty} \frac{\alpha^n}{n!} \frac{d^{n-1}}{da^{n-1}} \{\phi'(a)[h(a)]^n\}$$

which converges for all α subject to the condition that

$$|\alpha h(z)| < |z - a|$$

on C.

Proof. Consider the two functions $f(z) = z - a$ and $g(z) = -\alpha h(z)$. We assume that $|g(z)| < |f(z)|$ on the boundary C of a circle of radius ρ centered at a. The function $f(z)$ is clearly analytic in and on the boundary C so that $f(z)$ and $g(z)$ satisfy the criteria for Rouchè's theorem. Hence if we consider the zeros of the functions

$$f(z) = z - a \tag{1}$$

$$F(z) = f(z) + g(z) = z - a - \alpha h(z) \tag{2}$$

inside C, since $f(z)$ contains but one root inside C, then $F(z)$ also contains a single root inside C, and this is at ζ because it is assumed that $\zeta \neq a$ satisfies (A.3.1). [This means that $h(a)$ cannot be zero.] Thus $F(z) = (z - \zeta) R(z)$, where $R(z)$ does not have any zeros in C.

Let $B(z)$ be a function analytic in, and continuous on, the boundary of C. Then the function $Q(z)$ defined by

$$Q(z) \equiv \frac{B(z)}{F(z)} = \frac{B(z)}{(z - \zeta) R(z)} \tag{A.3.2}$$

has a single pole at $z = \zeta$ interior to C. The residue r of $Q(z)$ at $z = \zeta$ is

$$\begin{aligned} r &= \lim_{z\to\zeta} [(z - \zeta) Q(z)] \\ &= \lim_{z\to\zeta} \left[\frac{(z - \zeta) B(z)}{(z - \zeta) R(z)}\right] = \frac{B(\zeta)}{R(\zeta)} \end{aligned}$$

Since $Q(z)$ has but a single pole inside C, we can write

$$\frac{B(\zeta)}{R(\zeta)} = r = \frac{1}{2\pi i}\int_C Q(z)\,dz = \frac{1}{2\pi i}\int_C \frac{B(z)}{F(z)}\,dz \tag{A.3.3}$$

Now

$$F(z) = (z - \zeta)\,R(z)$$

and

$$F'(z) = R(z) + (z - \zeta)\,R'(z)$$

so that

$$F'(\zeta) = R(\zeta)$$

and (A.3.3) can be written

$$\frac{B(\zeta)}{F'(\zeta)} = \frac{1}{2\pi i}\int_C \frac{B(z)}{F(z)}\,dz = \frac{1}{2\pi i}\int_C \frac{B(z)\,dz}{z - a - \alpha h(z)} \tag{A.3.4}$$

The denominator in the integrand in (A.3.4) can be written in the form

$$\frac{1}{z - a - \alpha h(z)} = \frac{(1 - \xi)^{-1}}{(z - a)} \qquad \left[\xi \equiv \frac{\alpha h(z)}{z - a}\right]$$

Carrying out the long division up to the remainder of degree $n + 1$ in $z - a$, we have

$$\frac{1}{z - a - \alpha h(z)} = \frac{1}{z - a} + \frac{\alpha h(z)}{(z - a)^2} + \cdots + \frac{[\alpha h(z)]^n}{(z - a)^{n+1}}$$
$$+ \frac{1}{z - a - \alpha h(z)}\left[\frac{\alpha h(z)}{z - a}\right]^{n+1}$$

When this expansion is substituted into (A.3.4), we obtain

$$\frac{B(\zeta)}{F'(\zeta)} = \frac{1}{2\pi i}\int_C \frac{B(z)}{z - a}\,dz + \frac{\alpha}{2\pi i}\int_C \frac{B(z)\,h(z)}{(z - a)^2}\,dz + \cdots$$
$$+ \frac{\alpha^n}{2\pi i}\int_C \frac{B(z)\,h(z)^n}{(z - a)^{n+1}} + \frac{1}{2\pi i}\int_C \frac{B(z)}{z - a - \alpha h(z)}\left[\frac{\alpha h(z)}{z - a}\right]^{n+1} dz$$

If the generalized Cauchy integral formula is used, this last equation can be written in the form

$$\frac{B(\zeta)}{F'(\zeta)} = B(a) + \alpha\frac{d}{da}[B(a)\,h(a)] + \frac{\alpha^2}{2!}\frac{d^2}{da^2}[B(a)\,h(a)^2] + \cdots$$
$$+ \frac{\alpha^n}{n!}\frac{d^n}{da^n}\{B(a)[h(a)]^n\} + R_{n+1} \tag{A.3.5}$$

where

$$R_{n+1} = \frac{1}{2\pi i} \int_C \frac{B(z)}{z - a - \alpha h(z)} \left[\frac{\alpha h(z)}{z - a} \right]^{n+1} dz$$

To show that R_{n+1} vanishes as $n \to \infty$, let m and M be the maximum values of $| \alpha h(z)|$ and $| B(z)|$ respectively, along C. By hypothesis, $| \alpha h(z)| < | z - a | = \rho$. Thus we can write

$$\begin{aligned} | R_{n+1} | &= \left| \frac{1}{2\pi i} \int_C \frac{B(z)}{z - a - \alpha h(z)} \left[\frac{\alpha h(z)}{z - a} \right]^{n+1} dz \right| \\ &\leqslant \frac{1}{2\pi} \int_C \frac{| B(z)|}{| z - a - \alpha h(z)|} \left| \frac{\alpha h(z)}{z - a} \right|^{n+1} | dz | \\ &\leqslant \frac{1}{2\pi} \frac{M}{\rho - m} \left(\frac{m}{\rho} \right)^{n+1} \int_C | dz | = \frac{1}{2\pi} \frac{M}{\rho - m} \left(\frac{m}{\rho} \right)^{n+1} 2\pi\rho \end{aligned}$$

Since $m/\rho < 1$, it is clear that

$$\lim_{n \to \infty} | R_{n+1} | = 0$$

Thus the convergent series

$$\frac{B(\zeta)}{F'(\zeta)} = B(a) + \sum_{n=1}^{\infty} \frac{\alpha^n}{n!} \frac{d^n}{da^n} \{B(a)[h(a)]^n\} \tag{A.3.6}$$

is obtained if $| \alpha h(z)| < | z - a |$ on C.

It may be noted that the function $F(z)$ has a derivative

$$F'(z) = 1 - \alpha h'(z)$$

which is clearly analytic in, an on, C, with $F'(\zeta) = 1 - \alpha h'(\zeta)$. Thus we are led to choose

$$B(z) = \phi(z)[1 - \alpha h'(z)] = \phi(z) F'(z)$$

where $\phi(z)$ is to be analytic in, and continuous on, C, so that $B(z)$ is analytic in, and continuous on, C. We can then write

$$\phi(\zeta) = \frac{B(\zeta)}{F'(\zeta)}$$

If $B(\zeta)/F'(\zeta)$ is replaced by $\phi(\zeta)$ in (A.3.6), we can write

$$\phi(\zeta) = \phi(a) - \alpha\phi(a)\, h'(a) + \sum_{n=1}^{\infty} \frac{\alpha^n}{n!} \frac{d^n}{da^n} \{\phi(a)[1 - \alpha h'(a)][h(a)]^n\} \tag{A.3.7}$$

Equation (A.3.7) can be cast in a somewhat tidier form by noting that, for any given n, there are two terms of degree n in α because of the presence of the factor α which multiplies $h'(a)$. Thus, from the first of the two terms for n and the last of the two terms for $n-1$, we pick up two terms of degree n in α. Specifically, we have the four successive terms

$$\frac{\alpha^{n-1}}{(n-1)!}\frac{d^{n-1}}{da^{n-1}}\{\phi(a)[h(a)]^{n-1}\} - \frac{\alpha^n}{(n-1)!}\frac{d^{n-1}}{da^{n-1}}\{\phi(a)\,h'(a)[h(a)]^{n-1}\}$$

$$+\frac{\alpha^n}{n!}\frac{d^n}{da^n}\{\phi(a)[h(a)]^n\} - \frac{\alpha^{n+1}}{n!}\frac{d^n}{da^n}\{\phi(a)\,h'(a)[h(a)]^n\}$$

Collecting the coefficient of α^n, we can then write

$$\frac{\alpha^n}{n!}\frac{d^n}{da^n}\{\phi(a)[h(a)]^n\} - \frac{\alpha^n}{(n-1)!}\frac{d^{n-1}}{da^{n-1}}\{\phi(a)\,h'(a)[h(a)]^{n-1}\}$$

$$= \frac{\alpha^n}{n!}\frac{d^{n-1}}{da^{n-1}}\{\phi'(a)[h(a)^n] + n\phi(a)[h(a)]^{n-1}\,h'(a) - n\phi(a)[h(a)]^{n-1}\,h'(a)\}$$

$$= \frac{\alpha^n}{n!}\frac{d^{n-1}}{da^{n-1}}\{\phi(a)[h(a)]^n\} \tag{A.3.8}$$

Substituting this result into (A.3.7), we obtain the Lagrange expansion formula

$$\phi(\zeta) = \phi(a) + \sum_{n=1}^{\infty}\frac{\alpha^n}{n!}\frac{d^{n-1}}{da^{n-1}}\{\phi'(a)[h(a)]^n\} \tag{A.3.9}$$

Appendix 4

Table of First Six Bessel Coefficients and Derivatives

$$\frac{2}{e} J_1(e) = 1 - \frac{e^2}{8} + \frac{e^4}{192} - \frac{e^6}{9216} + \cdots$$

$$\frac{2}{e} J_2(2e) = e\left(1 - \frac{e^2}{3} + \frac{e^4}{24} - \cdots\right)$$

$$\frac{2}{e} J_3(3e) = \frac{9e^2}{8}\left(1 - \frac{9e^2}{16} + \frac{81e^4}{640} - \cdots\right)$$

$$\frac{2}{e} J_4(4e) = \frac{4e^3}{3}\left(1 - \frac{4e^2}{5} + \cdots\right)$$

$$\frac{2}{e} J_5(5e) = \frac{625e^4}{384}\left(1 - \frac{25e^2}{24} + \cdots\right)$$

$$\frac{2}{e} J_6(6e) = \frac{81e^5}{40}$$

. . . .

$$2J_1'(e) = 1 - \frac{3e^2}{8} + \frac{5e^4}{192} - \frac{7e^6}{9216} + \cdots$$

$$2J_2'(2e) = e\left(1 - \frac{2e^2}{3} + \frac{e^4}{8} - \frac{e^6}{90} + \cdots\right)$$

$$2J_3'(3e) = \frac{9e^2}{8}\left(1 - \frac{15e^2}{16} + \frac{189e^4}{640} - \cdots\right)$$

$$2J_4'(4e) = \frac{4e^3}{3}\left(1 - \frac{6e^2}{5} + \frac{8e^4}{15} - \cdots\right)$$

$$2J_5'(5e) = \frac{625e^4}{384}\left(1 - \frac{35e^2}{24} + \cdots\right)$$

$$2J_6'(6e) = \frac{81e^5}{40}\left(1 - \frac{12e^2}{7} + \cdots\right)$$

References

1. S. Newcomb, Tables of the Motion of the Earth on Its Axis and Around the Sun, *Astron. Papers Am. Ephemeris Nautical Almanac* **6**, 7 (1898).
2. R. Kurth, "The Mechanics of the Solar System." Macmillan, New York, 1959.
3. S. Newcomb, A New Determination of the Precessional Constant with the Resulting Precessional Motion, *Astron. Papers Am. Ephemeris Nautical Almanac* **8**, 73 (1898).
4. S. Newcomb, "Compendium of Spherical Astronomy." Macmillan, New York, 1906.
5. "Planetary Coordinates for the Years 1960–1980." H. M. Stationery Office, London, 1958.
6. F. R. Moulton, "An Introduction to Celestial Mechanics." Macmillan, New York, 1914.
7. H. C. Plummer, "An Introductory Treatise on Dynamical Astronomy." Cambridge Univ. Press, New York and London, 1918.
8. R. F. Arenstorf, A New Method of Perturbation Theory and Its Application to the Satellite Problem of Celestial Mechanics, *J. Reine Angew. Math.* **221**, 1966.
9. A. N. Kolmogorov, General Theory of Dynamical Systems and Classical Mechanics, *Proc. Intern. Congr. Math. 1954*, Vol. 1, pp. 315–333. North-Holland. Amsterdam, 1957.
10. J. Moser, On Invariant Curves of Area-Preserving Mappings of an Annulus, *Nachr. Akad. Wissenschaften Göttingen, Math. Phys. Kl.* No. 1 (1962).
11. V. I. Arnold, Generation of Almost Periodic Motion from a Family of Periodic Motions, *Dokl. Akad. Nauk SSSR* **138**, 13 (1961).
12. C. L. Siegel, "Vorlesungen über Himmelsmechanik." Springer, Berlin, 1956.

12a. G. Hori, Theory of General Perturbations with Unspecified Canonical Variables, *Publ. Astron. Soc. Japan*, **18**, 287 (1966).

12b. A. Deprit, Canonical Transformations Depending on a Small Parameter, *Boeing Scientific Research Labs. document No. D1-82-0755*, The Boeing Company, Seattle, Washington, 1968 .

13. F. H. Miller, "Partial Differential Equations." Wiley, New York, 1941.
14. R. V. Churchill, "Fourier Series and Boundary Value Problems." McGraw-Hill, New York, 1963.

14a. Y. Kozai, "Seasonal Variations of the Geopotential Inferred from Satellite Observations," *SAO Special Rept.* in preparation. Smithsonian Inst. Astrophys. Obs., Cambridge, Massachusetts, 1970.

15. W. M. Kaula, Determination of the Earth's Gravitational Field, *Rev. Geophys.* **1**, 507 (1963).

16. C. A. Lundquist and G, Veis (eds.), "Geodetic Parameters for a 1966 Smithsonian Institution Standard Earth," *SAO Special Rept. No. 200*, Vols. 1, 2. Smithsonian Inst. Astrophys. Obs., Cambridge, Massachusetts, 1966.
17. W. M. Kaula, Tesseral Harmonics of the Gravitational Field and Geodetic Datum Shifts Derived from Camera Observations of Satellites, *J. Geophys. Res.* **68**, 473 (1963).
18. O. Callandreau, Sur le Développement en Série du Potentiel, *J. Polytech.* (1889).
19. G. Birkhoff and S. MacLane, "A Survey of Modern Algebra." Macmillan, New York, 1949.
20. J. L. Arsenault, J. D. Enright, and C. Purcell, "General Perturbation Techniques for Satellite Orbit Prediction Study," *Tech. Document Rept. No. AL-TDR-64-70*, Vols. 1, 2. AF Avionics Lab., Wright-Patterson AFB, Ohio, 1964.
21. T. E. Sterne, "An Introduction to Celestial Mechanics." Wiley (Interscience), New York, 1960.
22. D. Brouwer and G. Hori, Theoretical Evaluation of Atmospheric Drag Effects in the Motion of an Artificial Satellite, *Astron. J.* **66**, 193 (1961).
23. M. H. Lane, The Development of an Artificial Satellite Theory Using a Power-Law Atmospheric Density Representation, *AIAA Paper No. 65–35*. AIAA 2d Aerospace Sci. Mtg., New York, 1965.
23a. K. Aksnes, A Second-Order Solution for the Motion of an Artificial Earth Satellite Based on an Intermediate Orbit, Dissertation, Dept. of Astron., Yale Univ., New Haven, Connecticut, 1969.
24. L. G. Jacchia, Properties of the Upper Atmosphere Determined from Satellite Orbits, *Phil. Trans. Roy. Soc. London* **262**, 157 (1967).
25. M. H. Lane, P. M. Fitzpatrick, and J. J. Murphy, "On the Representation of Air Density in Satellite Deceleration Equations by Power Functions with Integral Exponents," *Tech. Documentary Rept. No. APGC–TDR–62–15*. Air Proving Ground Center, Eglin AFB, Florida, 1962.
26. D. King-Hele, "Theory of Satellite Orbits in an Atmosphere." Butterworths, London, 1964.
27. D. Brouwer, Solution of the Problem of Artificial Satellite Theory without Drag, *Astron. J.* **64**, 378 (1959).
28. H. von Zeipel, Recherches sur le Mouvement des Petites Planètes, *Arkiv. Mat. Astron. Fys.* **11**, No. 1 (1916a).
29. H. von Zeipel, Recherches sur le Mouvement des Petites Planètes, *Arkiv. Mat. Astron. Fys.* **11**, No. 7 (1916b).
30. H. von Zeipel, Recherches sur le Mouvement des Petites Planètes, *Arkiv. Mat. Astron. Fys.* **12**, No. 9 (1917).
31. R. H. Lyddane, Small Eccentricities or Inclinations in the Brouwer Theory of the Artificial Satellite, *Astron. J.* **68**, 555 (1963).
32. P. Davenport, "Modification of Brouwer's Solution for Artificial Satellites to Include Small Eccentricities and Inclinations," *Document X–543–65–23*. Goddard Space Flight Center, Greenbelt, Maryland, 1965.
32a. M. H. Lane and K. H. Cranford, An Improved Analytical Drag Theory for the Artificial Satellite Problem, *AIAA Paper No. 69–925*. AIAA/AAS Astrodynamics Conference, Princeton, New Jersey, 1969.
33. P. M. Fitzpatrick and G. Findley, "Criteria for Schroedinger Separability Applied to Satellite Orbit Theory," *Tech. Documentary Rept. No. APGC–TDR–62–67*. Air Proving Ground Center, Eglin AFB, Florida, 1962.
34. P. M. Fitzpatrick, J. Cochran, G. Harmon, and J. Liu, "A Study of Orbital Perturb-

ing Forces," Interim Rept. April 1, 1969, NASA Contract NAS 8-20175, Dept. of Math., Auburn Univ., Auburn, Alabama, 1969.

35. K. H. Cranford and P. M. Fitzpatrick, "On the Influence of Gravity on the Rotational Motion of Artificial Earth Satellites," Environment Induced Orbital Dynamics Seminar Program, June 6–7, 1967, Vol. 2, NASA Marshall Space Flight Center, Huntsville, Alabama, 1967.
36. J. W. Crenshaw and P. M. Fitzpatrick, Gravity Effects on the Rotational Motion of a Uniaxial Artificial Satellite, *AIAA J.* **6**, 2140 (1968).
37. R. L. Holland, A First-Order Theory and Applications for the Rotational Motion of a Triaxial Rigid Body Orbiting an Oblate Primary, Thesis, Dept. of Math., Univ. of Alabama, Tuscaloosa, Alabama, 1967.
38. R. L. Holland and H. J. Sperling, A First-Order Theory for the Rotational Motion of a Triaxial Rigid Body Orbiting an Oblate Primary, *Astron. J.* **74**, 490 (1969).
39. V. V. Beletsky, Motion of an Artificial Earth Satellite about Its Center of Mass, *in* "Artificial Earth Satellites," Vol. 1, pp. 30–54, Plenum, New York, 1960.
40. V. V. Beletsky, Some Problems of Motion of Artificial Satellites about the Center of Mass, *in* "Dynamics of Satellites," pp. 158–167, Academic Press, New York, 1963.
41. G. Colombo, The Motion of Satellite 1958 Epsilon around Its Center of Mass, *in* "Research in Space Science, Smithsonian Contributions to Astrophysics," Vol. 6, p. 149. Smithsonian Inst. Astrophys. Obs., Cambridge, Massachusetts, 1963.
42. G. Colombo, "On the Motion of Explorer XI around Its Center of Mass," *SAO Special Report No. 70*. Smithsonian Inst. Astrophys. Obs., Cambridge, Massachusetts, 1962.
43. G. Colombo, Cassini's Second and Third Laws, *Astron. J.* **71**, 891 (1966).
44. V. V. Beletsky, "Motion of an Artificial Satellite about Its Center of Mass" (translated from Russian by Z. Lerman, Israel Program for Scientific Translations Ltd., Jerusalem 1966; available from U.S. Dept. of Commerce, Clearinghouse, Springfield, Virginia 22151).
45. P. F. Byrd and M. D. Friedman, "Handbook of Elliptic Integrals for Engineers and Physicists." Springer, Berlin, 1954.
46. J. W. Crenshaw, The Rotational Motion of a Uniaxial Artificial Satellite, Dissertation, Dept. of Phys., Auburn Univ., Auburn, Alabama, 1968.
47. A. E. Danese, "Advanced Calculus," Vol. 1. Allyn and Bacon, Boston, 1965.
48. D. L. Hitzl and J. V. Breakwell, Resonant and Nonresonant Gravity-Gradient Perturbations of a Tumbling Triaxial Satellite, *AIAA Paper No. 69–922*. AIAA/AAS Astrodynamics Conference, Princeton, New Jersey, 1969.
49. J. E. Cochran, B. D. Tapley, and P. M. Fitzpatrick, Gravity Effects on the Rotational Motion of a Triaxial Rigid Satellite, Paper presented to Eight Annual Seminar on Current Problems in Celestial Mechanics, The Univ. of Texas at Austin, Austin, Texas, Jan. 1970. Abstract to appear in *Bull. Am. Astron. Soc.* (in preparation), 1970.

Index

B

C

D

H

I

J

K

L

M

N

O

P

R

S